Livestock: Genetic Improvement Techniques

Livestock: Genetic Improvement Techniques

Edited by Carlos Hassey

SYRAWOOD
PUBLISHING HOUSE

New York

Published by Syrawood Publishing House,
750 Third Avenue, 9th Floor,
New York, NY 10017, USA
www.syrawoodpublishinghouse.com

Livestock: Genetic Improvement Techniques
Edited by Carlos Hassey

Cataloging-in-Publication Data

Livestock : genetic improvement techniques / edited by Carlos Hassey.
      p. cm.
Includes bibliographical references and index.
ISBN 978-1-68286-659-7
1. Livestock--Genetics. 2. Livestock--Genetic engineering. 3. Livestock--Breeding.
I. Hassey, Carlos.
SF756.5 .L58 2019
636.082 1--dc23

TABLE OF CONTENTS

PREFACE

Livestock refers to those domesticated animals which are used to produce commodities like meat, milk, leather, eggs, etc. They can also be used for the purpose of labor. Livestock animals are raised in an agricultural setting like a farm. Cows and goats are the most commonly reared livestock animals. In modern times, a number of technologically advanced techniques such as artificial insemination and selective breeding are being practiced to improve the genetic characteristics of livestock. This book elucidates the concepts and innovative models around prospective developments with respect to genetic manipulation and improvement of livestock. It consists of contributions made by international experts. This book will prove immensely beneficial to researchers and students in this field.

The information contained in this book is the result of intensive hard work done by researchers in this field. All due efforts have been made to make this book serve as a complete guiding source for students and researchers. The topics in this book have been comprehensively explained to help readers understand the growing trends in the field.

I would like to thank the entire group of writers who made sincere efforts in this book and my family who supported me in my efforts of working on this book. I take this opportunity to thank all those who have been a guiding force throughout my life.

Editor

Assessing accuracy of imputation using different SNP panel densities in a multi-breed sheep population

Ricardo V. Ventura[1,2], Stephen P. Miller[1,3]*, Ken G. Dodds[3], Benoit Auvray[4], Michael Lee[4], Matthew Bixley[3], Shannon M. Clarke[3] and John C. McEwan[3]

Abstract

Background: Genotype imputation is a key element of the implementation of genomic selection within the New Zealand sheep industry, but many factors can influence imputation accuracy. Our objective was to provide practical directions on the implementation of imputation strategies in a multi-breed sheep population genotyped with three single nucleotide polymorphism (SNP) panels: 5K, 50K and HD (600K SNPs).

Results: Imputation from 5K to HD was slightly better (0.6 %) than imputation from 5K to 50K. Two-step imputation from 5K to 50K and then from 50K to HD outperformed direct imputation from 5K to HD. A slight loss in imputation accuracy was observed when a large fixed reference population was used compared to a smaller within-breed reference (including all 50K genotypes on animals from different breeds excluding those in the validation set i.e. to be imputed), but only for a few animals across all imputation scenarios from 5K to 50K. However, a major gain in imputation accuracy for a large proportion of animals (purebred and crossbred), justified the use of a fixed and large reference dataset for all situations. This study also investigated the loss in imputation accuracy specifically for SNPs located at the ends of each chromosome, and showed that only chromosome 26 had an overall imputation (5K to 50K) accuracy for 100 SNPs at each end higher than 60 % (r^2). Most of the chromosomes displayed reduced imputation accuracy at least at one of their ends. Prediction of imputation accuracy based on the relatedness of low-density genotypes to those of the reference dataset, before imputation (without running an imputation software) was also investigated. FIMPUTE V2.2 outperformed BEAGLE 3.3.2 across all imputation scenarios.

Conclusions: Imputation accuracy in sheep breeds can be improved by following a set of recommendations on SNP panels, software, strategies of imputation (one- or two-step imputation), and choice of the animals to be genotyped using both high- and low-density SNP panels. We present a method that predicts imputation accuracy for individual animals at the low-density level, before running imputation, which can be used to restrict genomic prediction only to the animals that can be imputed with sufficient accuracy.

Background

Imputation refers to a statistical approach that is able to infer single nucleotide polymorphism (SNP) genotypes, which are not obtained from a low-density panel, by using information from a group of animals that are genotyped with higher density panels [1–3]. Widespread

implementation of genomic selection [4] in dairy cattle quickly followed the development of the Illumina SNP50 Genotyping beadchip [5]. The technology was subsequently launched for sheep [6] and beef cattle [7] as reference datasets of genotyped animals with a suitable size became available, as well as SNP panels (http://support.illumina.com/array/array_kits/). The next advancement in the technology was the use of lower density panels, which are available at a lower cost compared to the higher density panels required for genomic selection, and

*Correspondence: miller@uoguelph.ca
[1] Centre for Genetic Improvement of Livestock, University of Guelph, Guelph, ON N1G2W1, Canada
Full list of author information is available at the end of the article

can be imputed to higher densities with high accuracy in cattle [1, 8–10]. Imputation is also a key strategy for the implementation of genomic selection within the New Zealand sheep industries [6].

Several studies have investigated accuracy of genotype imputation and its impact on the accuracy of genomic selection in dairy and beef cattle through the adoption of high-density SNP panels, and more recently, whole-sequence data [1, 11–17]. Several panels that vary in the number of SNPs they include are currently available on the market and the number of genotyped individuals is rapidly growing in livestock sectors due to the reduction in costs and the development of new genotyping tools [9]. Although the imputation efficiency of each SNP panel is well documented [1, 18, 19], few articles evaluated imputation accuracy across different panels using both crossbred and purebred populations [20, 21] and, more specifically, strategies for the prediction of imputation accuracy are scarce.

Imputation is a robust tool to minimize costs of genotyping, but many factors can influence imputation accuracy, which provide opportunities for further improvements and optimal implementation of this technology. For some animal populations, missing SNPs cannot be inferred with high accuracy and this depends on the structure of the reference population (i.e. the group of animals genotyped with high-density SNPs) and the marker density of both reference and imputed populations. Gains in imputation accuracy are closely associated with the level of relationship between the animals to be imputed and the reference population, the number of animals in the reference population, the position of the SNPs on the chromosome, the density of the SNP panel used for the reference population, and the breed composition [1, 9, 13, 22].

Imputation of rare alleles is a particularly difficult task that is directly associated with minor allele frequencies (MAF); it can influence accuracy of genomic selection because of the potential influence of such alleles on the genetic expression of the trait under study [9, 23]. For example, for a chromosomal region that contains SNPs with a low MAF, association methods can generate spurious results due to genotyping errors [24]. Variants with a MAF lower than 0.05 could be under selection or in a related process that removes them from the population. According to Sargolzaei et al. [9], such variants with a low MAF tend to be recent mutations and are more likely to be identified after detecting long haplotypes. The same study [9] reported gains in imputation accuracy by using information on relatives, which can also optimize the imputation of rare alleles compared with other algorithms. Different measures of accuracy have been implemented, which depend on the methods used to compare

the original and imputed genotypes, and the output generated from each software/method [12, 15]. Calus et al. [13] evaluated different measures of correctness of genotype imputation in the context of genomic prediction and suggested that correlation between imputed and true genotypes is the most useful and unbiased measure of imputation accuracy and is suitable for comparisons across loci regardless of the MAF of SNPs [13]. The same authors suggested that individual specific imputation accuracies should be computed from genotypes that are centered and scaled. We did not apply this approach in our investigation but plan to evaluate it in future studies.

Hayes et al. [14] evaluated the accuracy of genotype imputation from low-density to 50K panels in sheep breeds by comparing fastPHASE [25] and BEAGLE [26] software programs. Recently, a new approach for efficient genotype imputation was reported by Sargolzaei et al. [9] and is implemented in the newest version of the FIM-PUTE software. Ventura et al. [1] assessed the impact of the reference population on accuracy of imputation from 6K and 50K SNP chips in purebred and crossbred beef cattle. These authors showed that IMPUTE2 and FIM-PUTE imputed almost all the individuals more accurately than BEAGLE by testing several scenarios and that they were also very efficient in terms of run time.

The objective of our study was to provide practical directions on the implementation of imputation strategies in a multi-breed sheep population that was genotyped with three SNP panels: 5K, 50K and HD (600K SNPs), and to compare these strategies with the current implementation of imputation that is carried out in practice for genomic selection in the New Zealand sheep industry. We evaluated: (1) composition of the reference population; (2) SNP density; (3) imputation of rare variants; (4) imputation software; (5) measures of imputation accuracy; and (6) prediction of imputation accuracy.

Methods

Population imputation was implemented using BEA-GLE 3.3.2 [26] and FIMPUTE 2.2 software [9] and several scenarios were generated by alternating the animals that were included in the reference population and in the set of animals to be imputed. The reference population consisted of animals that were genotyped with the Illumina OvineSNP50 Genotyping BeadChip (53,903 SNPs) (http://www.illumina.com/products/ovinesnp50_dna_analysis_kit.html) and/or the Ovine Infinium® HD SNP BeadChip (603,350 SNPs). Only autosomal SNPs were included in this study.

Data

A dataset including 2409 animals that were genotyped with the Ovine Infinium® HD and 17,176 animals that were genotyped with the Illumina OvineSNP50 were used to evaluate imputation accuracy. Before describing the imputation scenarios that were used to evaluate issues such as relatedness, multi- versus one-breed reference population and SNP density, we present the multi-breed populations according to the density of the SNP panel used to genotype animals and to the proportion of the main breed that composes the population. Animals in this dataset were primarily sires from breeders' flocks along with a group of animals of both sexes from research flocks. Average breed composition as deduced from pedigree information is described here for the two groups of animals that were genotyped with the 50K and HD panels:

1. 50K animals: 37 % Romney (30 % purebred Romney), 19 % Coopworth (8 % purebred), 4 % Texel (1 % purebred), 6 % Perendale (5 % purebred), 5 % Primera (composite of terminal sire breeds http://www.focus-genetics.com/sheep/sheep-breeding-programme/primera/) and other breeds with less than 3 % each.
2. HD animals: 33 % Romney (30 % purebred Romney), 10 % Coopworth (7 % purebred), 12 % Texel (1 % purebred), 9 % Perendale (6 % purebred), 11 % Primera (8 % purebred) and for the remaining animals, the breed was not identified (this set of individuals was not incorporated in any of our imputation scenarios). The distribution of the animals per breed/group is in Fig. 1a. This information was used to guide the choice of the most suitable imputation scenario since it is mainly influenced by factors such as number of breeds/groups available for investigation and number of individuals genotyped at each density. Animals that were genotyped with the HD panel but with an unknown breed composition were excluded from our investigation since they were not connected with the groups of animals analyzed, as determined by cluster analysis. The distribution of the genotyped animals for each panel density (50K or HD) according to birth year is in Fig. 1b.

Genotype conversion and quality control

Animals were genotyped with the Illumina OvineSNP50 and the Ovine Infinium® HD panels. Genotypes were coded as 0, 1, or 2 for AA, AB and BB genotypes, A and B being the two alleles of an SNP. Quality controls included removal of SNPs that (1) did not have defined positions on the ovine genome, (2) had a minor allele frequency (MAF) lower than 0.0005, (3) had a call rate lower than 95 % or (4) deviated from Hardy–Weinberg equilibrium (threshold p value: 1×10^{-5}). Finally, 48,241 and 568,569 autosomal SNPs (from the original 50K and HD panels, respectively) were retained for the analyses. In addition,

genotyped animals were excluded if their average genotype call rate was lower than 95 %.

Design of the low-density SNP panel

Two low-density SNP panels (5K and 50K) were simulated to test imputation by deleting part of the SNPs from the 50K and HD panels, i.e.:

1. only SNPs that were shared between the Illumina Ovine 5K SNP chip (http://www.illumina.com/documents/products/datasheets/) that is used commercially for genomic selection in New Zealand sheep [27] and the 50K original panel were retained, which resulted in 5095 SNPs (5K)
2. only SNPs that were shared between the Illumina OvineSNP50 and the Ovine Infinium® HD panels were retained, which resulted in 41,708 SNPs (50K).

Genomic relationships between animals from different breeds were determined by clustering

Relatedness is one of the key factors that affect the success of any imputation process. The genomic relationship matrix (**G** matrix) was calculated as follows and used for clustering analysis to verify the genetic connectivity (based on SNPs) among individuals from different breeds. In order to verify the connection of the genotyped animals among different breeds/groups and to better define the imputation scenarios, 100 animals from each breed or group were randomly selected to derive the **G** matrix and a cluster analysis was implemented by using the multidimensional scaling (MDS) approach, which is part of the package ggplot2 in R language. The **G** matrix was calculated as:

$$\mathbf{G} = \frac{\mathbf{X}\mathbf{X}'}{2 \sum p_i(1 - p_i)} \; [28],$$

where p_i is the allele frequency of the i-th SNP and $\mathbf{X}$ is the incidence matrix for SNPs.

Imputation scenarios

Thirty-one imputation scenarios were considered and animals in the reference population were selected based on the following criteria: density of the SNP panel (50K or HD), birth year (older animals), breed composition (multi- versus one-breed) and level of genomic relationship with imputed animals, as described in Tables 1, 2 and 3. For most of the 31 scenarios, the set of animals with imputed genotypes was composed of younger animals, which had their HD or 50K genotypes masked back to 50K or 5K genotypes, respectively.

The ten scenarios that are listed in Table 1 were designed to investigate different SNP densities and imputation of purebred Romney animals using alternate

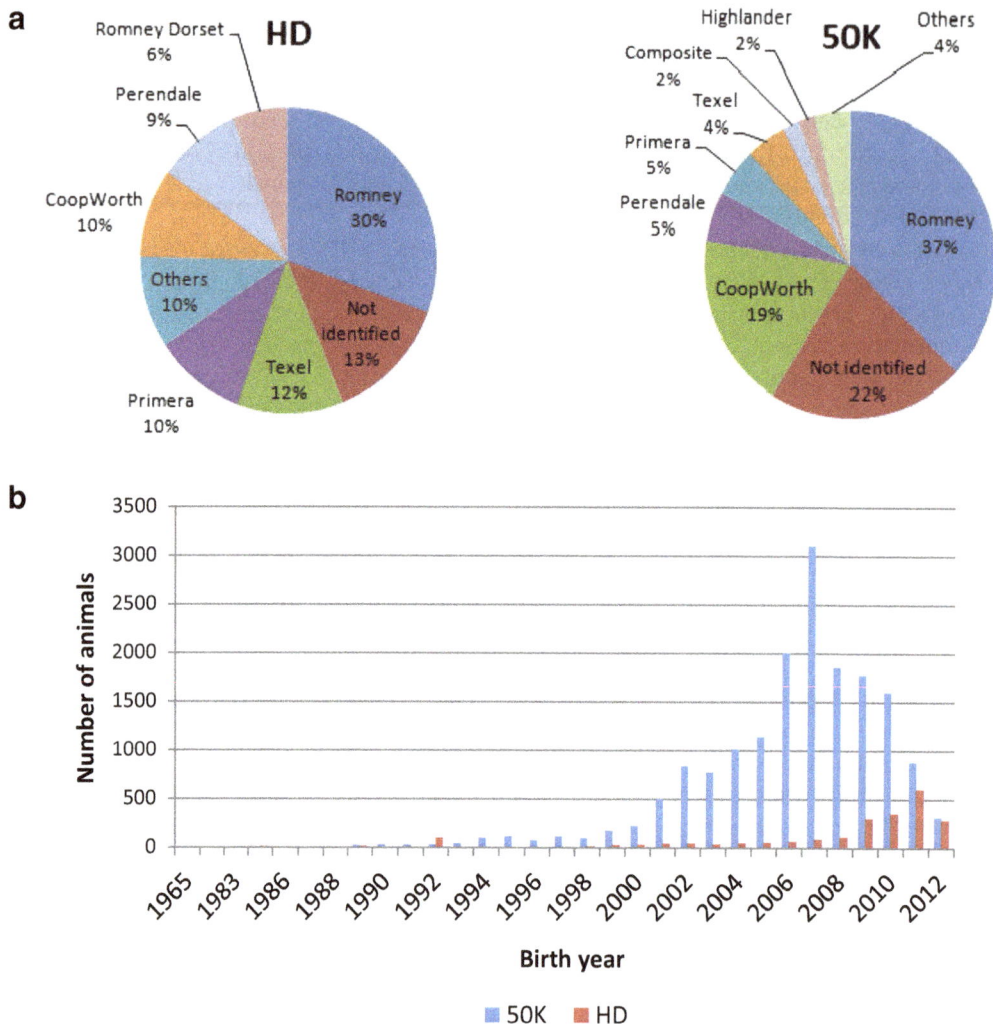

Fig. 1 Distribution of animals genotyped with 50K and HD. According to **a** main breed composition and **b** birth year

reference sets. These scenarios consisted in the imputation of the 116 youngest Romney animals using the oldest 500 Romney animals as reference population (except for the 2STEP Scenario, which included 17,000 animals that were genotyped with the 50K panel and constituted the reference set during the first step of imputation from 5K to 50K).

One-step versus two-step with a larger intermediate density reference set

In Scenario 1B_5KHD_2STEP, imputation from 5K to HD was done by using a two-step approach: from 5K to 50K and then from 50K to HD. This scenario allowed us to determine if a larger dataset that included animals genotyped with the 50K panel would improve haplotype reconstruction and hence imputation accuracy.

Relatedness, and impact of the size and breed composition of the reference population

In Scenarios 3, 3B and 4, 31 animals were excluded from the reference population because their relationship with at least one animal from the group of animals with imputed genotypes resulted in a relationship coefficient (based on the **G** matrix) that was higher than 0.45 (defined after parentage testing). In Scenarios 5, 5B and 6, randomly selected animals from another related breed (Perendale) were added to the reference population.

The scenarios that are listed in Table 2 evaluated the efficiency of imputation from 5K to 50K for Romney, composite, Primera terminal composite group (http://www.focusgenetics.com/sheep/sheep-breeding-programme/primera) and Coopworth animals (genotypes were obtained with the 50K Illumina panel and not with

Table 1 Imputation scenarios with HD genotypes using different groups of purebred and crossbred animals

Scenario[a]	Number of reference animals[b]	Number of imputed animals	Description of reference animals	Density of reference panel[c]	Imputed group breed[d]	Density of panel of imputed animals
1_5K50K	500	116	Romney	50K	Romney	5K
1B_5KHD_1STEP	500	116	Romney	HD	Romney	5K
1B_5KHD_2STEP	17,000 + 500	116	Romney	HD	Romney	5K
2_50KHD	500	116	Romney	HD	Romney	50K
3_5K50K	469	116	Romney-31 animals related with the imputed group	50K	Romney	5K
3B_5KHD	469	116	Romney-31 animals related with the imputed group	HD	Romney	5K
4_50KHD	469	116	Romney-31 animals related with the imputed group	HD	Romney	50K
5_5K50K	500 (R) + 100 (P)	116	Romney + Perendale	50K	Romney	5K
5B_5KHD	500 (R) + 100 (P)	116	Romney + Perendale	HD	Romney	5K
6_50KHD	500 (R) + 100 (P)	116	Romney + Perendale	HD	Romney	50K

[a] Imputation scenarios were from 5K to 50K (50K was a subset of the HD panel), 5K to HD and 50K to HD

[b] 2-Step imputation: from 5K to 50K using all genotyped animals as reference population (N = 17,000) and from 50K imputed to HD using 500 animals as the reference population

[c] The oldest animals in each scenario were used as reference population

[d] The youngest animals in each scenario were imputed

Table 2 Imputation scenarios with 50K genotypes using different groups of purebred and crossbred animals

Scenario[a]	Number of reference animals[b]	Number of imputed animals	Description of reference animals	Imputed group breed[c]
7_5K50K	466	500	Romney	Romney
8_5K50K	933	500	Romney	Romney
9_5K50K	1860	500	Romney	Romney
10_5K50K	2860	500	Romney	Romney
11_5K50K	4862	500	Romney	Romney
12_5K50K	933	200	Romney	Composite
13_5K50K	1000 (R) + 893 (C)	200	Romney + Coopworth	Composite
14_5K50K	1000 (R) + 893 (C) + 500 (P) + 500 (T)	200	Romney + Coopworth + Perendale + Texel	Composite
15_5K50K	710	500	Primera	Romney
16_5K50K	710 (P) + 933 (R)	500	Primera + Romney Scenario 8	Romney
17_5K50K	710 (P) + 1860 (R)	500	Primera + Romney Scenario 9	Romney
18_5K50K	350	200	Primera	Primera
19_5K50K	506	200	Primera	Primera
20_5K50K	350 (P) + 77 (S,PD)	200	Primera + Suffolk + Poll Dorset	Primera
21_5K50K	506 (P) + 77 (S,PD)	200	Primera + Suffolk + Poll Dorset	Primera
22_5K50K	470	300	Coopworth	Coopworth
23_5K50K	951	300	Coopworth	Coopworth
24_5K50K	951 (C) + 933 (R)	300	Coopworth + Romney	Coopworth

[a] Imputation scenarios were from 5K to 50K (original 50K panel)

[b] The oldest animals in each scenario were used as the reference population

[c] The youngest animals in each scenario were imputed

a subset from the HD panel). Combining 50K genotypes and subsets of genotypes obtained with the HD panel resulted in a larger number of animals available for the analyses.

For Scenarios 7_5K50K to 11_5K50K, within-breed imputation of 510 Romney animals was performed by enlarging the reference set (n = 466, 933, 1860, 2862 and 4862, respectively), i.e. by sorting the animals according to birth year and then by selecting them randomly within year groups.

For Scenarios 15_5K50K to 17_5K50K, the Primera set was first used as reference population (N = 710) to impute Romney animals (N = 500, Scenario 15_5K50K). Scenarios 16_5K50K and 17_5K50K were performed to check the effect of including Romney animals (same group of animals as in Scenarios 8_5K50K and 9_5K50K) to compose a multi-breed reference population. Scenarios 18_5K50K to 21_5K50K were used to evaluate the imputation of Primera animals (N = 200) by enlarging the reference population (N = 350 and 506) and combining animals from breeds that were used to create the Primera terminal composite group (Suffolk and Poll Dorset, N = 77). The last three scenarios in Table 2 (Scenarios 22_5K50K, 23_5K50K and 24_5K50K) aimed at investigating the imputation of Coopworth animals (N = 300) after doubling the size of the reference population (from 470 to 951, Scenarios 22_5K50K and 23_5K50K, respectively) and the impact of adding Romney animals in the reference population (Scenario 24_5K50K, N = 934).

Imputation of composite animals by expanding related breeds in the reference population

Scenarios 12_5K50K, 13_5K50K and 14_5K50K were used to evaluate imputation of composite animals by (1) using only Romney animals in the reference population (Scenario 12_5K50K), (2) adding Coopworth animals (Scenario 13_5K50K), and (3) including Perendale and Texel animals in the reference population (Scenario 14_5K50K). In New Zealand, much of the genetic background of commercial ewes used as dual-purpose sheep as studied here, originates from the Romney breed and both the Coopworth and Perendale breeds have a Romney origin. Texel is a breed that has recently been used

in composite dual-purpose meat sheep to increase lean yield [6, 27].

Within-group imputation or use of a fixed reference population that includes animals from all breeds with HD genotypes

Table 3 describes Scenarios 25_5K50K to 31_5K50K that aimed at assessing imputation accuracy of Romney (25_5K50K and 26_5K50K), Coopworth (28_5K50K and 29_5K50K), Perendale (30_5K50K) and composite (31_5K50K) animals; two different reference populations were used for each scenario: (1) a fixed reference population that included a large group of animals from all breeds (N = 15,443) and (2) a within-breed reference population. Romney and Coopworth imputed animals were also divided into two subgroups each, according to breed proportion: 100 % Romney or < 65 % (Scenarios 25_5K50K and 26_5K50K, respectively) and 100 % or < 70 % Coopworth (Scenarios 28_5K50K and 29_5K50K, respectively.

Imputation of rare alleles and accuracy of imputation for SNPs located at the ends of chromosomes

Scenario 27_5K50K was specifically designed to investigate within-breed imputation of Romney animals for rare alleles and to verify regions with reduced imputation accuracy using the squared Pearson correlation coefficient as a measure of accuracy. This scenario had the largest number of imputed animals and was deemed best to test imputation accuracy of rare variants.

Prediction of imputation accuracy before imputing missing genotypes

Based on SNP data, the relatedness among animals from the imputed and reference populations was investigated for each scenario, as the genomic relationship

Table 3 Imputation scenarios from 5K to 50K (50K original) using two types of reference population

Scenario	Number of reference animals	Number of imputed animals	Description of reference animals[d]	Imputed group breed[b]
25_5K50K	15,443[a] and 4564[b]	218	All breeds/Romney	Romney 100 %
26_5K50K	15,443[a] and 4326[b]	142	All breeds/Romney	Romney < 65 %
27_5K50K	4256	1000[c]	Romney	Romney
28_5K50K	15,443[a] and 2324[b]	250	All breeds/Coopworth	Coopworth 100 %
29_5K50K	15,443[a] and 2279[b]	250	All breeds/Coopworth	Coopworth < 70 %
30_5K50K	15,443[a] and 640[b]	250	All breeds/Perendale	Perendale > 95 %
31_5K50K	15,443[a] and 138[b]	172	All breeds/Composites	Composites > 50 % < 95 %

[a] Fixed reference population that included 15,443 animals from all breeds with genotyped animals

[b] Within-breed/group reference population: some groups contained a small number of genotyped animals

[c] 1000 animals defined as the imputed set to optimize the calculation of the r^2 imputation accuracy per SNP

[d] Two types of reference population were used: (1) a fixed reference population that included a large number of animals from all breeds and (2) a within-group reference population

average value (extracted from the **G** matrix) between each imputed animal and the 10 most related individuals from the reference population. The minimum and maximum top 10 relationships (upper and lower value for each group of the 10 most related animals) for each scenario were also calculated to compare the estimated accuracies of imputation. Another measure of relatedness was also investigated to predict imputation accuracy before running the imputation process: Mendelian inconsistency (MI), which is the average number of Mendelian inconsistencies between an imputed animal and the top 10 related individuals from the reference group, where MI reflects the number of opposing homozygotes between two individuals. Two individuals that have high MI values after genotype comparison are likely to share fewer haplotypes than individuals that have a low MI value.

Comparison of imputation software packages

We compared two software packages: BEAGLE and FIMPUTE. We do acknowledge that changes to BEAGLE software are now available (Version 4) and that this new version should be evaluated in future studies, along with any other available updates of these software packages, to determine if there are advantages for the New Zealand sheep industry. BEAGLE exploits linkage disequilibrium between SNPs and implements a population imputation method that assumes that all animals are unrelated. This software uses a hidden Markov model (HMM) and a localized haplotype clustering method to infer genotypes as described by Browning et al. [26]. All analyses using BEAGLE were carried out by setting default parameters. The FIMPUTE software uses a deterministic approach that combines family and population imputation methods. The population imputation method is based on the assumption that all individuals have some degree of relationship and share haplotypes that may differ in frequency and length depending on the relationships. FIMPUTE is a two-step procedure, i.e. first it searches for long haplotypes by applying a family imputation method, and second, it identifies short segments (two SNPs) by applying a population imputation method that analyzes overlapping sliding windows. BEAGLE analyses that were not complete within 1 week of computing time or failed at least twice during the process (the cause of failure could not be determined) were excluded and are not presented in this paper (13 occurrences).

Determination of imputation accuracy

Imputation accuracy (per individual and per SNP) was determined with two different measurements: (1) allelic squared Pearson correlation coefficient (r^2) as an appropriate approach to minimize the dependency on allele frequency and (2) concordance rate: proportion of correctly called SNP genotypes versus all called SNPs. Both values were determined by comparing imputed and true genotypes. Since imputation accuracy of specific SNPs was useful for Scenario 27_5K50K, which investigated imputation of rare variants, r^2 per SNP was calculated.

Run-time comparison (overall computing time)

FIMPUTE analyzes a set of chromosomes simultaneously by implementing parallel computing. For each software package, the total length of running time (overall computing time) was measured for all scenarios but comparison of values between BEAGLE 3.3.2 and FIMPUTE was not possible. Due to the long computation time required with BEAGLE, these analyses were carried out using the Condor server located at the University of Wisconsin (Linux server (fedora core 16) with dual Intel Xeon X5690@3.47 GHz CPUs). FIMPUTE analyses were performed on a local server, located at the Invermay Agricultural Centre, Agresearch (Linux server (CentOS 6.5) with 48 AMD Opteron 6176SE @2.3 GHz CPUs). Ten parallel jobs were implemented for BEAGLE and FIMPUTE, for comparison among scenarios (within software).

Results

In this paper, tables are used to report the concordance rate (CR) and r^2 measures of imputation accuracy, and figures show the variation in imputation efficiency for all animals genotyped with the low-density panel in each scenario. All figures provide imputation accuracy per animal in terms of CR. Tables 4, 5, 6, and 7 and Figs. 2, 3, 4, 5, 6 and 7 report the results for Scenarios 1–31 that are defined in Tables 1, 2 and 3.

First, we assessed imputation accuracy using two population imputation methods (BEAGLE and FIMPUTE) applied to HD genotypes of purebred Romney animals.

One- versus two-step imputation

The two-step imputation scenario (Scenario 1B_5KHD_2STEP) that imputed animals first from 5K to 50K and then, from 50K imputed to HD, was compared to the one-step imputation scenario from 5K to HD (Scenario 1B_5KHD_1STEP), which showed that the two-step procedure increased imputation accuracy by 5.67 % (CR) and 8.87 % (r^2). Based on Fig. 2d, animals for which imputation accuracy (CR) was lower than 95.1 % using the one-step approach, inference of missing genotypes was more efficient with the two-step procedure.

Imputation from a medium-density panel (50K) to HD (Scenarios 2_50KHD, 4_50KHD and 6_50KHD) resulted in the highest imputation accuracies i.e. higher than 97.25 % (CR) (see Table 4).

Table 4 Accuracy of genotype imputation and computing time for BEAGLE and FIMPUTE algorithms

Scenario	CR_F[a]	r²_F[b]	Run Time_F m:s	CR_B[c]	r²_B[d]	Run Time_ B h:m:s	Mean Top10[e]	Min Top10[e]	Max Top10[e]
1_5K50K	86.98	78.75	00:57	83.80	73.80	02:16:25	0.115	0.034	0.234
1B_5KHD_1STEP	87.61	80.73	06:51	84.10	74.00	23:12:35	0.115	0.034	0.234
1B_5KHD_2STEP	93.28	89.6	–	NA	NA	NA	0.115	0.034	0.234
2_50KHD	97.56	96.2	07:42	96.98	95.42	21:55:35	0.115	0.034	0.234
3_5K50K	84.35	74.15	00:53	82.12	70.94	03:15:10	0.090	0.033	0.179
3B_5KHD	85.3	76.85	06:43	82.23	71.12	27:17:35	0.090	0.033	0.179
4_50KHD	97.25	95.71	07:11	96.63	94.91	12:33:02	0.090	0.033	0.179
5_5K50K	87.19	78.98	01:08	83.58	73.37	03:18:52	0.097	0.037	0.252
5B_5KHD	87.68	80.81	08:45	83.99	76.00	25:16:22	0.097	0.037	0.252
6_50KHD	98.06	97.01	09:14	Failed	Failed	Failed	0.097	0.037	0.252

[a] CR_F = concordance rate using the FIMPUTE software

[b] r²_F = Squared Pearson correlation using the FIMPUTE software

[c] CR_B = concordance rate using the BEAGLE software

[d] r²_B = Squared Pearson correlation using the BEAGLE software

[e] Mean Top10, Min Top10 and Max Top10 = mean, min and max relationship among the 10 most related animals between the reference and imputed sets

Table 5 Accuracy of genotype imputation from 5K to 50K and computing time when using the FIMPUTE software

Scenario	CR_F[a]	r²_F[b]	Run time_F m:s	Mean Top10[c]	Min Top10[c]	Max Top10[c]
7_5K50K	74.82	57.79	01:15	0.058	0.011	0.178
8_5K50K	77.10	61.64	02:14	0.076	0.036	0.210
9_5K50K	84.42	74.05	03:33	0.135	0.054	0.310
10_5K50K	87.55	79.29	05:47	0.152	0.052	0.394
11_5K50K	91.06	85.38	09:08	0.177	0.054	0.398
12_5K50K	60.93	35.25	02:04	0.085	0.055	0.168
13_5K50K	66.69	44.25	03:32	0.095	0.056	0.338
14_5K50K	72.12	52.44	05:47	0.123	0.056	0.349
15_5K50K	51.82	17.89	01:28	0.004	0.003	0.006
16_5K50K	75.18	58.25	03:07	0.117	0.052	0.259
17_5K50K	84.07	73.41	05:04	0.153	0.058	0.335
18_5K50K	92.21	86.78	00:44	0.140	0.091	0.183
19_5K50K	95.10	91.90	01:01	0.042	0.001	0.270
20_5K50K	92.8	87.77	00:55	0.045	0.001	0.187
21_5K50K	95.32	92.26	01:11	0.066	0.002	0.270
22_5K50K	77.53	62.36	01:07	0.070	0.022	0.211
23_5K50K	88.46	80.92	02:05	0.167	0.023	0.370
24_5K50K	87.99	80.14	03:34	0.204	0.055	0.417

[a] CR_F = concordance rate when using the FIMPUTE software

[b] r²_F = Squared Pearson correlation when using the FIMPUTE software

[c] Mean Top10, Min Top10 and Max Top10 = mean, min and max relationship among the 10 most related animals between the reference and imputed sets

Imputation from 5K to both 50K and HD panels using one or more breeds in the reference population and impact of relatedness on imputation accuracy

Table 4 shows the accuracy of genotype imputation from 5K to 50K and HD, and from 5K to 50K. All SNP panels represented a subset of the HD panel. The highest CR (87.19 %) and r² (78.98 %) values (Table 4) for imputation from 5K to 50K were obtained when Romney and Perendale animals were combined in the reference population (Scenario 5_5K50K; Table 1). The difference in

Table 6 Accuracy of genotype imputation with the FIMPUTE software using two types of reference population

Scenario[a]	CRAll[c]	r²All[c]	CRW[d]	r²W[d]	MeanA[e]	MinA[e]	MaxA[e]	MeanW[e]	MinW[e]	MaxW[e]
25_5K50K	93.39	89.38	89.16	82.17	0.023	0.079	0.467	0.145	0.049	0.376
26_5K50K	95.45	92.10	82.05	70.47	0.267	0.096	0.432	0.185	0.077	0.355
27_5K50K[b]	89.07	82.06	–	–	0.180	0.055	0.401	–	–	–
28_5K50K	89.94	84.01	89.80	83.27	0.250	0.100	0.427	0.200	0.050	0.384
29_5K50K	96.24	93.12	87.55	79.76	0.283	0.085	0.426	0.201	0.075	0.387
30_5K50K	87.89	81.23	88.32	80.55	0.215	0.100	0.535	0.162	0.061	0.310
31_5K50K	90.16	82.17	65.05	41.57	0.243	0.109	0.413	0.03	0.001	0.260

[a] Genotype imputation was from 5K to 50K using two types of reference population: (i) fixed reference population containing a large number of animals from all breeds and (ii) within-group reference population

[b] Scenario defined for the calculation of SNP r² using 1000 animals as imputed

[c] CRAll and r²All = concordance rate and squared Pearson correlation, respectively, using the FIMPUTE software when a large set of animals from all breeds was defined as the reference population

[d] CRW and r²W = concordance rate and squared Pearson correlation, respectively, using the FIMPUTE software when the within-group population was defined as the reference population

[e] MeanA, MinA, MaxA, MeanW, MinW and MaxW = mean, min and max relationship among the 10 most related animals between the reference and imputed sets (all animals (A) or within-group (W))

Table 7 Rare allele imputation accuracy (r²) for different ranges of MAF

MAF	Number of SNPs	r²[a]
0 < MAF = 0.0001	35	0
0.0001 < MAF = 0.001	96	6.6
0.001 < MAF = 0.01	625	38.9
0.01 < MAF = 0.05	2360	57.8

[a] Allelic imputation accuracy (r²) for Scenario 27_5K50K where 1000 Romney animals were imputed using a within-breed reference set that included 4256 animals

overall average imputation accuracy between Scenarios 1_5K50K and 5_5K50K was very small (0.21 and 0.23 % for CR and r², respectively). Figure 2b shows that a small improvement in imputation accuracy for imputation to

50K and HD was observed for some animals for which CR accuracy was lower than 70 % (imputation to 50K) in Scenario 1_5K50K and imputation of genotypes was improved by adding Perendale animals in the training dataset.

On average, CR accuracy and r² decreased by 2.63 and 4.60 %, respectively, when 31 Romney animals, which were highly related with the animals that had imputed genotypes, were removed from the training dataset (comparison of Scenarios 1_5K50K and 3_5K50K). As also shown by Fig. 2a, the removal of these 31 animals caused a decrease in imputation accuracy for imputation from low-density to 50K for several animals in all ranges of accuracy, except for the seven animals that showed the lowest imputation efficiencies (CR < 70 %). For this set of animals, average MI values were higher than 3000, which

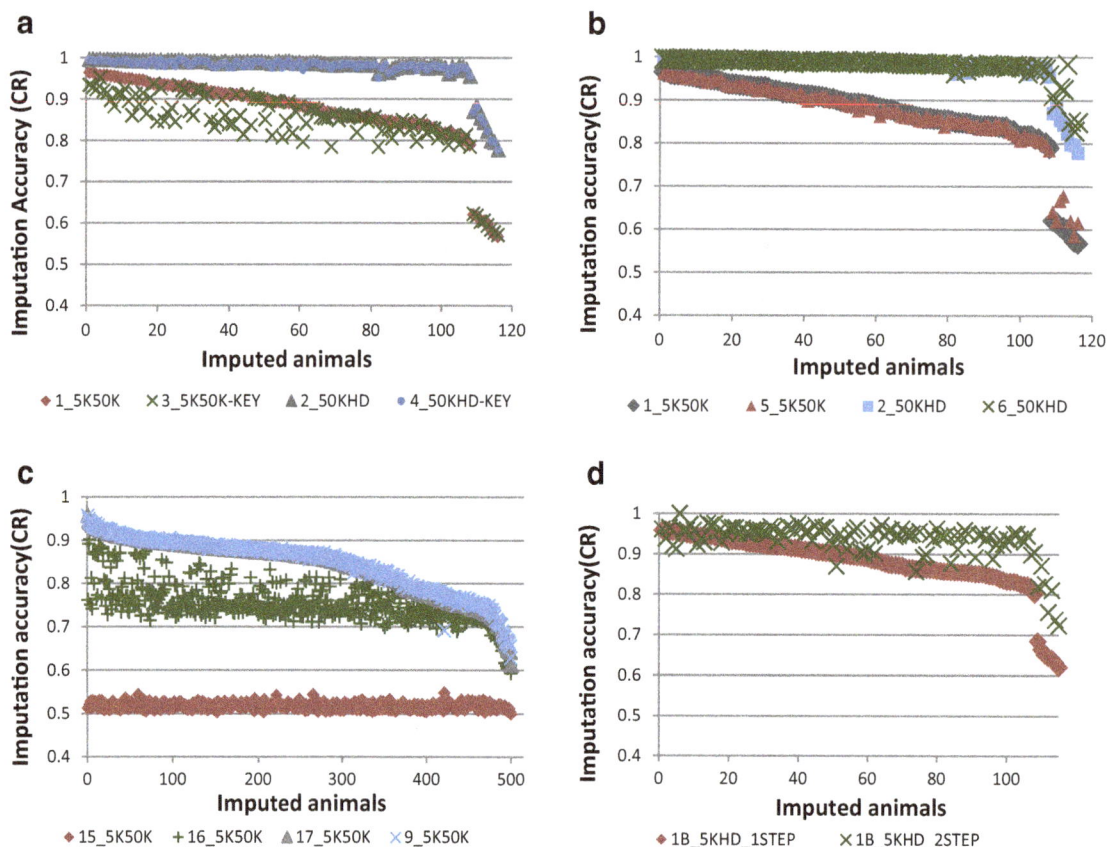

Fig. 2 Imputation accuracy assessed by alternative approaches. **a** Imputation from 5K to 50K (Scenarios 1_5K50K and 3_5K50K) and from 50K to HD (Scenarios 2_50KHD and 4_50KHD) using the FIMPUTE software for purebred Romney animals (The suffix "KEY" refers to the 31 animals that are highly related with the group of imputed animals). **b** Imputation from 5K to 50K (Scenarios 1_5K50K and 5_5K50K) and from 50K to HD (Scenarios 2_50KHD and 6_50KHD) using the FIMPUTE software for purebred Romney animals after including Perendale animals in the reference set. **c** Imputation from 5K to 50K using the FIMPUTE software for purebred Romney animals after using the Primera group as reference set (Scenario 15_5K50K) and inclusion of Romney animals in the reference set (Scenarios 16_5K50K and 17_5K50K). Scenario 9 was included in this plot for comparison with within-breed imputation. **d** Imputation from 5 K to HD using the FIMPUTE software for purebred Romney animals by one- or two-step imputation (Scenarios 1B_5KHD_1STEP and 1B_5KHD_2STEP). The x-axis represents the number of imputed individuals sorted from the highest to the lowest accuracy value

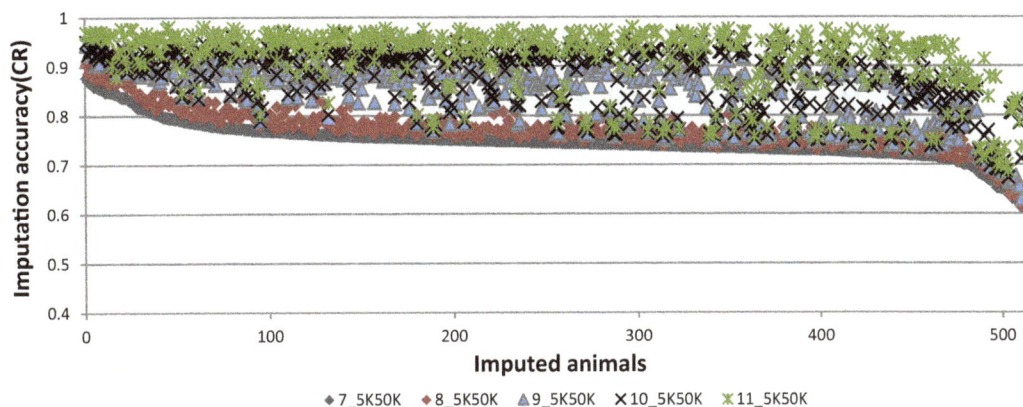

Fig. 3 Imputation from 5K to 50K using the FIMPUTE software for purebred Romney animals. Scenarios 7_5K50K to 11_5K50K. The x-axis represents the number of imputed individuals sorted from the highest to the lowest accuracy value

indicates a low level of relationship compared to the animals in the reference population. Imputation from 50K to HD (Scenario 4_50KHD) was not affected by removing the key related animals from the reference population. Imputation from 5K to HD (3B_5KHD) was also performed in this study and imputation accuracies (CR using FIMPUTE) were on average slightly higher (0.95 %) than for imputation from 5K to 50K (3_5K50K). Imputation accuracy from 5K to HD ranged from 82.23 to 87.68 % (CR) and from 71.12 to 80.81 % (r^2) for Scenarios 1B_5KHD, 3B_5KHD and 5B_5KHD (see Table 4).

Size of the reference set

The first five scenarios (Scenarios 7_5K50K to 11_5K50K) were used to evaluate the within-breed accuracy of imputation for 510 Romney animals by enlarging the reference population from 466 to almost 5000 animals. CR (and r^2) accuracies ranged from 74.82 % (57.79 %) for Scenario 7_5K50K to 91.06 % (85.38 %) for Scenario 11_5K50K, respectively. The highest accuracy was reached when 4862 animals (the largest set of Romney animals) were included in the reference population (Scenario 11_5K50K). Figure 3 shows imputation accuracy (CR) per animal for the same set of results presented above. A large average gain in accuracy (16.24 %) was obtained by increasing the reference population by tenfold.

Imputation of composite animals, multi- versus one-breed reference population and use of a single reference population for all imputed animals

The overall average imputation accuracy of composite animals using different reference populations that consisted of Romney animals and additional individuals from other groups (Coopworth, Perendale and Texel) ranged from 60.93 to 72.12 % (CR) and from 35.25 to 52.44 % (r^2) (Scenarios 12_5K50K to 14_5K50K). As shown in Fig. 4c, gains in imputation accuracy per animal were obtained by adding animals from different breeds to the reference population.

Accuracies of imputation of Romney animals using a reference population that comprised animals from another breed (Primera) were close to those of imputation by chance (i.e. replacing a missing genotype by the allele of higher frequency), also defined as random imputation (CR = 51.82 % and r^2 = 17.89 % (Scenarios 15_5K50K to 17_5K50K). Addition of Romney animals to the reference population (Scenarios 16_5K50K and 17_5K50K) increased imputation accuracy to values that were similar to those obtained for within-breed imputation (Scenario 9_5K50K; Fig. 2c). Overall, average gains in accuracy of 2.89 % in CR and 5.12 % in r^2 were observed by enlarging the reference Primera population (Scenarios 18_5K50K and 19_5K50K) with animals related to those that were at the origin of this group (Suffolk and Poll Dorset) (0.22 % in CR and 0.36 % in r^2). Only 6 % of the animals from the imputed set showed little overall gain in accuracy (2.3 %) by including animals from the two additional breeds (Fig. 4a). A slight decrease (0.47 % in CR and 0.78 % in r^2) in imputation accuracy was observed when Romney animals were included in the scenario for which Coopworth individuals were used in both the reference and imputed sets (Scenarios 23_5K50K to 24_5K50K). A near two-fold reduction in reference population size decreased imputation accuracy more than the addition of a second breed in the reference population, which resulted in a very slight decrease in accuracy (Fig. 4b). With FIMPUTE software and across all scenarios (see Table 5), the shortest

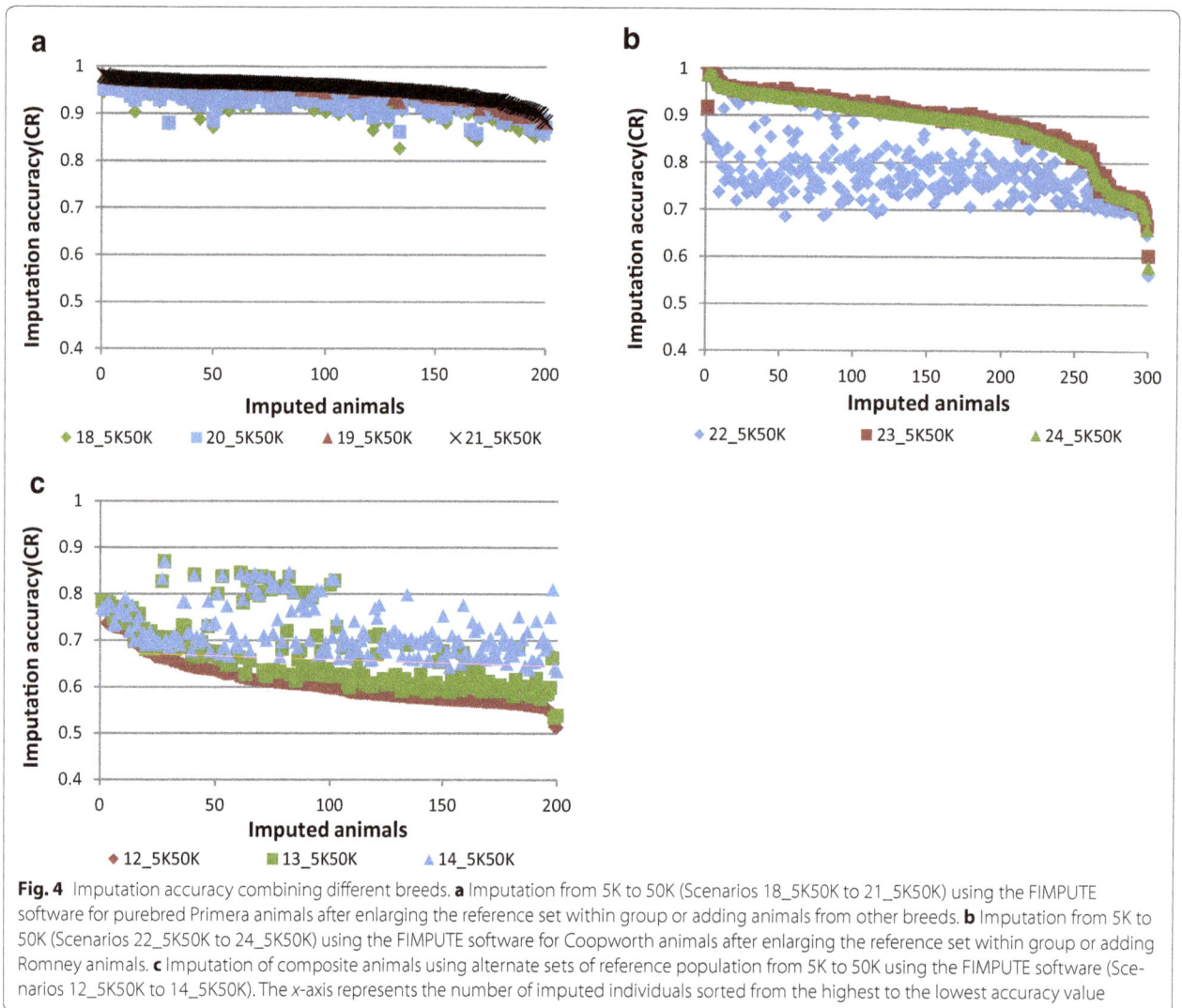

Fig. 4 Imputation accuracy combining different breeds. **a** Imputation from 5K to 50K (Scenarios 18_5K50K to 21_5K50K) using the FIMPUTE software for purebred Primera animals after enlarging the reference set within group or adding animals from other breeds. **b** Imputation from 5K to 50K (Scenarios 22_5K50K to 24_5K50K) using the FIMPUTE software for Coopworth animals after enlarging the reference set within group or adding Romney animals. **c** Imputation of composite animals using alternate sets of reference population from 5K to 50K using the FIMPUTE software (Scenarios 12_5K50K to 14_5K50K). The x-axis represents the number of imputed individuals sorted from the highest to the lowest accuracy value

and longest computing times were observed for Scenario 18_5K50K (44 s) and Scenario 9_5K50K (5 min and 47 s), respectively.

Overall, average gains in accuracy of 8.52 % in CR and 14.03 % in r^2 were obtained for all scenarios that compared a within-group reference population versus a fixed and large reference population that comprised animals from all groups (Table 6, Scenarios 25_5K50K to 31_5K50K).

The highest gain (25.11 % in CR and 40.6 % in r^2) was obtained for Scenario 31_5K50K for which the reference population of 138 animals (within-group reference set) that was used to impute composite animals was replaced by a larger set consisting of 15,443 animals (see Table 6, fixed reference population for all scenarios). Figure 5 shows imputation accuracies per animal for imputation from 5K to 50K with a reference population composed of animals from the same group as those to be imputed

(within-breed imputation): they are sorted from the highest to the lowest CR accuracy.

Imputation of Romney animals with different breed proportions (<100 % and <65 %), Coopworth (<70 %), and of composite animals, benefited from using a unique large reference population that included animals from all breeds/groups. Imputation of animals 100 % Coopworth and Perendale did not benefit substantially by including animals from all breeds/groups in the reference population compared to a within-breed reference population, with only a slight change in imputation accuracy observed for a few animals (see Fig. 5, Scenarios 28 and 30).

Comparison of BEAGLE and FIMPUTE

Accuracies of imputation and corresponding computing times for FIMPUTE and BEAGLE are provided in Table 4. FIMPUTE outperformed BEAGLE across all

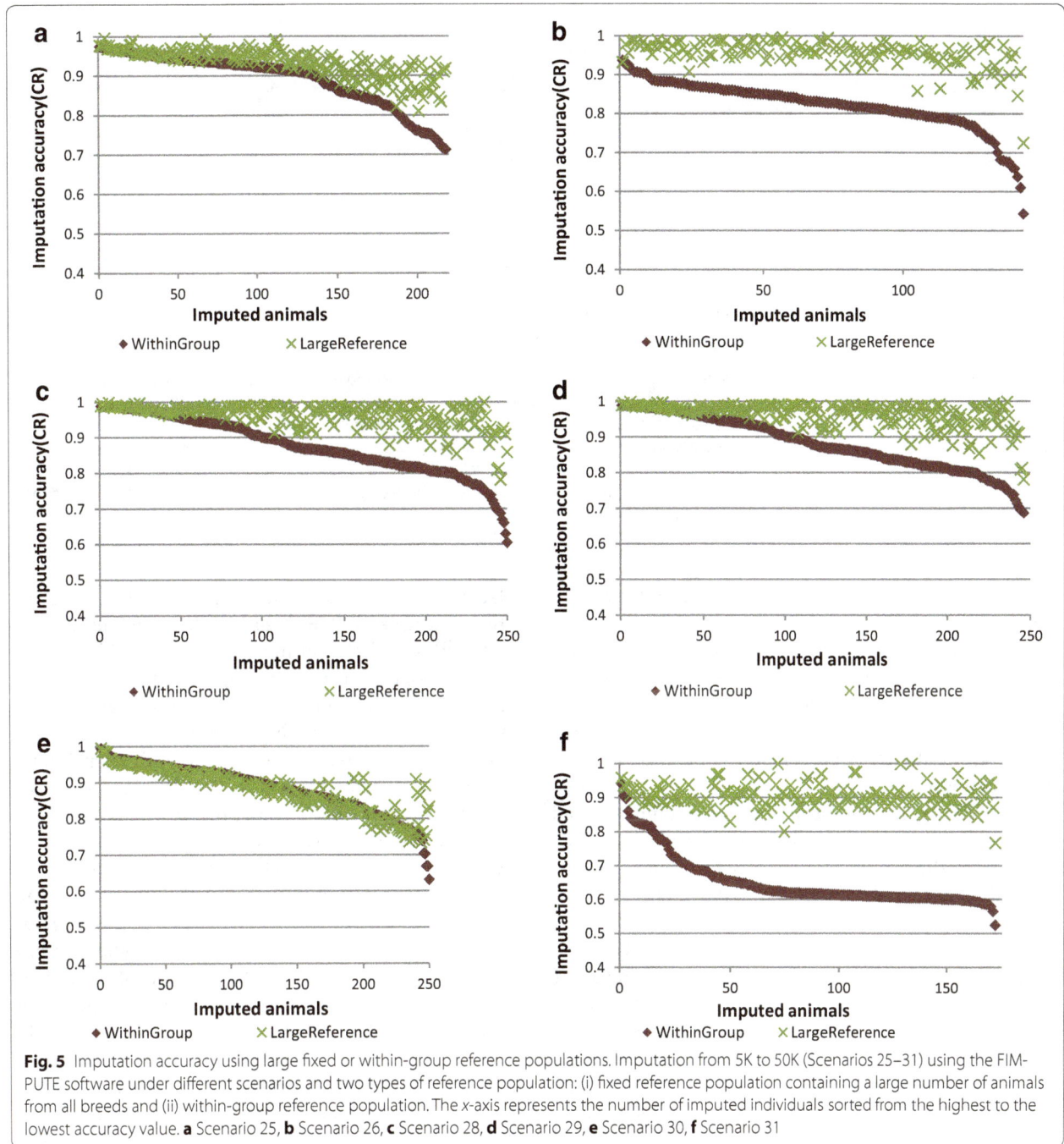

Fig. 5 Imputation accuracy using large fixed or within-group reference populations. Imputation from 5K to 50K (Scenarios 25–31) using the FIM-PUTE software under different scenarios and two types of reference population: (i) fixed reference population containing a large number of animals from all breeds and (ii) within-group reference population. The x-axis represents the number of imputed individuals sorted from the highest to the lowest accuracy value. **a** Scenario 25, **b** Scenario 26, **c** Scenario 28, **d** Scenario 29, **e** Scenario 30, **f** Scenario 31

scenarios. Overall, average decreases in accuracy of 3.06 % (CR) and 4.59 % (r^2) for imputation from 5K to 50K and of 3.42 % (CR) and (r^2) for imputation from 5K to HD were found with BEAGLE compared to FIMPUTE. Computation time was shortest in Scenario 1_5K50K for both software packages: 57 s with FIMPUTE and over 2 h with BEAGLE. Twenty GB of RAM (random-access memory) were allocated for both algorithms. For some

analyses that failed using BEAGLE, the RAM threshold had to be increased to 100 GB for the computation of scenarios that investigated imputation to HD genotypes. Scenarios that used BEAGLE and were not completed within 5 days or failed twice are not presented in this paper and the cause of these failures was not determined. Imputation with BEAGLE in all Scenarios from 7_5K50K to 31_5K50K (results are only presented for FIMPUTE

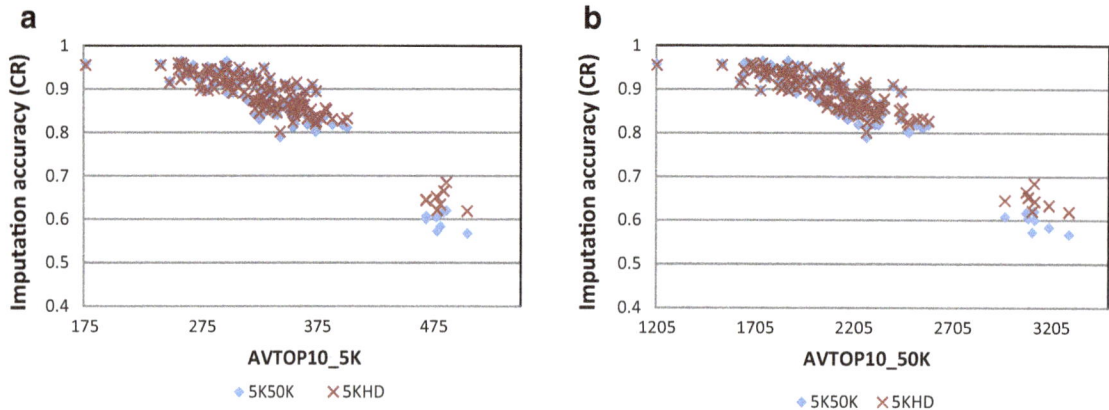

Fig. 6 Imputation accuracy and its relation with the connectivity between each imputed animal and the reference set. Average numbers of Mendelian inconsistencies (AVTOP10_5K and AVTOP10_50K) between each animal in the imputed set and all animals from the reference set were calculated and are presented for each imputed animal as the average of 10 pairs of animals (one from the reference set and one from the imputed set) with the lowest Mendelian inconsistency. Imputation from 5K to 50K (Scenario 1_5K50K) and 5K to HD (Scenario 1B_5KHD_1STEP) using the FIMPUTE software for purebred Romney animals is also compared with the value defined above. **a** AVTOP10_5K calculated using the 5K panel before imputation. **b** AVTOP10_50K calculated using the 50K panel. The x-axis represents the average number of Mendelian inconsistencies and the y-axis the imputation accuracy per animal measured by concordance rate (CR)

in Tables 5, 6 and 7) was not feasible and is not reported here.

Table 5 shows the accuracy of genotype imputation from 5K to 50K that was reached with FIMPUTE for Scenarios 7_5K50K to 24_5K50K.

Predicting imputation accuracy before imputation and relatedness

Figure 6 shows imputation accuracy per animal across two scenarios measured by concordance rate (CR) according to the average number of Mendelian inconsistencies (MI) observed with 5K (Fig. 6a) and 50K (Fig. 6b) panels: a similar trend is observed in both plots.

The highest imputation accuracy (98.7 % in CR) was obtained for an individual for which the average MI between itself and the top 10 most related animals in the reference population was equal to 176.9 when the 5K panel was used and 1208.7 when the 50K panel was used. The lowest imputation accuracy was found for an animal for which MI was equal to 504.2 and 3297.9 when the 5K and 50K panels were used, respectively.

Tables 4, 5 and 6 also show the top 10 relationships between animals from the imputed and reference populations. The mean, minimum and maximum average top 10 values across all scenarios were equal to 0.129, 0.041 and 0.296, respectively. Scenario 15_5K50K (imputation of Romney animals using the Primera group as the reference population) resulted in the lowest values of relatedness [0.004 (mean), 0.003 (min) and 0.006 (max)]. Imputation of Coopworth animals using all the other animals as the reference population resulted in the highest average

relatedness value (0.283) and in one of the highest imputation accuracies (CR = 96.24 %). After carefully examining the classes of relationship among the individuals in the reference population and imputed set (results not shown), we found that, in most cases, the most highly related animal was a half-sib, with genetic relatedness dropping quickly, where the relationship for the 10th animal in the top 10 most related set was close to 0.03 (Min Top10 stats in Tables 4 and 5). This indicates that in the scenarios that were designed for this study, the number of highly related animals (for example, family members that are shared between imputed and reference sets) was quite small. This is confirmed by the comparison of Scenario 1_5K50K (Table 5) with Scenario 3_5K50K, for which the reference population was enlarged by the addition of 31 animals that were highly related with animals in the imputed set; in this case the Max Top10 statistics did not exceed 0.234.

Imputation of chromosome tails and rare alleles

Figure 7a shows imputation accuracies (r^2) per SNP for the 26 autosomal sheep chromosomes for the animals described in Scenario 27_5K50K, in which 1000 animals were used as the imputed set. In general, imputation accuracies for the SNPs that were located at each end of each chromosome were lower than for those in other chromosomal regions. Figure 7b shows that 14 out of the 26 autosomes had at least one of their extremities covered by 100 SNPs with an average imputation accuracy lower than 40 % (r^2).

Chromosome 4 shows the best marker coverage at the proximal end (average $r^2 = 80.40$ %) whereas the lowest imputation accuracy was found for the 100 SNPs

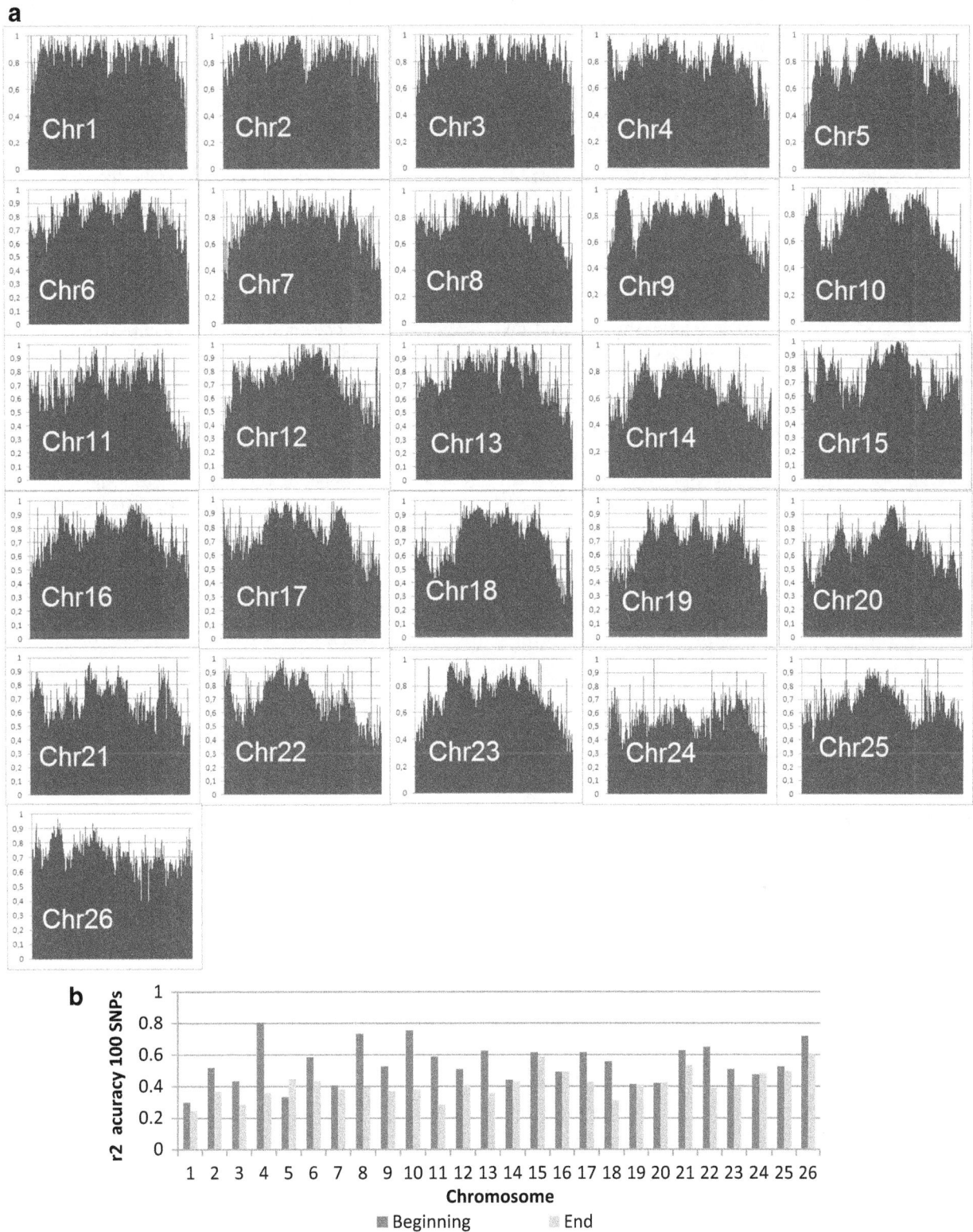

Fig. 7 Imputation accuracy per chromosome and at both chromosome ends. **a** Squared Pearson correlation measure of imputation accuracy (r^2) across different chromosomes after imputation from 5K to 50K for Romney sheep using the FIMPUTE software (Scenario 27_5K50K). **b** Squared Pearson correlation measure of imputation accuracy (r^2) for both ends of each chromosome (each chromosome end is covered by 100 markers); imputation accuracy defined as the average r^2 value for the 100 markers

located at the telomeric end of chromosome 1 (average $r^2 = 29.89$ %).

Imputation accuracies of rare alleles as measured by r^2 and grouped into four categories according to the MAF of each imputed SNP allele $(0 < MAF < 0.05)$ are in Table 7. Thirty-five SNPs were reported in the first category $(0 < MAF < 0.0001)$ and their r^2 was equal to 0. The overall average imputation accuracies (r^2) for the MAF groups $(0.0001 < MAF < 0.001; 0.001 < MAF < 0.01;$ and $0.01 < MAF = 0.05)$ were equal to 6.6, 38.9 and 57.8 %, respectively.

Genetic relationships among breeds based on MDS cluster

Figure 8 illustrates the genetic relationships (based on genomic distances estimated from SNPs) between animals of each group or breed. Primera and Texel groups showed reduced connectivity with other breeds (Romney, Coopworth, composites and Perendale). This plot was used to determine the most relevant imputation scenarios and for the description of population structure.

Discussion

We used a 50K SNP subset that was extracted from the HD panel to compare the imputation accuracy from 5K to 50K, 5K to HD using a one- or two-step procedure, and from 50K to HD. Animals genotyped with the HD panel were not re-genotyped with the 50K panel, but the 50K panel was derived as a subset of the HD genotypes. The large number of animals (17,176) that were genotyped with the Illumina OvineSNP50 (50K) panel allowed us to investigate the use of alternate reference populations, i.e. that comprised samples of animals of various sizes and breed composition, the impact of removing animals that were closely related to the reference population and also to identify the chromosomal regions that are not imputed efficiently in Romney animals, for which a large imputed set (N = 1000) was used to reduce the bias in r^2 imputation results.

Impact of reference population on the imputation of purebred and crossbred animals

Imputation accuracies that were obtained in our study were on average higher than those reported by Hayes et al. [14] for Australian sheep. These authors investigated different breeds and smaller populations. Imputation accuracy depends on several factors, including the number of immediate ancestors in the reference population, size of reference population and density of the SNP panel used for both imputed and reference sets [13]. Scenarios 7 to 11 in our study resulted in a substantial gain in accuracy by enlarging the reference population used for the within-group imputation of Romney animals

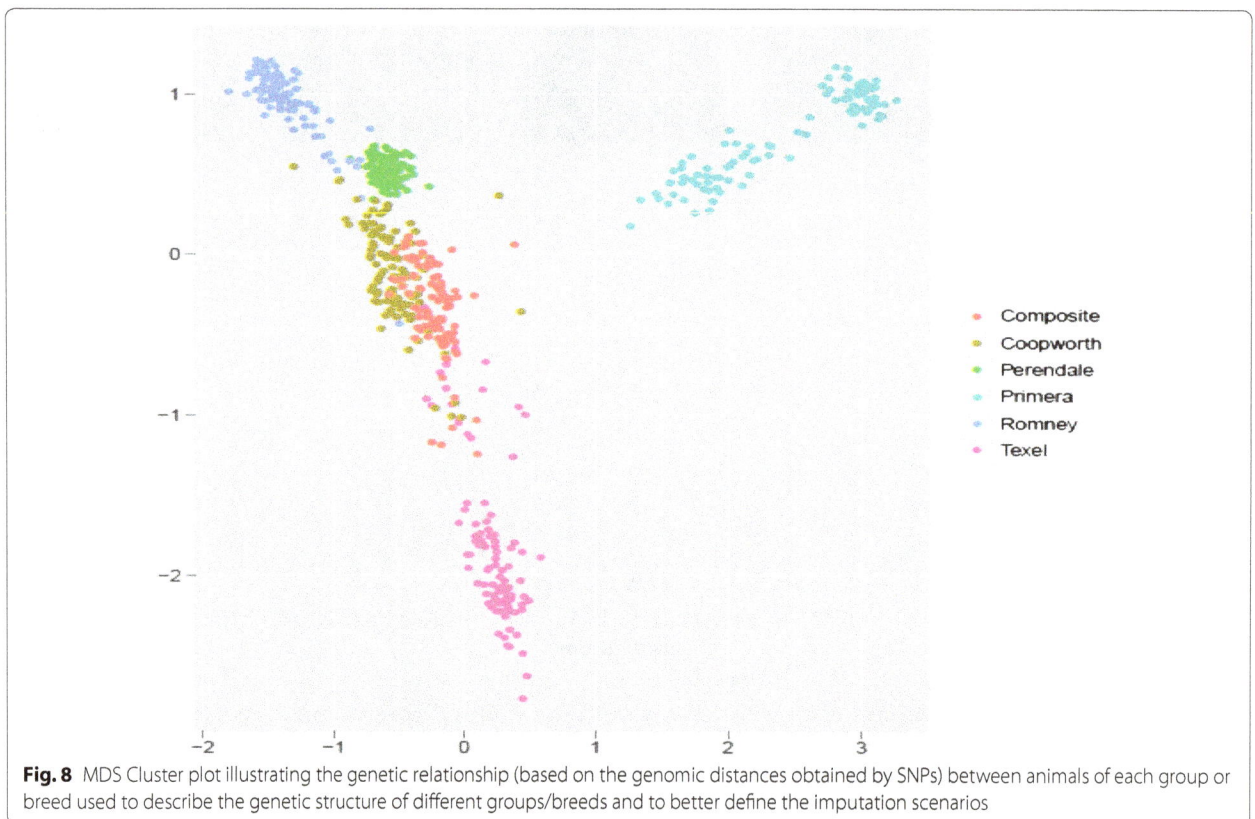

Fig. 8 MDS Cluster plot illustrating the genetic relationship (based on the genomic distances obtained by SNPs) between animals of each group or breed used to describe the genetic structure of different groups/breeds and to better define the imputation scenarios

which agrees with the findings of [13]. Ventura et al. [1] investigated the accuracy of imputation from 5K to 50K in a multi-breed beef cattle population and reported higher CR accuracies when closely-related individuals to the imputed group along with a representation of the breed composition of the imputed group were included in the reference population. These authors also showed that adding another purebred population in the reference population did not improve the within-breed imputation for imputation from low- to medium-density panels. Sargolzaei et al. [9] reported that imputation from denser panels (i.e. from 50K to HD) depended less on the size of the reference population than that from sparser panels (i.e. from 5K to 50K). The existence of strong relationships between animals in the reference and imputed sets, helps to better detect long haplotypes that are used to infer missing SNPs. Hayes et al. [14] cited problems of pedigree structure and small family sizes in sheep breeds, which affect the imputation process if a population imputation method is not applied. McRae et al. [29] reported that, in sheep, the linkage disequilibrium between SNPs that are separated by less than 10 cM is lower than that for SNPs separated by similar distances in the dairy cattle population, thus reducing the power of the population imputation method which depends on linkage disequilibrium. The number of haplotypes shared between breeds is small and a large reference population is required to capture haplotype diversity for different sheep breeds [1]. Imputation accuracies were higher for almost all the scenarios for which a fixed and large reference population was used and this was consistent with the above studies. Across all scenarios (FIMPUTE was the only software used) for imputation from 5K to 50K, a slight loss in accuracy when using the fixed and large reference population was observed only for a few animals. In addition, a large gain in accuracy for a large proportion of animals (purebred and crossbred) in the imputed set, justifies the use of a fixed and large reference population for all situations. This may be associated with the complexity of the breed composition of each animal considered in some cases as purebred.

The top 10 measures of relatedness demonstrated that accuracy of imputation was strongly associated with the level of relationships between animals in the imputed and reference sets and that it increased as the average top 10 relationships increased. The relationship between imputation accuracy and top relationships was also demonstrated by Bolormaa et al. [22].

Imputation from 5K to both 50K (HD subset) and HD panels

Imputation from 5K to HD was slightly better (0.6 %) than imputation from 5K to 50K. This result is not consistent with previous studies in other species. A study

on Hereford cattle by Picolli et al. [8] showed that imputation accuracies were higher for imputation from 5K to 50K (CR = 94.60 %) than for imputation to HD (CR = 89.80 %). This implies that longer chromosome segments need to be inferred if the targeted SNP density for imputation is the HD panel, when the number of SNPs in the low-density panel is fixed (5K). The fact that there are more misplaced SNPs in the medium-density panel (50K), compared to the HD panel may cause more problems when imputing to 50K from the same low-density panel. Further studies with other datasets are necessary to check this issue. Imputation of animals that are highly-related to individuals in the reference population can benefit from the identification of long haplotype blocks and thus could lead to smaller differences in imputation accuracies for imputation to 50K and HD panels from the same low-density panel. The difference in overall imputation accuracy between imputations to these two panels is reduced from less than 1 to 0.2 % if animals with lower CR than 80 % are not considered in the statistics (animals that are weakly related to those in the reference population). Further investigations on this topic are also necessary. Individuals for which the 5K SNPs were imputed to 50K with an imputation accuracy lower than 70 % (Fig. 2a, b) had an overall average gain in CR accuracy of 21.32 % after imputation from 50K to HD panel. According to Sargolzaei et al. [9], closely-related animals share long haplotypes that usually occur at a low frequency in the population, while less related individuals may share short haplotypes that occur at higher frequencies in the population. Based on these results, it is likely that these short haplotypes were captured by increasing both panel densities (i.e. imputation from 50K to HD compared to 5 to 50K) and the effect was largest for the animals for which imputation accuracies were lowest in the imputation using the low-density panels.

Imputation from low- and medium-density panels to the HD panel

The two-step imputation from 5K to HD (5K to 50K and then from 50K to HD) outperformed the one-step imputation from 5K to HD (+5.67 % in CR). A comparison of one- and two-step imputation approaches in Canadian dairy breeds (Ayrshire, Guernsey and Holstein) reported by Larmer et al. [10], also showed that the two-step procedure resulted in higher accuracies. A similar study on Braford and Hereford beef cattle in Brazil [8] reported a gain in CR of 8.06 % with a two-step imputation procedure. These authors suggested that the gain in accuracy can be attributed to the larger number of SNPs present in the low-density (50K) panel used in the second step of imputation.

Imputation of rare variants

Imputation of rare variants was recently investigated on human data [30, 31]. Kreiner-Møller et al. [32] proposed a new approach to improve imputation accuracy of rare alleles that was based on a two-step imputation procedure, i.e. (step 1) genotyping many additional individuals only for the rare variants to constitute a specific reference population for the rare segments and (step 2) imputation to the highest density panel as usual. Using data on a purebred dairy cattle population, Sargolzaei et al. [9] showed the importance of having information on closely-related animals for the efficiency of the imputation of rare variants and reported gains in accuracy relative to the increase in reference population size and panel density [9]. These authors showed that rare variants tend to be recent events and are directly associated with longer haplotypes. They reported imputation accuracies for rare variants using various sizes of reference populations and found that they were higher than 80 % for a reference population size similar to that described in this study (N > 4000). This pronounced disparity in imputation accuracies between our study (58 %) and the study of Sargolzaei et al. [9] in dairy cattle (at least 80 %) is mainly due to differences between the structure of dairy and sheep populations. Population structure will also directly affect the number of closely-related animals that will positively influence the imputation of rare variants. The imputation accuracies (r^2) of 0 (N = 35) for SNPs with MAF lower than 0.0001 that were obtained in our study are likely due to genotyping errors or the absence of variation for this specific set of SNPs, which directly impacts the correlation calculation.

Software comparison

We chose the version 3.3.2 of BEAGLE for our study because it is implemented in practice for genomic selection in New Zealand sheep at the industry level [20]. Computation run-time and efficiency of BEAGLE and FIMPUTE software packages have been reported by several authors in other species [1, 8, 9]. Our results corroborate the findings from those authors and show that FIMPUTE V2.2 outperformed BEAGLE 3.3.2 across all imputation scenarios. Since FIMPUTE is able to parallelize chromosomes on multi-core systems [9], it will become an important tool for imputation of thousands of animals genotyped with a variety of panel densities.

r^2 and concordance rate measures of imputation accuracy

Concordance rates as a measure of imputation accuracy have been reported by several authors, including for imputation in sheep [14]. Sargolzaei et al. [9] used allelic r^2 (squared correlation between imputed and true genotypes)

as a measure of imputation accuracy that minimizes the dependency of SNP allele frequencies. The r^2 calculation can be carried out on a SNP or animal basis. Unlike the calculation on an animal basis that uses the large number of SNP genotypes per animal for the calculation of r^2, the calculation of r^2 per SNP requires a large number of animals to compose the imputed set, in order to obtain an unbiased estimate of the correlation. For this reason, Scenario 27_5K50K was considered the most appropriate (for which the imputed set included 1000 animals) to estimate r^2 accuracy per SNP. Our study used three different measures of imputation accuracy depending on the scenarios: concordance rate (plots reporting imputation accuracy per individual in Figs. 2, 3, 4, 5 and 6, reported as an average value in Tables 4, 5 and 6) and r^2, both per animal (Tables 4, 5 and 6), and per SNP (Scenario 27_5K50K), used to investigate regions that were imputed less accurately.

Prediction of imputation accuracy before imputation

Imputation accuracy can be determined only after masking chromosome segments from the individual's genotype and by comparing the true and masked genotypes to the imputed genotype. According to Calus et al. [13], imputation accuracy depends mainly on the ability of identifying the correct haplotype for a specific SNP and on the number of genotyped immediate ancestors. In this paper, we report a novel and efficient approach to identify, prior to imputation, the animals for which regions in the genome are less likely to be inferred efficiently. Imputation from 5K to 50K and HD SNP panels was investigated and we found that there was a clear trend relating the resulting imputation accuracy with the number of MI at the 5K genotype level (before imputing). The same trend was observed using the 50K genotypes (original and not masked genotypes). MI values (average value between an imputed animal and the top 10 related individuals from the reference group) higher than 400 (measured at the 5K level) or 3000 (at the 50K level) were obtained for individuals for which imputation accuracy was lower than 80 %. Further analyses are necessary on other populations with a different structure to better evaluate this method. If the imputation process is evaluated for denser or sparser panels, a similar investigation with different SNP densities is required.

Imputation efficiency per chromosome region in Romney animals

According to Picolli et al. [8], in beef cattle, imputation accuracy is associated with chromosome length. They reported that CR accuracies were highest for bovine chromosome 1 and lowest for chromosome 28, which is consistent with our results. However, little is known on the imputation accuracy of proximal and telomeric regions for each chromosome in sheep. We showed that only ovine chromosome 26 had an overall imputation accuracy over 100 SNPs at each end higher than 60 % (r^2). Most of the ovine chromosomes had problems at least at one of the ends. If a trait is affected by a locus located in one of these regions, association studies will be impacted or biased if the genotypes investigated are imputed. Incorporation of additional SNPs located in these regions in the low-density panel may improve imputation accuracy.

Conclusions

In this study, we identified several critical factors that influence imputation accuracy and that need to be taken into account for the implementation of genomic selection in industry breeding programs for New Zealand dual-purpose sheep breeds. These factors include the SNP panels and software used, both of which should be carefully evaluated when new technologies are presented. Strategies of imputation (one- or two-step) and the choice of the animals to be genotyped using both high- and low-density panels are important since we highlighted the influence of the presence of closely-related animals in the reference population as well as the improved imputation accuracy reached when a subset of more closely-related animals is added to the reference population compared to a larger reference population that includes all the animals. Incorporation of additional SNPs in the lowest density panel (5K) increases imputation accuracy furthermore. Since it is not possible to have a high imputation accuracy for all the animals, we present a method that allows imputation accuracy to be predicted based on the low-density genotypes, which can then be used to restrict genomic prediction only to animals that can be imputed with sufficient accuracy. Imputation of rare alleles is a difficult task that needs to be better investigated in future studies, especially for regions under selection pressure and for scenarios for which the size of the reference set is limited.

Authors' contributions
RVV participated in the design of the study, performed the statistical analyses, and drafted the manuscript. SPM, KGD, BA, ML, MB, SMC and JCM participated in the design of the study and helped to draft the manuscript. All authors read and approved the final manuscript.

Author details
[1] Centre for Genetic Improvement of Livestock, University of Guelph, Guelph, ON N1G2W1, Canada. [2] Beef Improvement Opportunities, Guelph, ON N1K1E5, Canada. [3] Invermay Agricultural Centre, AgResearch Limited, Mosgiel 9053, New Zealand. [4] Department of Mathematics and Statistics, University of Otago, Dunedin 9016, New Zealand.

Acknowledgements
AgResearch core funding and Beef and Lamb New Zealand Genetics are acknowledged for financial support. We are also grateful to Dr. Mehdi Sargolzaei (L'Alliance Boviteq and CGIL University of Guelph) for sharing the FIMPUTE software. The data used in this study came from Beef and Lamb New Zealand Genetics (formerly Ovita) and FarmIQ funded research and the HD chip was created by FarmIQ in conjunction with the International Sheep Genomics Consortium. The support and encouragement of many New Zealand ram breeders are acknowledged as well as access to their flocks and DNA samples and the laboratory and field staff of the AgResearch Animal Genomics team responsible for genotyping the animals.

Competing interests
The authors declare that they have no competing interests.

References
1. Ventura RV, Lu D, Schenkel S, Wang Z, Li C, Miller SP. Impact of reference population on accuracy of imputation from 6K to 50K single nucleotide polymorphism chips in purebred and crossbreed beef cattle. J Anim Sci. 2014;92:1433–44.
2. Roberts A, McMillan L, Wang W, Parker J, Rusyn I, Threadgill D. Inferring missing genotypes in large SNP panels using fast nearest-neighbor searches over sliding windows. Bioinformatics. 2007;23:i401–7.
3. Su SY, White J, Balding DJ, Coin LJM. Inference of haplotypic phase and missing genotypes in polyploid organisms and variable copy number genomic regions. BMC Bioinformatics. 2008;9:513.
4. Pryce JE, Wales WJ, de Haas Y, Veerkamp RF, Hayes BJ. Genomic selection for feed efficiency in dairy cattle. Animal. 2014;8:1–10.
5. Matukumalli LK, Lawley CT, Schnabel RD, Taylor JF, Allan MF, Heaton MP, et al. Development and characterization of a high density SNP genotyping assay for cattle. PLoS One. 2009;4:e5350.
6. Dodds KG, Auvray B, Newman S-AN, McEwan JC. Genomic breed prediction in New Zealand sheep. BMC Genet. 2014;15:92.
7. Saatchi M, McClure MC, McKay SD, Rolf MM, Kim J, Decker JE, et al. Accuracies of genomic breeding values in American Angus beef cattle using K-means clustering for cross-validation. Genet Sel Evol. 2011;43:40.
8. Piccoli M, Braccini J, Cardoso F, Sargolzaei M, Schenkel F. Impact of imputation in Braford and Hereford beef cattle. BMC Genet. 2014;15:157.
9. Sargolzaei M, Chesnais JP, Schenkel FS. A new approach for efficient genotype imputation using information from relatives. BMC Genomics. 2014;15:478.
10. Larmer SG, Sargolzaei M, Schenkel FS. Extent of linkage disequilibrium, consistency of gametic phase, and imputation accuracy within and across Canadian dairy breeds. J Dairy Sci. 2014;97:3128–41.
11. Erbe M, Hayes BJ, Matukumalli LK, Goswami S, Bowman PJ, Reich CM, et al. Improving accuracy of genomic predictions within and between dairy cattle breeds with imputed high-density single nucleotide polymorphism panels. J Dairy Sci. 2012;95:4114–29.
12. Hickey JM, Crossa J, Babu R, de los Campos G. Factors affecting the accuracy of genotype imputation in populations from several maize breeding programs. Crop Sci. 2012;52:654–63.
13. Calus MPL, Bouwman AC, Hickey JM, Veerkamp RF, Mulder HA. Evaluation of measures of correctness of genotype imputation in the context of genomic prediction: a review of livestock applications. Animal. 2014;8:1743–53.
14. Hayes BJ, Bowman PJ, Daetwyler HD, Kijas JW, van der Werf JHJ. Accuracy of genotype imputation in sheep breeds. Anim Genet. 2012;43:72–80.
15. Pei YF, Li J, Zhang L, Papasian CJ, Deng HW. Analyses and comparison of accuracy of different genotype imputation methods. PLoS One. 2008;3:e3551.
16. Brøndum RF, Guldbrandtsen B, Sahana G, Lund MS, Su G. Strategies for imputation to whole genome sequence using a single or multi-breed reference population in cattle. BMC Genomics. 2014;15:728.
17. van Binsbergen R, Bink MC, Calus MP, van Eeuwijk FA, Hayes BJ, Hulsegge I, et al. Accuracy of imputation to whole-genome sequence data in Holstein Friesian cattle. Genet Sel Evol. 2014;46:41.
18. Corbin LJ, Kranis A, Blott SC, Swinburne JE, Vaudin M, Bishop SC, et al. The utility of low-density genotyping for imputation in the Thoroughbred horse. Genet Sel Evol. 2014;46:9.

19. Pausch H, Aigner B, Emmerling R, Edel C, Götz KU, Fries R. Imputation of high-density genotypes in the Fleckvieh cattle population. Genet Sel Evol. 2013;45:3.

20. Cleveland MA, Hickey JM. Practical implementation of cost-effective genomic selection in commercial pig breeding using imputation. J Anim Sci. 2013;91:3583–92.

21. Moghaddar N, Gore KP, Daetwyler HD, Hayes BJ, van der Werf JHJ, Meuwissen T, et al. Accuracy of genotype imputation based on random and selected reference sets in purebred and crossbred sheep populations and its effect on accuracy of genomic prediction. Genet Sel Evol. 2015;47:97.

22. Bolormaa S, Gore K, van der Werf JHJ, Hayes BJ, Daetwyler HD. Design of a low-density SNP chip for the main Australian sheep breeds and its effect on imputation and genomic prediction accuracy. Anim Genet. 2015;46:544–56.

23. Martin AR, Tse G, Bustamante CD, Kenny EE. Imputation-based assessment of next generation rare exome variant arrays. Pac Symp Biocomput. 2014;2014:241–52.

24. Yuan M, Fang H, Zhang H. Correcting for differential genotyping error in genetic association analysis. J Hum Genet. 2013;58:657–66.

25. Scheet P, Stephens M. A fast and flexible statistical model for large-scale population genotype data: applications to inferring missing genotypes and haplotypic phase. Am J Hum Genet. 2006;78:629–44.

26. Browning BL, Browning SR. A unified approach to genotype imputation and haplotype-phase inference for large data sets of trios and unrelated individuals. Am J Hum Genet. 2009;84:210–23.

27. Dodds KG, Auvray B, Lee M, Newman S-A, McEwan JC. Genomic selection in New Zealand dual purpose sheep. In Proceedings of the 10th world congress on genetetics applied to livestock production: 17–22 August 2014; Vancouver. 2014. https://asas.org/docs/default-source/wcgalp-proceedings-oral/333_paper_10352_manuscript_1331_0.pdf?sfvrsn=2.

28. VanRaden PM. Efficient methods to compute genomic predictions. J Dairy Sci. 2008;91:4414–23.

29. McRae AF, McEwan JC, Dodds KG, Wilson T, Crawford AM, Slate J. Linkage disequilibrium in domestic sheep. Genetics. 2002;160:1113–22.

30. Pistis G, Porcu E, Vrieze SI, Sidore C, Steri M, Danjou F, et al. Rare variant genotype imputation with thousands of study-specific whole-genome sequences: implications for cost-effective study designs. Eur J Hum Genet. 2014;23:975–83.

31. Deelen P, Menelaou A, van Leeuwen EM, Kanterakis A, van Dijk F, Medina-Gomez C, et al. Improved imputation quality of low-frequency and rare variants in European samples using the "Genome of The Netherlands". Eur J Hum Genet. 2014;22:1321–6.

32. Kreiner-Møller E, Medina-Gomez C, Uitterlinden AG, Rivadeneira F, Estrada K. Improving accuracy of rare variant imputation with a two-step imputation approach. Eur J Hum Genet. 2014;23:395–400.

2

Genome-wide detection of genetic markers associated with growth and fatness in four pig populations using four approaches

Yuanmei Guo[†], Yixuan Huang[†], Lijuan Hou, Junwu Ma, Congying Chen, Huashui Ai, Lusheng Huang and Jun Ren[*] [iD]

Abstract

Background: Genome-wide association studies (GWAS) have been extensively used to identify genomic regions associated with a variety of phenotypic traits in pigs. Until now, most GWAS have explored single-trait association models. Here, we conducted both single- and multi-trait GWAS and a meta-analysis for nine fatness and growth traits on 2004 pigs from four diverse populations, including a White Duroc × Erhualian F_2 intercross population and Chinese Sutai, Laiwu and Erhualian populations.

Results: We identified 44 chromosomal regions that were associated with the nine traits, including four genome-wide significant single nucleotide polymorphisms (SNPs) on SSC2 (SSC for *Sus scrofa* chromosome), 4, 7 and X. Compared to the single-population GWAS, the meta-analysis was less powerful for the identification of SNPs with population-specific effects but more powerful for the detection of SNPs with population-shared effects. Multiple-trait analysis reduced the power to detect trait-specific SNPs but significantly enhanced the power to identify common SNPs across traits. The SNP on SSC7 had pleiotropic effects on the nine traits in the F_2 and Erhualian populations. Another pleiotropic SNP was observed on SSCX for these traits in the F_2 and Sutai populations. Both population-specific and shared SNPs were identified in this study, thus reflecting the complex genetic architecture of pig growth and fatness traits.

Conclusions: We demonstrate that the multi-trait method and the meta-analysis on multiple populations can be used to increase the power of GWAS. The two significant SNPs on SSC7 and X had pleiotropic effects in the F_2, Erhualian and Sutai populations.

Background

Growth and fatness traits are economically important and have been intensively selected in the global pig industry. Dissection of the genetic architecture of growth and fat deposition in pigs not only benefits the pig industry but also sheds insight into our understanding of human obesity, because pigs are more physiologically similar to humans than rodents and other model animals [1].

To understand the molecular basis of divergent phenotypes in pigs, researchers have established multiple

F_2 intercross populations using European and Chinese breeds as founders, and have mapped quantitative trait loci (QTL) for a list of phenotypic traits, including growth and fatness traits, using hundreds of microsatellite markers across the whole genome [2–6]. Until now, 1880 and 1070 QTL for growth and fatness traits have been deposited in the pig QTL database (http://www.animalgenome.org/cgi-bin/QTLdb/SS/index, Release 26, Apr 27, 2015), respectively. These findings have significantly advanced our understanding of the genetic architecture of porcine growth and fatness traits. Nevertheless, the resolution of traditional QTL mapping is relatively poor due to markers being sparse and insufficient recombination events in

*Correspondence: renjunjxau@hotmail.com
[†]Yuanmei Guo and Yixuan Huang contributed equally to this work
State Key Laboratory of Pig Genetic Improvement and Production Technology, Jiangxi Agricultural University, Nanchang 330045, China

the F_2 crosses. Thus, identification of causative mutations that underlie the identified QTL remains a big challenge.

With the availability of the Illumina Porcine SNP60 Beadchip, it has become feasible to exploit the association between high-density single nucleotide polymorphisms (SNPs) and phenotypic traits through genome-wide association studies (GWAS) [7]. Compared to the traditional QTL mapping approach, the GWAS approach allows the identification of SNPs that are significantly associated with traits. Nevertheless, large sample sizes are still required to identify SNPs that are weakly associated with target traits. A meta-analysis of GWAS can not only increase statistical power but also reduce the number of false positives by combining information from multiple independent studies [8]. Moreover, for a QTL with pleiotropic effects on multiple traits, a multi-trait analysis can improve the detection power of GWAS [9–11].

The aim of our study was to identify SNPs associated with nine traits related to growth and fatness across four pig populations, including a White Duroc × Erhualian F_2 intercross (referred hereafter as F_2), Sutai, Laiwu and Erhualian pigs by four GWAS methods: single-trait analysis on a single population (SS-GWAS), single-trait analysis on multiple populations (SM-GWAS), multi-trait analysis on a single population (MS-GWAS), and multi-trait analysis on multiple populations (MM-GWAS).

Methods
Ethics statement
All procedures used for this study and involving animals are in compliance with guidelines for the care and utility of experimental animals established by the Ministry of Agriculture of China. The ethics committee of Jiangxi Agricultural University specifically approved this study.

Animals and phenotypic measurements
A total of 2004 pigs were used in this study, including 925, 434, 331 and 314 individuals from the F_2, Sutai, Erhualian and Laiwu populations, respectively. The F_2 and Sutai populations were previously described in [12, 13]. Briefly, the F_2 animals originated from a cross between two White Duroc boars and 17 Erhualian sows [12]. In this population, nine F_1 boars and 59 F_1 sows were intercrossed by avoiding full-sib mating to produce 1912 F_2 pigs. A total of 925 F_2 pigs were randomly selected from all F_1 boar and sow families. These animals were slaughtered at the age of 240 ± 3 days and carcass and meat quality traits were measured; the remaining F_2 pigs were used to produce F_3 individuals or measure the reproductive traits. The Sutai pig is a Chinese synthetic breed that was derived from a cross between Western Duroc boars and Chinese Taihu (mainly Erhualian) sows. This breed has experienced directional selection for prolificacy and

growth for more than 18 generations. The 434 Sutai pigs used in the current study were offspring of four sires and 55 dams [13]. Erhualian and Laiwu are Chinese indigenous breeds. The former is known for its prolificacy, with a litter size that can exceed 15, and the latter is characterized by its exceptionally high intramuscular fat content (more than 9%) [14]. We obtained 334 Erhualian (168 sires and 166 dams) and 314 Laiwu (218 sires and 98 dams) pigs at the age of ~90 days from two national conservation farms of the two breeds in Jiangsu and Shandong provinces, respectively. The F_2, Sutai, Erhualian and Laiwu pigs were all raised in an experimental farm in Nanchang, Jiangxi province from 2001 to 2014. All 2004 pigs had ad libitum access to fresh water and consistent feed containing 16% crude protein, 3100 kJ of digestive energy, and 0.78% lysine during the fattening period. Each pig in the four populations was weighed at birth and at 210 and 240 days of age, and the average daily gains from 0 to 210 days of age (ADG_{0-210}) and from 210 to 240 days of age ($ADG_{210-240}$) were calculated. The F_2 and Sutai pigs were slaughtered in the same commercial abattoir at 240 ± 3 days of age and the Erhualian and Laiwu pigs at 300 ± 3 days of age. After slaughter, all pigs were measured for fatness traits, including backfat thickness at the shoulder (SBF), the first rib (FBF), the last rib (LBF), and at the hip (HBF), and weight of leaf fat (LFW), veil fat (VFW) and abdominal fat (AFW).

Genotyping and quality control
Genomic DNA was extracted from ear or tail tissues using a standard phenol/chloroform protocol, and was then quantified and adjusted to a final concentration of 50 ng/μl. All 2004 pigs were genotyped with the porcine 60 K SNP Beadchip on an iScan System (Illumina, USA) following the manufacturer's protocol. SNPs, including sex-linked SNPs, that had a call rate less than 95%, a minor allele frequency lower than 5%, or that strongly deviated from Hardy–Weinberg equilibrium ($P < 0.000001$) were discarded. Animals with a call rate less than 95% were also removed from further analyses. These quality controls were performed for each population separately to include as many qualified SNPs as possible for the GWAS in each population. A common set of 15,429 qualified SNPs across the four populations was used in the meta-analysis of GWAS.

Statistical methods
Descriptive statistics of phenotypic traits were calculated by the MEANS procedure of SAS9.0 (SAS Institute Inc., USA) and phenotypic differences between sexes were tested by the TTEST procedure. The MIXED procedure was used to determine the fixed effects and the covariates included in the GWAS model. Sex and fattening batch

were included as fixed effects in all GWAS models, and birth weight, body weight at 210 days, and carcass weight were treated as covariates for ADG_{0-210}, $ADG_{210-240}$ and for fatness traits, respectively, in the GWAS models. A polygenic effect for each animal with covariances based on genomic kinship, which was calculated based on the identity-by-state of the SNPs on autosomes [15, 16], was included as a random effect to account for the effect of population substructure. The P values of Bonferroni corrected thresholds for suggestive, 5 and 1% genome-wide significant levels were 1, 0.05 and 0.01, respectively, divided by the number of SNPs used in the GWAS. The suggestive level was first proposed by Lander and Kruglyak [17] and represents the threshold where, under the null hypothesis, one false positive is expected per genome scan.

GWAS for a single trait in a single population (SS-GWAS)
The two-stage approach implemented in the R package GenABEL was used to conduct SS-GWAS under an additive model. First, the following mixed model was used to calculate the phenotypic residual vector $\mathbf{e}^*$:

$$\mathbf{y} = \mathbf{Xb} + \mathbf{Zu} + \mathbf{e}^*,$$

where $\mathbf{y}$ is a vector of phenotypes; $\mathbf{b}$ is the estimator vector of fixed effects, including population mean μ; $\mathbf{u}$ is a vector of random polygenic additive effects that follows a normal distribution $N(\mathbf{0}, \mathbf{G}\sigma_u^2)$, with $\mathbf{G}$ being the genomic kinship matrix calculated from all autosomal SNPs based on identity-by-state [16], σ_u^2 is the polygenic additive variance, and $\mathbf{X}$ and $\mathbf{Z}$ are the incidence matrices for $\mathbf{b}$ and $\mathbf{u}$, respectively.

Then, a family-based score test was used to detect associations between SNPs and traits using the following simple regression model [18], one SNP at a time:

$$\mathbf{e}^* = \mathbf{Sa} + \mathbf{e},$$

where a is an estimator of the SNP allele substitution effect; $\mathbf{S}$ is the incidence vector for a (coded 0, 1, 2 based on allele dosage); and $\mathbf{e}$ is a vector of residual errors that follows a normal distribution $N(\mathbf{0}, \mathbf{I}\sigma_e^2)$, with $\mathbf{I}$ being an identity matrix and σ_e^2 the variance of the residual error. The P value of the association test was adjusted by the genomic control method to correct for residual inflation [19, 20].

GWAS for multiple traits in a single population (MS-GWAS)
MS-GWAS were conducted to detect pleiotropic SNPs using the method proposed by Bolormaa et al. [11]. In brief, a Chi square statistic, which approximately follows a Chi square distribution with the number of traits tested as the number of degrees of freedom, was calculated for each SNP using the following formula:

$$\chi^2_{multi-trait} = \mathbf{t}_i'\mathbf{V}^{-1}\mathbf{t}_i,$$

where $\mathbf{t}_i$ is a 9×1 vector of the signed t-values for the ith SNP from the SS-GWAS for the nine traits, $\mathbf{t}_i'$ is the transpose of $\mathbf{t}_i$, and $\mathbf{V}^{-1}$ is the inverse of the 9×9 correlation matrix between traits. The correlation between two traits was calculated by correlating the estimated effects (signed t-values) of the 15,429 qualified SNPs for the two traits.

Meta-analysis of GWAS for a single trait (SM-GWAS) and multiple traits (MM-GWAS)
We applied the inverse variance weighting method of [8], in which each population is weighted according to the inverse of its squared standard error, to perform SM-GWAS and MM-GWAS across the four populations based on the results of SS-GWAS and MS-GWAS, respectively. The weight (w_i) for the ith population was equal to the inverse of the square of the standard error (s_i) of the allele substitution effect in the ith population. Then, the pooled estimates of the allele effect (β) of a given SNP and its standard error (s) were calculated as follows:

$$\beta = \frac{\sum_{i=1}^n w_i\beta_i}{\sum_{i=1}^n w_i} \text{ and } s^2 = \frac{1}{\sum_{i=1}^n w_i},$$

where n is the population number; β_i is the allele effect in the ith population.

A statistic (Z score) of Z-test was calculated as follows:

$$Z = \frac{\beta}{s} = \frac{\sum_{i=1}^n w_i\beta_i}{\sqrt{\sum_{i=1}^n w_i}}.$$

An allele of an associated SNP may have a positive effect in some populations and a negative effect in the others. Such an inconsistent effect could significantly reduce the detection power of the meta-analysis of GWAS. To circumvent this, we used information on linkage disequilibrium and ignored information on phase, i.e., the absolute value of β_i was used to calculate the pooled β and Z values.

To determine if a SNP that was significantly associated with multiple traits was due to closely linked genes or pleiotropy, we conducted a conditional SS-GWAS by fixing the effect of the top SNPs identified by MS-GWAS in the statistic model.

Linkage and linkage disequilibrium analyses
To determine the approximate genomic positions of the unmapped significant SNPs, a two-point linkage analysis was used to detect linkage between the mapped and unmapped SNPs in the F_2 population [21]. The lower the recombination rate (θ) is, the tighter is the link

between the two SNPs. Haplotypes of the regions that were significantly associated with the target traits were inferred by Simwalk2.9 [22], and linkage disequilibrium blocks were defined using default parameters of Haploview4.2 [23]. The VennDiagram in R package was used to draw a Venn diagram that showed the loci that were

Table 1 Genome-wide significant loci for nine fatness and growth traits identified by four GWAS approaches in this study

QTL[a]	Chr[b]	Trait[c]	Population	Method	Top SNP	Pos Mb[d]	Effect ± SE[e]	P value[f]	N^g_{SNP}	Boundary SNPs	
1	2	HBF	Meta	SM	ss131211507	3.61	0.127 ± 0.026	1.28E−06*	1	–	–
2	4	Multi-trait	Meta	MM	ss131269439	81.71	–	7.88E−10**	8	ss131269439	ss478935222
2	4	FBF	Meta	SM	ss131269439	81.71	0.168 ± 0.032	1.20E−07**	1	–	–
2	4	HBF	Meta	SM	ss131269678	82.25	0.156 ± 0.032	1.08E−06*	1	–	–
2	4	SBF	Meta	SM	ss478935222	85.09	0.166 ± 0.035	1.53E−06**	1	–	–
3	6	ADG$_{210-240}$	Meta	SM	ss131029816	71.60	0.040 ± 0.008	1.38E−06*	1	–	–
4	7	ADG$_{210}$	F$_2$	SS	ss131342496	32.96	-0.026 ± 0.005	5.95E−07*	1	–	–
4	7	Multi-trait	Meta	MM	ss131343534	34.56	–	2.65E−18**	27	ss120018804	ss131348342
4	7	SBF	Meta	SM	ss131343534	34.56	0.255 ± 0.042	1.11E−09**	6	ss131341589	ss107890951
4	7	FBF	Meta	SM	ss131343534	34.56	0.315 ± 0.040	4.88E−15**	17	ss131341589	ss131348342
4	7	HBF	Meta	SM	ss131343534	34.56	0.379 ± 0.045	3.16E−17**	25	ss120018804	ss131348342
4	7	Multi-trait	F$_2$	MS	ss107837325	34.80	–	3.97E−33**	107	ss23131766	ss478941636
4	7	AFW	F$_2$	SS	ss107837325	34.80	0.163 ± 0.022	8.99E−11**	34	ss131066868	ss131345041
4	7	FBF	F$_2$	SS	ss107837325	34.80	0.691 ± 0.060	3.10E−17**	69	ss131341676	ss131348342
4	7	HBF	F$_2$	SS	ss107837325	34.80	0.773 ± 0.072	1.97E−14**	64	ss131341676	ss131347489
4	7	LBF	F$_2$	SS	ss107837325	34.80	0.571 ± 0.057	1.77E−14**	60	ss131341676	ss131348342
4	7	LFW	F$_2$	SS	ss107837325	34.80	0.650 ± 0.058	1.75E−18**	63	ss131341766	ss131347489
4	7	SBF	F$_2$	SS	ss107837325	34.80	0.487 ± 0.057	1.94E−12**	45	ss131341766	ss131346335
4	7	Multi-trait	Erhualian	MS	ss131343870	34.84	–	1.29E−15**	6	ss131342502	ss131347459
4	7	FBF	Erhualian	SS	ss131343870	34.84	0.585 ± 0.073	2.13E−13**	8	ss131336720	ss131347459
4	7	HBF	Erhualian	SS	ss131343870	34.84	0.502 ± 0.075	1.49E−10**	5	ss131343870	ss131347459
4	7	LBF	Erhualian	SS	ss131343870	34.84	0.523 ± 0.064	6.96E−14**	5	ss131343870	ss131347459
4	7	LFW	Erhualian	SS	ss131343870	34.84	0.453 ± 0.067	2.39E−10**	5	ss131343870	ss131347459
4	7	SBF	Erhualian	SS	ss131343870	34.84	0.453 ± 0.079	4.08E−08**	2	ss131343870	ss478941599
4	7	LFW	Meta	SM	ss131344553	36.20	0.140 ± 0.028	7.18E−07*	2	ss131344553	ss131347175
4	7	LBF	Meta	SM	ss131347175	40.85	0.155 ± 0.024	1.27E−10**	13	ss131337529	ss131348342
5	X	SBF	Meta	SM	ss131570179	46.75	0.124 ± 0.020	5.42E−10**	5	ss131036304	ss131562987
5	X	VFW	F$_2$	SS	ss107834496	51.70	0.091 ± 0.013	5.39E−10**	21	ss478944418	ss478935791
5	X	Multi-trait	F$_2$	MS	ss23131102	63.65	–	5.78E−35**	79	ss131067158	ss131563360
5	X	FBF	F$_2$	SS	ss23131102	63.65	0.164 ± 0.023	2.94E−07*	2	ss478943984	ss23131102
5	X	HBF	F$_2$	SS	ss23131102	63.65	0.278 ± 0.029	6.81E−12**	24	ss478936157	ss131562911
5	X	Multi-trait	Sutai	MS	ss478934917	78.58	–	3.19E−10**	38	ss478944000	ss131570171
5	X	Multi-trait	Meta	MM	ss131070541	106.48	–	1.61E−15**	14	ss478936157	ss131562987
5	X	FBF	Meta	SM	ss131070541	106.48	0.110 ± 0.018	1.33E−09**	10	ss131036304	ss131562987
5	X	HBF	Meta	SM	ss131070541	106.48	0.187 ± 0.021	9.06E−19**	18	ss131067158	ss131563051
5	X	LBF	Meta	SM	ss131070541	106.48	0.108 ± 0.018	1.46E−09**	1	–	–

[a] We operationally define two loci with the distance between their lead SNPs less than 5 Mb as the same QTL except the loci on chromosome X. All loci close to the recombination cold spot (more than 30 Mb, with an extremely low rate of recombination) in the middle of X chromosome was considered the same QTL because of too few SNPs in this region

[b] Chromosome

[c] Abbreviations of trait names are in Additional file 1: Table S1

[d] The position of the unmapped SNP (ss131029816) is deduced by its tightly linked SNP (ss131566312)

[e] The direction of SNP effect estimated by the meta-GWAS was not shown because the linkage phase may be inconsistent among the four populations, and the SNP effect cannot be estimated by the multi-trait GWAS

[f] ** 1% genome-wide significant; * 5% genome-wide significant

[g] Number of SNPs that surpass the significance level

in common among the four populations and the four methods.

Results

Descriptive statistics of the traits

The descriptive statistics for the nine growth and fatness traits in the F_2 and Sutai populations were reported in our previous publications [24, 25]. Additional file 1: Table S1 shows the means and standard errors for these nine traits as well as the phenotypic differences between females and males in the Erhualian and Laiwu populations. Growth rates were not significantly different between sexes in the two populations, except that Erhualian males grew faster ($P = 0.03$) than females from 210 to 240 days of age. Backfat thickness at the four localizations was significantly higher ($P < 10^{-6}$) in males than in females in the Erhualian population, while no significant difference was observed in the Laiwu population. Males deposited more ($P < 0.01$) leaf fat, while females stored more ($P < 10^{-10}$) fat in the abdomen in both populations. Veil fat was significantly heavier ($P = 6.77 \times 10^{-6}$) in males than in females in Erhualian pigs, while there was no significant difference in veil fat between sexes in Laiwu pigs.

Phenotypic correlation coefficients between traits

Raw phenotypic correlation coefficients between the measured traits in the two populations are in Additional file 2: Table S2. All correlation coefficients between the measured traits were significant ($P < 0.05$) in the Erhualian and Laiwu populations, except that between ADG_{0-210} and $ADG_{210-240}$ in the Laiwu population. In general, fatness traits were more significantly correlated with each other than growth traits in both populations. ADG_{0-210} had a moderate positive correlation with the other traits and $ADG_{210-240}$ showed a weak positive correlation with the other traits.

Qualified SNPs and animals in the GWAS

All genotyped animals passed quality control with SNP call rates higher than 0.95. A total of 34,636 (4825), 36,341 (4979), 24,602 (3492) and 32,058 (4527) mapped (unmapped) SNPs were qualified for GWAS in the F_2, Sutai, Erhualian and Laiwu populations, respectively. The P values of the 5% (suggestive) genome-wide significant threshold were equal to 1.27×10^{-6} (2.53×10^{-5}), 1.21×10^{-6} (2.42×10^{-5}), 1.78×10^{-6} (3.56×10^{-5}) and 1.37×10^{-6} (2.73×10^{-5}) in these four populations, respectively. Based on the pig genome assembly (Sscrofa10.2, http://www.ensembl.org/Sus_scrofa/Info/Index), the average physical distances between adjacent SNPs were 74.7, 71.2, 105.1 and 80.7 kb in these four populations, respectively. A common set of 15,429

qualified SNPs across the four populations was used in the GWAS meta-analysis, with average physical distance between adjacent SNPs of 167.7 kb. The P values of the 5% genome-wide and suggestive significant thresholds were equal to 3.24×10^{-6} and 6.48×10^{-5}, respectively, in the meta-analysis.

SNPs identified by single-trait GWAS

Table 1 shows the genome-wide significant regions for the nine fatness and growth traits identified by the four GWAS approaches in this study. The SS-GWAS and SM-GWAS identified 15 and 31 chromosomal regions (loci) associated with these nine traits (Table 1; Fig. 1 [see Additional file 3: Table S3, Additional file 4: Figure S1]),

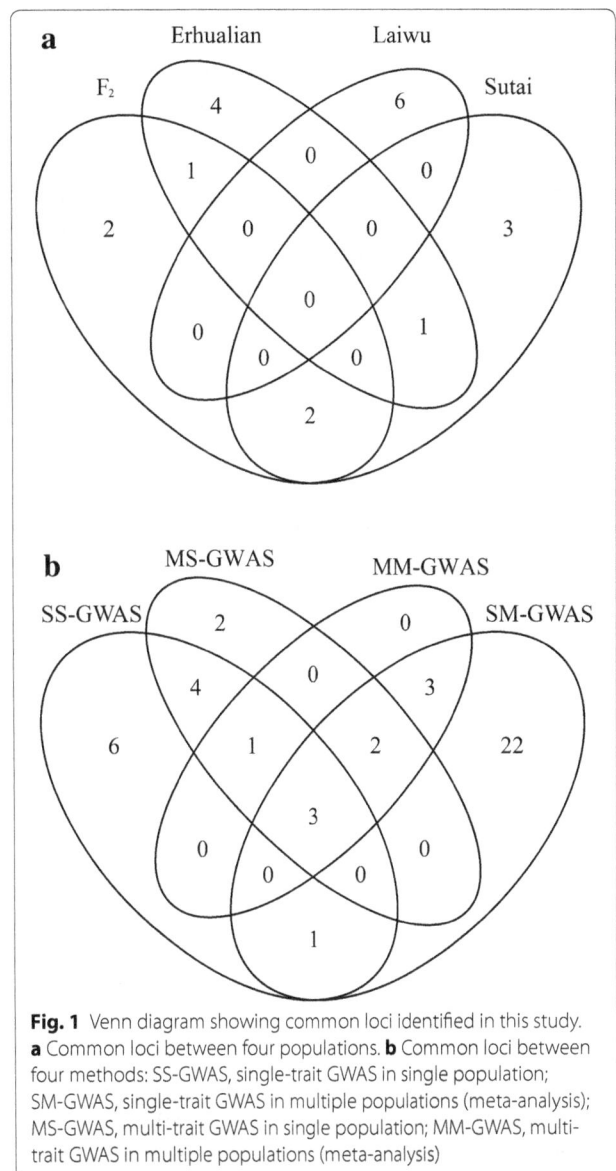

Fig. 1 Venn diagram showing common loci identified in this study. **a** Common loci between four populations. **b** Common loci between four methods: SS-GWAS, single-trait GWAS in single population; SM-GWAS, single-trait GWAS in multiple populations (meta-analysis); MS-GWAS, multi-trait GWAS in single population; MM-GWAS, multi-trait GWAS in multiple populations (meta-analysis)

respectively. Four loci were consistently detected by the two methods, one each on SSC1, 2, 7 and X.

SS-GWAS

In the F_2 population, the most promising locus was mapped at 34.8 Mb on SSC7 (Table 1). Another genome-wide significant locus was identified in the middle of chromosome X. Two suggestive loci were detected, one on SSC1 and one on SSC3 [see Additional file 3: Table S3]. Five and four suggestive loci were identified in the Sutai and Laiwu datasets, respectively. No locus was significant at the genome-wide level in these two datasets. The loci on SSC1, 2 and 8 were detected exclusively for the Sutai population [see Additional file 3: Table S3, Additional file 4: Figure S1]. A genome-wide locus was identified at 34.84 Mb on SSC7 for the Erhualian population. Another three suggestive loci were identified, including two population-specific loci, one each on SSC4 and 6 [see Additional file 3: Table S3, Additional file 4: Figure S1].

SM-GWAS

The SM-GWAS identified 31 loci that were distributed on all chromosomes except SSC11 and 13, including 22 loci that were not identified in the SS-GWAS and four genome-wide loci, one each on SSC2, 4, 7 and X, that were also identified in the SS-GWAS (Fig. 1; Table 2, and Table S3 [see Additional file 3: Table S3]). These four genome-wide loci were all associated with

backfat thickness at the four evaluated sites, and the loci on SSC4 and 7 were also associated with ADG and LFW, respectively.

Loci identified by multi-trait GWAS
MS-GWAS

The MS-GWAS revealed 12 loci associated with both fatness and growth traits, including two loci with genome-wide significance (Fig. 2; Table 1; Additional file 3: Table S3). Four loci were identified for the F_2 population, including two genome-wide significant loci, one at 34.80 Mb on SSC7 and one at 63.65 Mb on chromosome X. For the Sutai population, three loci were identified, including one genome-wide locus at 78.58 Mb on SSCX. For the Laiwu population, only four suggestive loci were identified. Four loci were detected for the Erhualian population, including one genome-wide locus at 34.84 Mb on SSC7 (Fig. 2).

MM-GWAS

Nine loci were identified in the MM-GWAS, including three genome-wide significant loci on SSC4, 7 and X (Fig. 2). All nine loci were identified by the SM-GWAS, except the suggestive locus at 10.53 Mb on SSC5 (Fig. 1).

Additive effects of top SNPs in the four populations

The additive effects of top SNPs at genome-wide and suggestive significant levels for the four populations are in Table 2 and see Additional file 5: Table S4, respectively.

Table 2 Genome-wide significant SNPs detected by the meta-analysis of single-trait GWAS common to the four populations, with corresponding additive effects

Chr[a]	Trait[b]	Top SNP	Pos Mb	Allele	Effect ± SE[c]			
					F_2	Sutai	Erhualian	Laiwu
2	HBF	ss131211507	3.61	A	0.149 ± 0.046**	0.146 ± 0.039**	0.051 ± 0.066	−0.042 ± 0.121
4	FBF	ss131269439	81.71	A	−0.148 ± 0.061*	−0.114 ± 0.060	−0.234 ± 0.068**	−0.198 ± 0.067**
4	HBF	ss131269678	82.25	A	−0.292 ± 0.076**	0.110 ± 0.046*	0.048 ± 0.077	−0.256 ± 0.079**
4	SBF	ss478935222	85.09	G	−0.176 ± 0.063**	−0.129 ± 0.050*	0.323 ± 0.136*	−0.196 ± 0.087*
7	SBF	ss131343534	34.56	G	0.420 ± 0.058**	0.094 ± 0.088	−0.083 ± 0.094	0.020 ± 0.169
7	FBF	ss131343534	34.56	G	0.565 ± 0.062**	0.097 ± 0.078	−0.194 ± 0.088*	−0.082 ± 0.125
7	HBF	ss131343534	34.56	G	0.655 ± 0.074**	0.266 ± 0.085**	−0.225 ± 0.089*	0.047 ± 0.145
7	LFW	ss131344553	36.20	G	0.325 ± 0.070**	0.119 ± 0.033**	0.296 ± 0.092**	0.146 ± 0.070*
7	LBF	ss131347175	40.85	A	−0.435 ± 0.062**	−0.050 ± 0.033	−0.315 ± 0.068**	0.119 ± 0.054*
X	SBF	ss131570179	46.75	A	−0.162 ± 0.030**	−0.084 ± 0.030**	−0.127 ± 0.072	−0.137 ± 0.088
X	HBF	ss131070541	106.48	G	−0.234 ± 0.026**	−0.028 ± 0.054	−0.145 ± 0.056**	−0.179 ± 0.080**
X	LBF	ss131070541	106.48	G	−0.141 ± 0.023**	0.029 ± 0.045	−0.058 ± 0.048	−0.091 ± 0.064
X	FBF	ss131070541	106.48	G	−0.131 ± 0.022**	−0.051 ± 0.052	−0.057 ± 0.055	−0.079 ± 0.070

[a] Chromosome

[b] Abbreviations of trait names are in Additional file 1: Table S1

[c] The significant level is a single point without Bonferroni correction: ** $P \leq 0.01$; * $P \leq 0.05$

Genome-wide detection of genetic markers associated with growth and fatness in four pig populations...

27

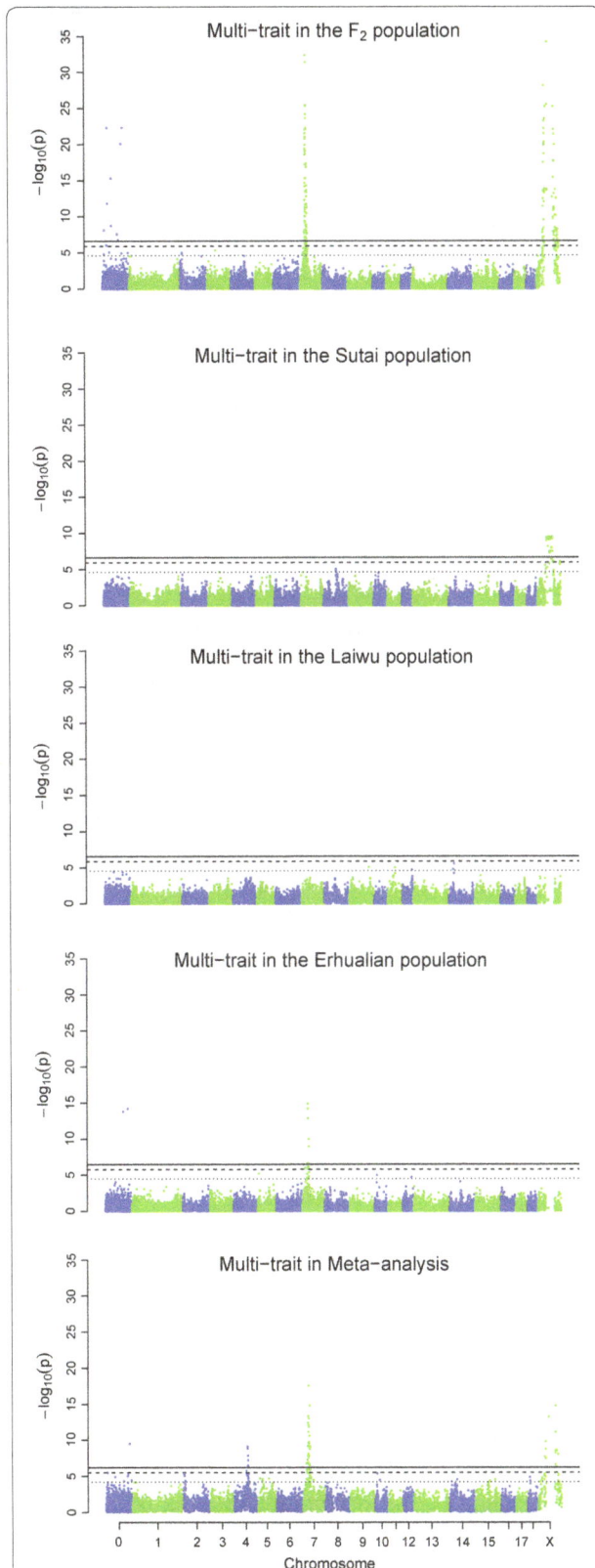

Fig. 2 Manhattan plots of the multi-trait GWAS and meta-analysis for fatness and growth traits in F_2, Sutai, Laiwu and Erhualian populations. The *solid*, *dashed* and *dotted horizontal lines* indicate the 1 and 5% genome-wide and suggestive significant threshold values, respectively. Unmapped SNPs are assigned on chromosome 0 and arbitrary ordered by their names

Thirteen SNPs were associated with fatness traits at the genome-wide significant level in the SM-GWAS. Among these 13 SNPs, five had a consistent direction of additive effects across the four populations, while five showed inconsistent effect directions. The other three SNPs just achieved the significance level for one population. We observed a similar pattern for suggestive SNPs: 12 consistent SNPs and nine inconsistent SNPs. This indicates that the linkage phases between these SNPs and the causative mutations that underlie the detected QTL are not always identical across populations.

Results of the conditional GWAS

We identified two loci associated with multiple traits at the genome-wide significant level by the MS-GWAS. One region was located on SSC7 (between 34.80 and 34.84 Mb) for the F_2 and Erhualian populations, and the other mapped to SSCX (between 63.65 and 66.25 Mb) for the F_2 and Sutai populations. After fixing the effects of the top SNPs (detected by the MS-GWAS) in the SS-GWAS model, we found that no SNP on SSC7 showed any association signal with multiple traits for both the F_2 and Erhualian populations (Fig. 3a, b). No SNP was significantly associated with the nine traits in the region between 60 and 90 Mb on SSCX for the F_2 and Sutai populations, respectively (Fig. 3c, d).

Discussion

Comparison of our results with previously reported loci

Previously, we had identified 15 significant loci on 11 chromosomes for the traits measured in the F_2 and Sutai populations that were also analysed here [25]. In the current study, we identified nine loci on six chromosomes for the same traits by single-trait and multi-trait GWAS in these two populations. Only four loci (one each on SSC2, 4, 7 and X, see Tables 1, 2) were common between both studies and have also been consistently reported in previous studies [2–4, 6, 24, 25]. The lower power in the current study can be attributed to the fact that residual inflation was not corrected in our previous study, but was corrected by genomic control here. To date, 1880 QTL for growth and 1070 for fatness traits have been reported in the pig QTL database [26]. Although we conducted GWAS on four divergent populations using four approaches, no novel loci were identified in the current study. The four GWAS methods yielded different results with few common results, which is likely due to the fact that each method has its own advantage to identify distinct loci. For example, the multi-trait analysis is suitable for detecting pleiotropic loci, the meta-analysis is sufficiently powerful to identify common loci across populations, and the single-population GWAS is an effective method to detect population-specific loci.

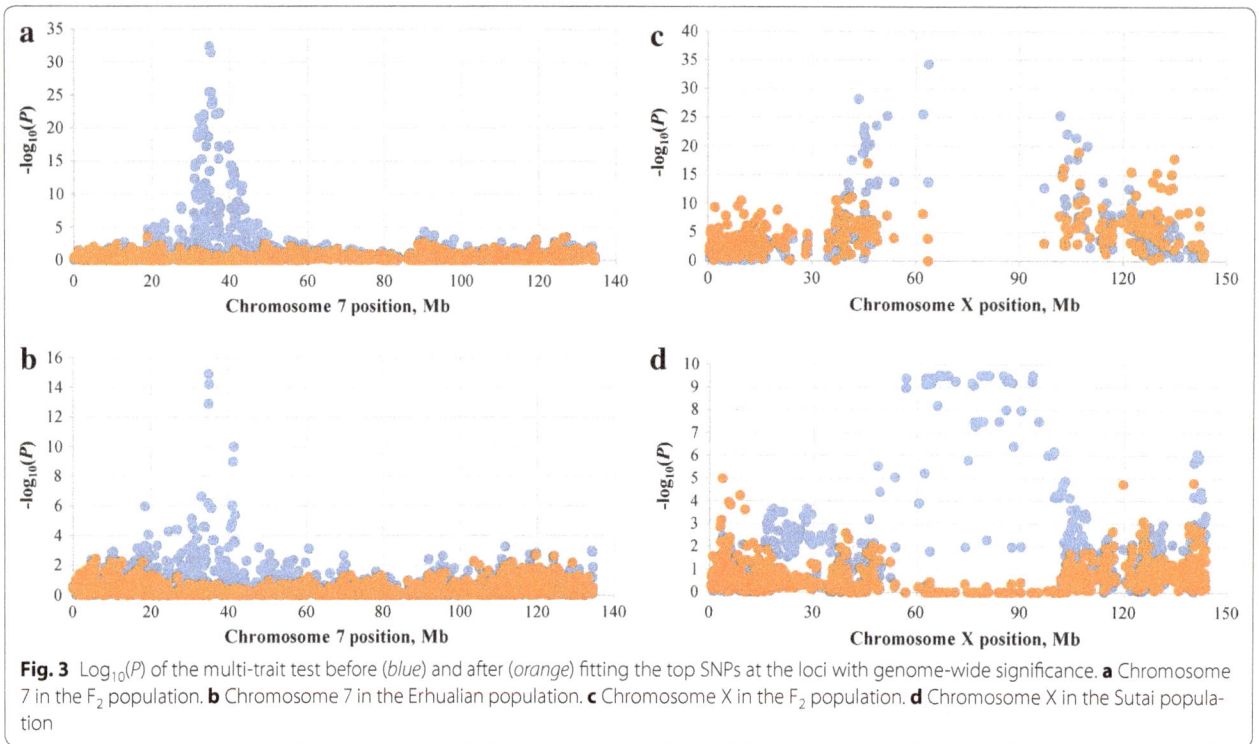

Fig. 3 $\log_{10}(P)$ of the multi-trait test before (*blue*) and after (*orange*) fitting the top SNPs at the loci with genome-wide significance. **a** Chromosome 7 in the F_2 population. **b** Chromosome 7 in the Erhualian population. **c** Chromosome X in the F_2 population. **d** Chromosome X in the Sutai population

Unmapped SNPs improve the power of association analyses

As shown in our previous studies [13, 27, 28], retaining the unmapped SNPs can improve the power of GWAS. In this study, we identified 18 unmapped SNPs that surpassed the suggestive or genome-wide significant level [see Additional file 4: Figure S1]. By applying linkage analysis in the F_2 population, we deduced the approximate genomic locations of 15 unmapped SNPs [see Additional file 6: Table S5], and the use of these SNPs improved the power of the current GWAS. For example, we did not identify any mapped SNPs associated with $ADG_{210-240}$ in the Sutai dataset. However, a SNP (ss131031851) that had two locations (65.29 and 65.38 Mb) on SSC10 displayed an association signal with $ADG_{210-240}$ [see Additional file 4: Figure S1, Additional file 6: Table S5]. The unmapped SNP ss131029816 is another example. We detected only one mapped SNP on SSC6 that was associated with $ADG_{210-240}$ at a suggestive significant level in the SM-GWAS, while the unmapped SNP ss131029816 showed a genome-wide significant association with this trait [see Additional file 4: Figure S1, Additional file 6: Table S5]. In a two-point linkage analysis on the F_2 population, ss131029816 was tightly linked ($\theta = 0.0022$) to ss131566312 at 71.6 Mb on SSC6. Therefore, a genome-wide significant locus for $ADG_{210-240}$ was identified at around 71.6 Mb on SSC6.

Single-population GWAS versus meta-analysis

The SS-GWAS showed that four loci were common to two populations (Fig. 1a). Two loci, one at the proximal end of SSC2 and one in the middle of SSCX, were shared between the F_2 and Sutai populations (Table 1 and Table S3 [see Additional file 3: Table S3]). The locus around 34.80 Mb on SSC7 was associated with fatness traits in both the F_2 and Erhualian datasets (Table 1). This locus may be segregating in the population of Laiwu pigs since we observed a weak signal around it (Fig. 1; see Additional file 4: Figure S1). Another locus in the middle of SSC12 was found to be shared between the Erhualian and Laiwu populations. The segregation of these common loci in multiple populations should allow us to efficiently fine map these loci by using higher-density chips, such as the 600 K SNP chip, since inter-population linkage disequilibrium usually extends over short distances ($r_{0.3}^2 = 10.5$ kb) in Chinese pigs [29].

We noted that there were no common loci across three or four populations in the SS-GWAS. However, the SM-GWAS identified 31 significant loci that were putative common loci across these four populations (Fig. 1b). One explanation for the discrepancy between the two GWAS methods is that the detection power of SS-GWAS is lower than that of SM-GWAS because of the small sizes of individual populations, which prevents the detection of these putative common loci. Another possibility is that some informative SNPs in the SS-GWAS were deleted during the filtering process in the SM-GWAS. Of the 31

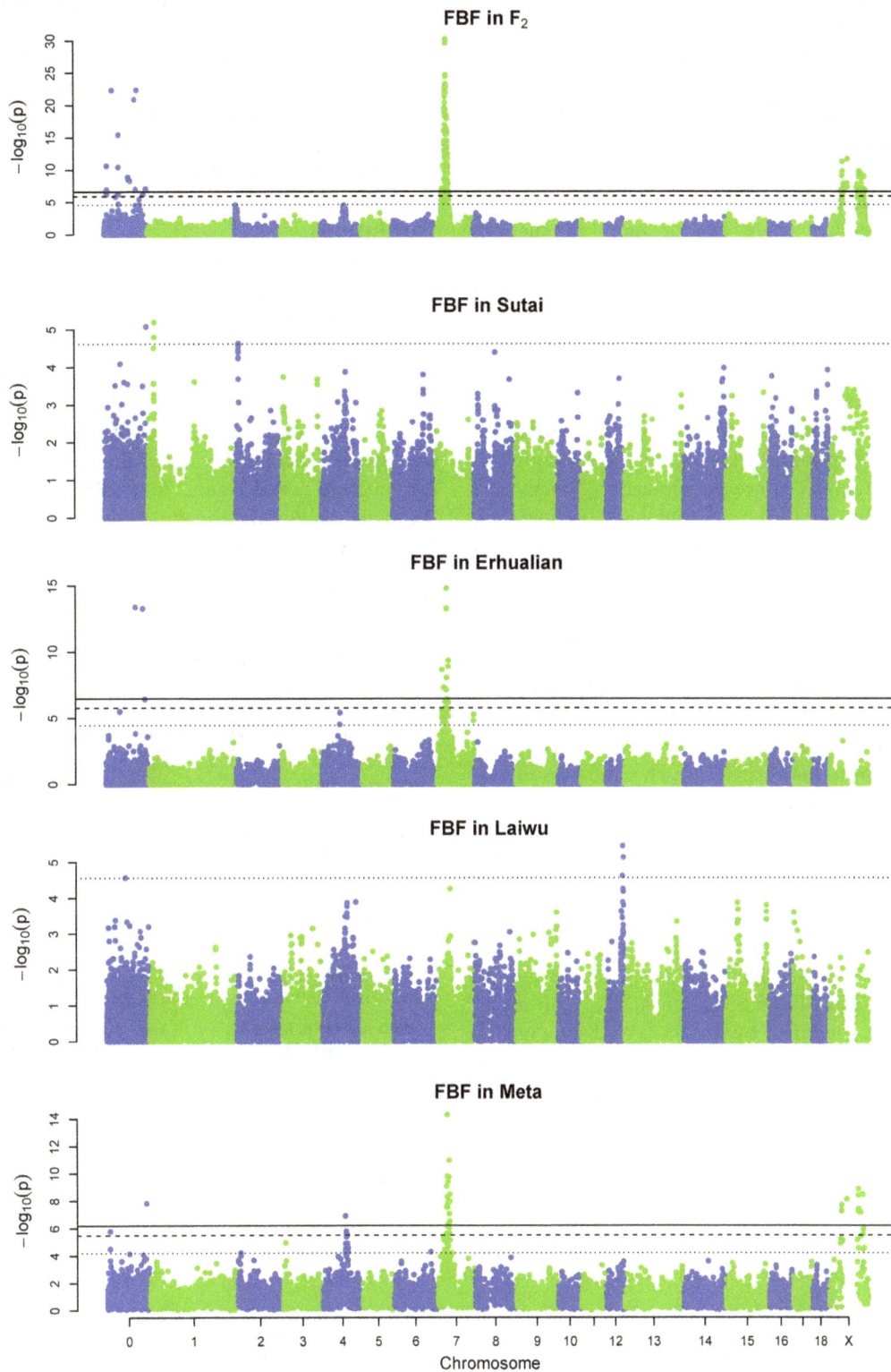

Fig. 4 Manhattan plots of the single-trait GWAS and meta-analysis for backfat thickness at first rib (FBF) in F_2, Sutai, Laiwu and Erhualian populations. The *solid*, *dashed* and *dotted horizontal lines* indicate the 1 and 5% genome-wide and suggestive significant threshold values, respectively. Unmapped SNPs are assigned on chromosome 0 and arbitrary ordered by their names

loci, 13 surpassed the genome-wide significant level. It is interesting that all 13 SNPs had 'single point significant' effects in the F_2 population, while only some of these SNPs showed such effects in the other three populations (Table 2). This is most likely due to the fact that the F_2 population has the largest population size and thus had greater impact on the results of the meta-analysis.

It is known that meta-analysis of GWAS can improve the detection power when two or more populations show consistent association signals with target traits. For example, we detected a suggestive locus for FBF at 65.72 Mb on SSC4 in the Erhualian population [see Additional file 3: Table S3], while this region had only a weak (non significant) association signal with FBF in the other three populations (Fig. 4). By performing a meta-analysis across the four populations, we observed an association signal at the 1% genome-wide significant level in this region (Fig. 4; Table 1). However, population-specific loci could not be detected by the meta-analysis. For instance, we identified an F_2-specific locus for $ADG_{210-240}$ at 53.06 Mb on SSC3. This locus was not detected by the meta-analysis.

Single-trait GWAS versus multi-trait GWAS

If a locus has pleiotropic effects on multiple traits, the multi-trait method can be used to enhance the power of GWAS. For example, the SS-GWAS identified 15 loci in this study, more than half of which were also identified by the MS-GWAS (Fig. 1b). In the SS-GWAS, the minimum P values at the locus on SSC7 for the F_2 and Erhualian populations were 1.75E−18 and 6.96E−14, respectively; whereas these two values decreased to 3.97E−33 and 1.29E−15 in the MS-GWAS, respectively. In contrast, if a locus has no pleiotropic effects, the MS-GWAS can decrease the signal. For instance, the suggestive locus at 303.92 Mb on SSC1 for ADG_{0-210} in the F_2 population was identified by the SS-GWAS but not by the MS-GWAS. Furthermore, it has been reported that multi-trait GWAS can map the locus more accurately than the single-trait method when the marker density is high [11]. It should be noted that the extent of linkage disequilibrium is usually large in F_2 populations, for which a multi-trait GWAS based on medium-density SNP chips, such as 60 K SNPs, is expected to have high detection power.

Linkage versus pleiotropy

By applying multi-trait GWAS, we detected three prominent loci on SSC4, 7 and X (Fig. 2), respectively. The detection of these loci could be caused by pleiotropy or closely-linked causal QTL. The SSC7 locus detected for the F_2 and Erhualian populations, as well as the SSCX locus for the Sutai pigs (Fig. 2), are most likely loci with pleiotropic effects, since we found no SNP that showed an association signal with the nine traits when the top SNPs were fixed in the MS-GWAS model (Fig. 3a, b, d). After correction for the effect of the top SNP on SSCX, several SNPs were still associated with the traits tested in F_2 pigs (Fig. 3c). This implies that the locus on SSCX could be caused by closely-linked causal variants in F_2 pigs.

Conclusions

In this study, we explored four GWAS approaches to identify genomic loci for nine growth and fatness traits in four pig populations. Compared to the single-trait analysis, the meta-analysis had less power to identify population-specific loci but more power to detect population-shared loci. Compared to the single-trait analysis, the multiple-trait analysis reduced the power to detect trait-specific loci but enhanced the power to identify the common loci across traits. Our findings demonstrate that the meta-analysis and the multi-trait method can be used to increase the power of GWAS.

Additional files

Additional file 1: Table S1. Descriptive statistics and differences between females and males for nine growth and fatness traits in the Erhualian and Laiwu pigs.

Additional file 2: Table S2. Phenotypic correlation coefficients between nine growth and fatness traits in the Laiwu and Erhualian pigs.

Additional file 3: Table S3. The suggestive significant loci for nine fatness and growth traits identified in this study.

Additional file 4: Fig. S1. Manhattan plots of the single-trait GWAS and meta-analysis for nine fatness and growth traits in the F_2, Sutai, Laiwu and Erhualian populations. The solid, dashed and dotted horizontal lines indicate the 1% and 5% genome-wide and suggestive significant threshold values, respectively. Unmapped SNPs are assigned on chromosome 0 and arbitrary ordered by their names.

Additional file 5: Table S4. The additive effects of suggestive significant SNPs detected by the meta-analysis in the four populations.

Additional file 6: Table S5. Tight linkage of unassigned significant SNPs with the mapped SNPs. This table shows approximate genomic positions of the unassigned significant SNPs by a two-point linkage analysis in the F_2 population.

Authors' contributions
LSH conceived and designed the experiment. YG and JR analyzed the data. YG, YH and LJH performed the experiment. JM, CC and HA contributed materials. JR and YG wrote the manuscript. All authors read and approved the final manuscript.

Acknowledgements
This research was supported by the National Natural Science Foundation of China (31525023, 31460590, 31660303) and Jiangxi province (20142BAB204017) to JR and YG.

Competing interests
The authors declare that they have no competing interests.

References

1. Miller ER, Ullrey DE. The pig as a model for human nutrition. Annu Rev Nutr. 1987;7:361–82.
2. Bidanel JP, Milan D, Iannuccelli N, Amigues Y, Boscher MY, Bourgeois F, et al. Detection of quantitative trait loci for growth and fatness in pigs. Genet Sel Evol. 2001;33:289–309.
3. Walling GA, Visscher PM, Andersson L, Rothschild MF, Wang L, Moser G, et al. Combined analyses of data from quantitative trait loci mapping studies. Chromosome 4 effects on porcine growth and fatness. Genetics. 2000;155:1369–78.
4. Rohrer GA. Identification of quantitative trait loci affecting birth characters and accumulation of backfat and weight in a Meishan-White Composite resource population. J Anim Sci. 2000;78:2547–53.
5. Sato S, Oyamada Y, Atsuji K, Nade T, Kobayashi E, Mitsuhashi T, et al. Quantitative trait loci analysis for growth and carcass traits in a Meishan × Duroc F2 resource population. J Anim Sci. 2003;81:2938–49.
6. Andersson L, Haley CS, Ellegren H, Knott SA, Johansson M, Andersson K, et al. Genetic mapping of quantitative trait loci for growth and fatness in pigs. Science. 1994;263:1771–4.
7. Ramos AM, Crooijmans RP, Affara NA, Amaral AJ, Archibald AL, Beever JE, et al. Design of a high density SNP genotyping assay in the pig using SNPs identified and characterized by next generation sequencing technology. PLoS One. 2009;4:e6524.
8. Evangelou E, Ioannidis JP. Meta-analysis methods for genome-wide association studies and beyond. Nat Rev Genet. 2013;14:379–89.
9. Korol AB, Ronin YI, Itskovich AM, Peng J, Nevo E. Enhanced efficiency of quantitative trait loci mapping analysis based on multivariate complexes of quantitative traits. Genetics. 2001;157:1789–803.
10. Knott SA, Haley CS. Multitrait least squares for quantitative trait loci detection. Genetics. 2000;156:899–911.
11. Bolormaa S, Pryce JE, Reverter A, Zhang Y, Barendse W, Kemper K, et al. A multi-trait, meta-analysis for detecting pleiotropic polymorphisms for stature, fatness and reproduction in beef cattle. PLoS Genet. 2014;10:e1004198.
12. Guo Y, Mao H, Ren J, Yan X, Duan Y, Yang G, et al. A linkage map of the porcine genome from a large-scale White Duroc × Erhualian resource population and evaluation of factors affecting recombination rates. Anim Genet. 2009;40:47–52.
13. Guo YM, Zhang XF, Ren J, Ai HS, Ma JW, Huang LS. A joint genome-wide association analysis of pig leg weakness and its related traits in an F2 population and a Sutai population. J Anim Sci. 2013;91:4060–8.
14. Wang L, Wang A, Wang L, Li K, Yang G, He R, et al. Animal genetic resources in China: pigs. Beijing: China Agricultural Press; 2011.
15. Astle W, Balding DJ. Population structure and cryptic relatedness in genetic association studies. Stat Sci. 2009;24:451–71.
16. Amin N, van Duijn CM, Aulchenko YS. A genomic background based method for association analysis in related individuals. PLoS One. 2007;2:e1274.
17. Lander E, Kruglyak L. Genetic dissection of complex traits: guidelines for interpreting and reporting linkage results. Nat Genet. 1995;11:241–7.
18. Chen WM, Abecasis GR. Family-based association tests for genomewide association scans. Am J Hum Genet. 2007;81:913–26.
19. Devlin B, Roeder K. Genomic control for association studies. Biometrics. 1999;55:997–1004.
20. Zheng G, Freidlin B, Li Z, Gastwirth JL. Genomic control for association studies under various genetic models. Biometrics. 2005;61:186–92.
21. Green P, Falls K, Crooks S. Cri-map Version 2.4. St. Louis: Washington University School of Medicine; 1994.
22. Sobel E, Lange K. Descent graphs in pedigree analysis: applications to haplotyping, location scores, and marker-sharing statistics. Am J Hum Genet. 1996;58:1323–37.
23. Barrett JC, Fry B, Maller J, Daly MJ. Haploview: analysis and visualization of LD and haplotype maps. Bioinformatics. 2005;21:263–5.
24. Ai H, Ren J, Zhang Z, Ma J, Guo Y, Yang B, et al. Detection of quantitative trait loci for growth- and fatness-related traits in a large-scale White Duroc × Erhualian intercross pig population. Anim Genet. 2012;43:383–91.
25. Qiao RM, Gao J, Zhang ZY, Lin L, Xie XH, Fan Y, et al. Genome-wide association analyses reveal significant loci and strong candidate genes for growth and fatness traits in two pig populations. Genet Sel Evol. 2015;47:17.
26. Hu ZL, Park CA, Wu XL, Reecy JM. Animal QTLdb: an improved database tool for livestock animal QTL/association data dissemination in the post-genome era. Nucleic Acids Res. 2013;41:D871–9.
27. Guo Y, Hou L, Zhang X, Huang M, Mao H, Chen H, et al. A meta analysis of genome-wide association studies for limb bone lengths in four pig populations. BMC Genet. 2015;16:95.
28. Guo YM, Zhang ZY, Ma JW, Ai HS, Ren J, Huang LS. A genomewide association study of feed efficiency and feeding behaviors at two fattening stages in a White Duroc × Erhualian F population. J Anim Sci. 2015;93:1481–9.
29. Ai H, Huang L, Ren J. Genetic diversity, linkage disequilibrium and selection signatures in chinese and Western pigs revealed by genome-wide SNP markers. PLoS One. 2013;8:e56001.

3

Genetic and economic benefits of selection based on performance recording and genotyping in lower tiers of multi-tiered sheep breeding schemes

Bruno F. S. Santos[1,2]* [ID], Julius H. J. van der Werf[2,3], John P. Gibson[2], Timothy J. Byrne[1] and Peter R. Amer[1]

Abstract

Background: Performance recording and genotyping in the multiplier tier of multi-tiered sheep breeding schemes could potentially reduce the difference in the average genetic merit between nucleus and commercial flocks, and create additional economic benefits for the breeding structure.

Methods: The genetic change in a multiple-trait breeding objective was predicted for various selection strategies that included performance recording, parentage testing and genomic selection. A deterministic simulation model was used to predict selection differentials and the flow of genetic superiority through the different tiers. Cumulative discounted economic benefits were calculated based on trait gains achieved in each of the tiers and considering the extra revenue and associated costs of applying recording, genotyping and selection practices in the multiplier tier of the breeding scheme.

Results: Performance recording combined with genomic or parentage information in the multiplier tier reduced the genetic lag between the nucleus and commercial flock by 2 to 3 years. The overall economic benefits of improved performance in the commercial tier offset the costs of recording the multiplier. However, it took more than 18 years before the cumulative net present value of benefits offset the costs at current test prices. Strategies in which recorded multiplier ewes were selected as replacements for the nucleus flock did modestly increase profitability when compared to a closed nucleus structure. Applying genomic selection is the most beneficial strategy if testing costs can be reduced or by genotyping only a proportion of the selection candidates. When the cost of genotyping was reduced, scenarios that combine performance recording with genomic selection were more profitable and reached breakeven point about 10 years earlier.

Conclusions: Economic benefits can be generated in multiplier flocks by implementing performance recording in conjunction with either DNA pedigree recording or genomic technology. These recording practices reduce the long genetic lag between the nucleus and commercial flocks in multi-tiered breeding programs. Under current genotyping costs, the time to breakeven was found to be generally very long, although this varied between strategies. Strategies using either genomic selection or DNA pedigree verification were found to be economically viable provided the price paid for the tests is lower than current prices, in the long-term.

*Correspondence: bsantos@abacusbio.co.nz
[1] AbacusBio Limited, PO Box 5585, Dunedin 9058, New Zealand
Full list of author information is available at the end of the article

Background

In most commercial sheep production systems, improvement of genetic merit is limited to outside sire purchases. In general, selection of ewes in commercial flocks is driven by conformation traits such as soundness and constitution, and sometimes, based on the size of the litter in which the replacement candidate was born. While there is limited scope for selection among ewes within commercial sheep flocks due to limited opportunities to undertake voluntary culling, there may be value in selecting future commercial sires and dams in the multiplier tier of multi-tiered breeding structures composed of nucleus, multiplier and commercial tiers. Performance recording of candidates in the multiplier tier could potentially reduce the difference in the average genetic merit between nucleus and commercial tiers, normally referred to as genetic lag, creating additional economic benefits for the breeding structure. In a typical multiplier flock, selection based on performance brings complexity and can involve substantial costs [1]. However, in commercially integrated multi-tiered breeding structures, sufficient value may be captured to offset these costs, particularly if performance recording in large-scale flocks is facilitated by technology such as electronic identification tools (EID), which ensure reliable individual identification and allow automation for accurate performance data recording [2].

There is also opportunity to exploit recent advances in molecular genetics technology. For instance, DNA parentage testing allows the combination of information from an individual's relatives and its own phenotypic records to provide the basis for prediction of genetic merit [3]. Meuwissen et al. [4] proposed genomic selection (GS), which enables selection decisions to be made early in the life of animals [5], with highest benefits for traits that are more difficult to measure and have low heritability, or are recorded late in life [6]. For GS to be accurate, it is necessary to record large amounts of phenotypic data on genotyped animals [7]. Traits recorded in the multiplier tier could contribute to the reference population needed for genomic prediction, increasing the overall accuracy of selection within the breeding scheme.

Previous studies have estimated the potential benefits of DNA testing in nucleus sheep breeding [6, 8–10]. However, these studies did not model the impact of implementation of performance recording combined with DNA testing in the multiplier tier of multi-tiered breeding schemes. Lack of estimates of the economic benefit of such technologies prevents the identification of a price point for DNA testing at which implementation becomes profitable in multi-tiered breeding schemes.

In the absence of selection in lower tiers, the genetic lag between the nucleus and multiplier, and also between the multiplier and commercial tiers is typically two generations of genetic progress [11–13]. This assumes that all rams used in the lower tiers are obtained from random progeny selected in the tier immediately above, and that no ewe transfer occurs between the tiers. Supplying improved breeding males to the commercial flock by recording the multiplier tier could consequently reduce the genetic lag relative to the nucleus. The hypothesis of this study is that benefits arise from the selection differential created in the multiplier tier through selection based on breeding values and/or genomic prediction when selecting rams for transfer to the commercial tier, and also when selecting replacement ewes from the multiplier to the nucleus tier.

The objective of this study was to quantify production and economic benefits obtained in the commercial tier of a multi-tiered breeding scheme after the introduction of performance recording and DNA parentage testing or genomic selection in the multiplier tier. A deterministic model was developed to evaluate different selection decisions based on combinations of trait recording, genotyping strategies and ewe replacement policies.

Methods
Overview

In this study, we calculated the additional economic benefits from improved performance of commercial animals due to recording and DNA testing in the multiplier tier, relative to a breeding structure without performance recording in the multiplier tier. We developed a simulation model for an integrated three-tier structure to estimate the overall benefits of genetic gain over time. Selection differentials were calculated for traits within a multiple trait breeding objective using selection index theory. The dissemination of this genetic superiority throughout the population was predicted based on gene flow methodology, as developed by Bichard [11]. We developed a simulation model which predicts trait-specific estimates of genetic merit in each cohort (*tier × sex × age group*) and calculates discounted genetic expression (DGE) coefficients used to quantify the timing and frequency of expressions of the genetic superiority that flows through to the commercial tier. Trait heritabilities, phenotypic and genetic variances and correlations, along with descriptions of the numbers of records on individual selection candidates and their relatives were used in the selection index model developed by van der Werf [14]. The total extra revenue and associated costs of applying recording and selection practices in the multiplier tier of the breeding scheme were calculated based on trait DGE coefficients and economic values. The marginal benefits of performance recording and genotyping for various scenarios were then compared with a base scenario, where there was no performance recording, parentage assignment, or genomic selection in the multiplier tier of the breeding scheme.

Structure of the breeding scheme

The study was based on the structure of the sheep industry in New Zealand. The breeding scheme underlying most of these systems is a multi-tiered structure, which normally involves the nucleus (ram breeder) and the commercial tier, and in some situations, a multiplier tier. Table 1 presents key parameters related to the modelled breeding scheme that supports a large commercial tier of 180,000 ewes.

Recursive prediction of genetic merit in the nucleus and multiplier tiers

Recursive equations were used to calculate the average genetic merit in the nucleus and multiplier tiers. The resulting equations were functions of genetic merit of animals born in previous years and corresponding selection differentials. The average genetic merit of offspring born in year y for animals in tier T of the breeding structure, for trait j, was calculated as:

$$O^j_{y,T} = \frac{S^j_{y,T} + D^j_{y,T}}{2},$$

where $S^j_{y,T}$ and $D^j_{y,T}$ are the average genetic merit of the sires and dams of the offspring, respectively. These in turn were calculated as:

Table 1 Parameters describing the structure and performance of different tiers within the three-tiered breeding scheme

Parameter	Unit	Tier within the breeding program		
		Nucleus	Multiplier	Commercial
Flock breeding ewes	Head	826	7000	180,000
Flock breeding rams	Head	10	70	1800
Ewes mated to terminal sires	%	0	0	20
Ewes/ram	Head	80	100	100
Ewe replacement rate	%	35	35	30
Mixed age ewe lambing rate[a]	%	210	190	165
Ewe lamb lambing rate	%	100	90	80
Lamb survival	%	79	79	82
Weaning rate	%	166	150	135
Lambs sold as stores[b]	%	–	–	20

[a] Lambing percentage of ewes of 2 years old or older, per ewe mated

[b] Slaughter lambs sold to be finished off farm

$$S^j_{y,T} = \phi_{y,T=N}$$
$$\cdot \left[\sum_{t=1}^{6} \rho_{t,T=N,s=m} \cdot \left(O^j_{y-t,T=N} + \Delta^j_{y-t,T=N,s=m} \right) \right]$$
$$+ \left(1 - \phi_{y,T=N} \right)$$
$$\cdot \left[\sum_{t=1}^{6} \rho_{t,T=M,s=m} \cdot \left(O^j_{y-t,T=M} + \Delta^j_{y-t,T=M,S=m} \right) \right]$$

and

$$D^j_{y,T} = \phi_{y,T=N}$$
$$\cdot \left[\sum_{t=1}^{7} \rho_{t,T=N,s=f} \cdot \left(O^j_{y-t,T=N} + \Delta^j_{y-t,T=N,s=f} \right) \right]$$
$$+ \left(1 - \phi_{y,T=N} \right)$$
$$\cdot \left[\sum_{t=1}^{7} \rho_{t,T=M,s=f} \cdot \left(O^j_{y-t,T=M} + \Delta^j_{y-t,T=M,s=f} \right) \right]$$

where $\phi_{y,T}$ is the proportion of sires and dams of lambs born in year y that originate from either the nucleus tier ($T = N$) or the multiplier tier ($T = M$), and $\rho_{t,T,s}$ is the proportion of sires ($s = m$) and dams ($s = f$) coming from age group t in tier T. The selection differentials $\Delta_{y-t,T,s}$ give the superiority of animals of sex s in age group t that were born in year $y - t$, selected to become parents in tier T over all animals born that year, and are calculated using selection index theory as described in a later section. For the initial years ($y - t < 0$ where 0 is the initial year of recording and selection in the multiplier tier), when dam and sire genetic merit values would be required to be derived from offspring that have not yet been generated, the merit of the missing offspring was calculated by assuming a constant linear rate of genetic progress among age groups. The model does not currently allow optimised selection across age cohorts as it was assumed that the age profile of selected rams is often predetermined due to the need to have a number of mature rams to mate younger ewes, and to provide genetic connectedness among year classes of lambs. In addition, there is an implicit assumption that those rams and ewes mated in any year subsequent to their first mating were kept at random from the ewe and ram lambs originally selected.

Elite rams required for the nucleus and multiplier tiers were sourced from within the nucleus. Ram lambs born and selected in the multiplier tier were used as sires of lambs in the commercial tier. The recorded multiplier tier produced its own ewe replacements where young female candidates were selected based on genetic merit or on their phenotypic performance depending on the scenario under consideration. Replacement nucleus ewes were sourced from within either the nucleus or multiplier tiers, depending on the scenario. The commercial tier produced its own ewe replacements based on traditional non-recording methods.

Flow of multiplier ram's genes into the commercial tier

Discounted genetic expressions (DGE) coefficients predict the proportion of genetic superiority that is transmitted to an individual's descendants through transfer of genes [15]. In the current study, DGE coefficients model the flow of genes from multiplier rams once they enter the commercial flock for mating. We followed the methodology first proposed by McClintock and Cunningham [16], also used by Amer [17] and Berry et al. [18], to predict the timing and frequency of genetic expressions which deliver the ultimate economic benefits at commercial flock level. The expressions of genes in different age groups were calculated using five distinct matrices by adapting the methodology described by Amer [17]. Tables 2 and 3 present the age distribution and survival in the different tiers, assuming a constant age distribution structure of breeding ewes and rams in all tiers.

These parameters, used in the calculations of DGE coefficients in the gene-flow model, were obtained from the commercial breeding scheme representing a typical set of farmers within the New Zealand sheep industry.

The five expression matrices $\mathbf{D}$, $\mathbf{E}$, $\mathbf{F}$, $\mathbf{G}$ and $\mathbf{H}$, account for the probability ($\mathbf{a}_i$) of ewes of all ages surviving to the next age, across the different age groups in time, quantifying the flow of genes from parents to the other age groups. The matrix $\mathbf{D}$ has dimension $h \times h$, where h is defined in years equivalent to the number of age groups, i.e. $h = 7$. Matrix $\mathbf{D}$ maps the probability ($\mathbf{a}_i$) of a ewe surviving to the next age group in successive years and is defined as:

$$\mathbf{D}_{i,j} = \begin{cases} \mathbf{a}_{i-j}, & \text{for} \ldots j < i-1 \ldots \text{and} \ldots i-j \leq c \\ 0 & \ldots \text{otherwise} \end{cases},$$

where i and $j = 1, \ldots h$, and c is a cull for age threshold. The vectors of increments of genetic superiority, by year k of expression for each generation ($\mathbf{g}_k$) for the seven age groups across all cohorts, account for the ewe's genetic contribution to the progeny and is calculated as:

$$\mathbf{g}_k = \frac{1}{2} \cdot f \cdot \mathbf{D} \cdot \mathbf{g}_{k-1},$$

Table 2 Ewe age structure in different tiers of the breeding scheme supporting the commercial tier

Age of ewes (years)	Proportions of ewe age groups in the flock ($\rho_{s=f}$)			Probability of ewe survival to age group i given presence in age group 2 ($\mathbf{a}_i$)	Probability of a ewe dying or being culled at age i ($\mathbf{d}_i$)	Prolificacy by age group (Lrp$_i$)
	Nucl	Mult	Comm	Comm	Comm	Comm
1[a]	0.09	0.00	0.00	1.00	0.00	0.80
2	0.32	0.35	0.30	1.00	0.23	1.49
3	0.25	0.23	0.23	0.77	0.10	1.65
4	0.18	0.18	0.20	0.67	0.17	1.65
5	0.14	0.13	0.15	0.50	0.17	1.65
6	0.01	0.09	0.10	0.33	0.28	1.65
7	0.00	0.02	0.02	0.06	0.05	1.65

Nucl is the nucleus tier of the breeding scheme, Mult is the multiplier tier and Comm is the commercial tier

[a] This refers to the proportion of ewes mated in the first year of age as ewe lambs

Table 3 Ram age structure in different tiers of the breeding scheme supporting the commercial tier

Age of rams (years)	Proportions of ram age groups in the flock ($\rho_{s=f}$)			Probability of a ram surviving to age group i given presence in age group 1 ($\mathbf{a}_i$)		
	Nucl	Mult	Comm	Nucl	Mult	Comm
1	0.35	0.28	0.28	1.00	1.00	1.00
2	0.30	0.27	0.27	0.86	0.95	0.95
3	0.15	0.25	0.25	0.43	0.90	0.90
4	0.10	0.20	0.20	0.29	0.72	0.72
5	0.05	0.00	0.00	0.14	0.00	0.00
6	0.05	0.00	0.00	0.14	0.00	0.00

Nucl is the nucleus tier of the breeding scheme, Mult is the multiplier tier and Comm is the commercial tier

where f is the number of ewe lambs required as replacements per ewe lambing per year. Aggregate yearly genetic expressions accumulated over the generations are calculated as the sum of vectors, i.e. $g_{sum} = \sum_{k=1}^{7} \mathbf{g}_k$.

Additional expression matrices $\mathbf{E}$, $\mathbf{F}$, $\mathbf{G}$ and $\mathbf{H}$ (all with $h \times h$ dimension) are used to map the occurrence of genes at the birth of a new generation that expresses specific categories of traits over the years of the lives of different classes of animals in each generation. The (i, j) elements of matrix $\mathbf{E}$ represent the number of lambs produced for slaughter per ewe within each age group repeated as columns within matrix $\mathbf{E}$, but with elements shifted down by one row for successive columns, and are calculated as:

$$\mathbf{E}_{i,j} = \begin{cases} \mathbf{a}_{i-j} \cdot (\mathbf{v}_i - f), & \text{for} \ldots j < i - 1 \ldots \text{and} \ldots i - j \leq c \\ 0 & \ldots \text{otherwise} \end{cases},$$

where $\mathbf{v}_i$ is the number of lambs weaned at age i in years, c is a cull for age threshold. The elements of the lambing expressions matrix $\mathbf{F}$ represent the number of lambs born within each age group, such that:

$$\mathbf{F}_{i,j} = \begin{cases} \mathbf{a}_{i-j} \cdot \mathbf{v}_{i-j}, & \text{for} \ldots j < i - 1 \ldots \text{and} \ldots i - j \leq c \\ 0 & \ldots \text{otherwise} \end{cases},$$

while $\mathbf{G}$ contains elements representing proportions of ewes dying or being culled in the different ages $(\mathbf{d}_i)$ as:

$$\mathbf{G}_{i,j} = \begin{cases} \mathbf{d}_{i-j}, & \text{for} \ldots j < i - 1 \ldots \text{and} \ldots i - j \leq c + 1 \\ 0 & \ldots \text{otherwise} \end{cases}$$

and $\mathbf{H}_{i,j}$ describes the expressions of replacement ewes (18 months old), with elements of 1 for $i = 2$ and $j = i + 1$, or 0 otherwise.

For breeding rams that are transferred from the multiplier tier to the commercial tier, genetic expressions transmitted via replacement daughters were obtained by multiplying the cumulative genetic superiority that is expressed in each age group (g_{sum}) by the relevant expression matrix of each trait group. These traits were grouped in four vectors denoted $\mathbf{w}$ which count the number of expressions of the genes of a ewe replacement that enters the flock by itself, and her descendants. The rows of the vector $\mathbf{w}$ correspond to the year following the birth of the replacement female. Vectors $\mathbf{w}$ are superscripted for traits that are expressed annually by breeding ewes ($\mathbf{w}^A$), in hogget replacement ewes ($\mathbf{w}^H$), at birth by lambs ($\mathbf{w}^L$), and at slaughter by lambs ($\mathbf{w}^S$), and were calculated as:

$$\mathbf{w}^A = \mathbf{D} \cdot g_{sum}, \quad \mathbf{w}^H = \mathbf{H} \cdot g_{sum},$$

$$\mathbf{w}^L = \mathbf{F} \cdot \frac{1}{2 \cdot ls} g_{sum} + k_L \quad \text{and} \quad \mathbf{w}^S = \mathbf{E} \cdot \frac{1}{2 \cdot ls} g_{sum} + k_S$$

where ls is lamb survival from birth to slaughter. Because surplus lambs are generated in the process of breeding replacement ewes, and these lambs express traits at

both birth and slaughter, constant adjustment factors k_L (for lambing traits) and k_S (for slaughter traits), are incorporated into the equations for $\mathbf{w}^L$ and $\mathbf{w}^S$, respectively. These adjustment factors give the direct genetic expressions of a slaughter trait per replacement ewe kept, which is calculated by using an adaptation of formulae described in Amer [17], based on the proportion of ewe lambs surviving to slaughter age that are retained as ewe flock replacements (u) as:

$$k_L = \left(\frac{1}{ls}\right)\left(\frac{1}{2}u\right)^{-1} \quad \text{and} \quad k_L = \left(1 - \frac{1}{2}u\right)\left(\frac{1}{2}u\right)^{-1}.$$

It was necessary to link expressions by ewe replacements in the various $\mathbf{w}$ vectors to the number of sires mated in the commercial flock to generate replacements over multiple mating years. To do this, matrices $\mathbf{Z}^j$ for each trait type j (lambing, slaughter, annual ewe and replacement ewe traits) with columns made up of the lagged expression vectors for the corresponding trait type were applied. These matrices represent ram matings over successive years assuming they survive, but lagged down one row, for each successive potential year of re-mating. Thus, rows of matrices $\mathbf{Z}$ correspond to the year e following the first mating, and columns t correspond to the successive years of mating of the ram. Then, survival of the ram can be taken into account when computing a final vector of commercial trait expressions, and it was convenient to scale these expression vectors so that the sum of the elements in the expression vector equals 1. This procedure was applied to the different trait groups (j) and the calculation was:

$$\varepsilon_e^j = \frac{\sum_t \mathbf{Z}_{e,t}^j \cdot \alpha_t}{\sum_t \sum_e \mathbf{Z}_{e,t}^j \cdot \alpha_t},$$

where α_t is the probability of a breeding ram surviving in the commercial flock to year t after its first year of mating.

Thus, ε_e^j is the sum of discounted expressions of trait j by a ram from the multiplier tier in each year (e) of expression after its first mating in the commercial flock but expressed as a proportion of its total lifetime sum of expressions. The numerator adds up the expressions by year of expression (after first mating), and the denominator standardises the expressions into a proportion by year of expression relative to the overall expressions.

Selection differentials

This study applied selection index principles to quantify responses to selection based on a pre-determined multiple-trait breeding objective. The definition of the aggregate breeding value of selection candidates, across tiers, was calculated as the sum of the products of economic weights (ew_j) of the traits j composing the breeding objective (made up of n traits), and their respective breeding values (ebv_j), and computed by $H = \sum_1^n (ew_j \cdot ebv_j)$, as described by Hazel et al. [19].

Selection differentials were calculated for each of the component traits of the breeding objective described in Table 4. The selection differentials were computed based on deterministic selection index equations [14]. The selection index model predicted index weights, the index additive genetic variance, and the consequent regression coefficients of each component trait on the index. The selection differentials were obtained as the product of the index standard deviation ($\sigma_{T,s}$), the respective regression coefficients of traits on the index ($b^j_{T,s}$) and the selection intensities (i) corresponding to a year of the breeding program (y), in the tier from which the parents were selected (T) and the sex of the parents (s), as follows:

$$\Delta^j_{y,T,s} = \sigma_{T,s} \cdot b^j_{T,s} \cdot i_{y,T,s}.$$

The regression coefficients are calculated as $b^j_{T,s} = \frac{Cov(I_{T,s}, tbv_j)}{Var(I_{T,s})}$, where $I_{T,s}$ is the index derived by using

standard selection index theory and assuming a set of information sources appropriate for selection candidates in tier T and of sex s, and tbv is the true breeding value of trait j. The parameters required for the calculation of the regression coefficients and selection differentials are in Tables 4, 5 and 6. Table 4 presents trait economic values (EV_j) calculated as the marginal profit per unit change in the trait j per animal in the class where the trait is expressed [20]. Economic values were used to calculate benefits of genetic changes in animals that express the relevant trait, whereas economic weights (ew_j) were used to define optimal index weights underlying the selection index model, and therefore EV_j incorporated standard DGE coefficients [17] that are used in the national breeding index for dual purpose sheep in New Zealand. Trait accuracies (r) and genomic prediction accuracies (r_{GBV}) were also obtained from the New Zealand national genetic evaluation system, Sheep Improvement Limited (SIL). Trait accuracies represent the correlation between estimated breeding values (EBV) and true breeding values. Genomic prediction accuracies represent the correlation between pedigree-based EBV and the genomic breeding values (GBV) estimated based on genomic

Table 4 Heritability (h^2), genetic standard deviation (σ_g), accuracies (r), accuracy of genomic prediction (r_{GBV}), economic values (EV) and weights (ew) for various traits used in the simulation

DGE Trait group[a]	Trait (abbreviation)	Unit	h^2	σ_g	r	r_{GBV}	EV ($/unit)	ew ($/unit)
Slaughter	Carcase weight (CWT)	kg	0.30	1.10	0.60	0.50	2.60	3.74
	Weaning weight (WWT)	kg	0.20	1.57	0.58	0.48	0.93	1.36
Annual	Number of lambs born (NLB)	Lambs	0.10	0.18	0.25	0.54	22.31	22.31
	Ewe mature weight (EWT)	kg	0.45	0.99	0.30	0.50	−0.94	−1.49
	Ewe body condition score (BCS)	Score	0.18	0.30	0.30	–	12.93	12.93
	Survival maternal (SURm)	Lambs	0.05	0.09	0.16	–	52.20	83.78
	Weaning weight maternal (WWTm)	kg	0.10	1.11	0.25	0.47	1.02	1.21
Hogget	Stayability (Stay)	Binary	0.15	0.15	0.41	–	19.28	19.28
Lambing	Lamb survival (SUR)	Lambs	0.01	0.04	0.13	–	52.20	92.46

[a] Traits grouped by animal class that expresses the trait

Table 5 Genetic (below diagonal) and phenotypic (above diagonal) correlations between traits used in the selection index model

Trait (abbreviation)	WWT	WWTm	NLB	SUR	SURm	EWT	BCS	Stay	CWT
Weaning weight (WWT)	1.00	0.00	0.00	0.00	0.00	0.55	0.00	0.00	0.75
Weaning weight maternal (WWTm)	0.00	1.00	0.00	0.00	0.00	0.00	0.00	0.00	0.00
Number of lambs born (NLB)	0.00	0.00	1.00	0.00	0.00	0.00	0.00	0.00	0.00
Lamb survival (SUR)	0.00	0.00	0.00	1.00	0.00	0.00	0.00	0.00	0.00
Survival maternal (SURm)	0.00	0.00	0.00	0.00	1.00	0.00	0.00	0.00	0.00
Ewe mature weight (EWT)	0.55	0.00	0.00	0.00	0.00	1.00	0.50	0.20	0.75
Ewe body condition score (BCS)	0.00	0.00	0.00	0.00	0.00	0.50	1.00	0.20	0.00
Stayability (Stay)	0.00	0.00	0.00	0.00	0.00	0.20	0.20	1.00	0.00
Carcass weight (CWT)	0.75	0.00	0.00	0.00	0.00	0.75	0.00	0.00	1.00

Table 6 Selection proportions and resulting selection intensities for ewe and ram lambs in different tiers

Parameter	Nucleus		Multiplier		Commercial	
	Ewes	Rams	Ewes	Rams[b]	Ewes	Rams[c]
Selection proportion[a]	0.70	0.05	0.80	0.20	0.00	0.30
Selection intensity (i)	0.49	2.06	0.35	1.40	0.00	1.16

[a] Proportion of candidates selected to potentially become a replacement ewe or a breeding ram

[b] Rams selected in the nucleus to mate ewes in the multiplier tier

[c] Rams selected in the multiplier to mate ewes in the commercial tier

and phenotypic information from the national reference population [6, 7, 21]. According to Auvray et al. [21], the training set is made up of a mixture of pure and cross-bred animals, with Illumina OvineSNP50K BeadChip [50K single nucleotide polymorphism (SNP) chip] genotypes from 13,420 individuals to investigate BLUP with different genomic relationship matrices and SNPs and to predict the GBV of younger animals.

The methodology modelled genomic selection by defining GBV as additional traits, which are genetically and phenotypically correlated with the traits included in the selection index model (Table 5), similar to methodology described by Dekkers [22]. Because, in practice, GBV for specific traits are expected to be genetically correlated with other traits, the correlations between GBV traits and each other trait was calculated as, $r_{GBV_i,BV_j} = r_{BV_i,BV_j} \cdot r_{GBV_i,BV_i}$, where r_{BV_i,BV_j} is the genetic correlation between traits i and j, and r_{GBV_i,BV_i} is the accuracy of genomic prediction for trait i, presented in Table 4. The calculations assumed that GBV had a phenotypic standard deviation of 1.0, a heritability of 0.95 and an economic value of 0.

Table 6 presents selection proportions and selection intensities for each sex and tier. These result from replacement policy decisions and the age structure of the flock, which influence the proportions of ewe and ram lamb candidates selected in the different tiers.

Scenarios

The nucleus animals were assumed to be fully recorded, with phenotypes and parentage or genomic selection assigned to all individuals. In the nucleus tier, live weight at different ages, ultra-sound scanning, body condition score and maternal traits were routinely recorded. Parentage assignment was carried out through DNA testing in all lambs born. Alternatively, when genomic selection was assumed, selection candidates were tested on a 5K SNP chip to determine genomic relationships and SNP profiles. Sires used in the nucleus were assumed to be tested on the 50K SNP chip.

The base scenario assumed no performance recording or genotyping in the multiplier tier. A range of scenarios

were compared, which included different combinations of policies for phenotypic recording, DNA parentage, genomic selection, genotyping strategy, and nucleus replacement policy.

Two levels of phenotypic recording policies were evaluated. The "simple" performance recording policy was assigned for recording pregnancy scanning [as a proxy for number of lambs born (NLB)], live weight and body condition score on ewes of all ages, and only weights for lambs, without maternal information. In this case, selection could only take place based on individual performance information, since with no parentage assignment it is impossible to use information from relatives when evaluating candidates in the multiplier tier. The "complete" performance recording policy had the assumption that weaning, slaughter and carcass weights, body condition score, pregnancy scanning results, lambing and weaning rates were recorded on ewes of all ages and their lambs.

In the DNA parentage policy, it was assumed that lambs born in the multiplier tier were either assigned to their dams and sires by DNA testing (Yes), or that there was no DNA parentage assignment (No). On a commercial scale, the advantage of DNA parentage testing is that it is a more practical and accurate method of parentage determination than matching lambs to ewes at birth, which is labour intensive and requires single sire mating groups.

In the genomic selection policy, it was assumed that progeny born in the multiplier flock were genotyped on the 5K SNP chip (Yes), or GS was not applied (No). GS is different to pedigree allocation via DNA, in that it generates higher accuracy of selection than can be achieved through identification of an animal's relatives via pedigree information. GS is based on knowledge of the relationships between individuals, and between their genotypes (SNP) and phenotypes, established using reference populations.

The genotyping strategy policy also had two alternate levels, "All" and "Selective", when all lambs born in the multiplier tier were genotyped and when a subset of individuals were genotyped for GS, respectively. In the case where a "Selected" policy is implemented, only physically sound replacement ewes and ram candidates born as twins or triplets could be potentially selected, which resulted in effective genotyping of 47% of the ewe lambs and 28% of the ram lambs born in the multiplier. There are two situations where it may be necessary to genotype all progeny. First, in situations in which phenotypes of the ungenotyped animals may be important to avoid bias in genetic evaluation or to improve selection accuracy, or second, if the turn-around time for genotyping is too slow to allow genotype results to be back in the time period between culling of unsuitable candidates and the final selection of breeding animals.

Alternative replacement policies compared a "Closed" to an "Open" nucleus. In a closed nucleus, candidate ewe lambs were selected only from within the nucleus tier. The alternative policy selected part of the female nucleus replacements, based on truncation selection, from among the top ewes within the multiplier tier (open nucleus).

The different scenarios established the basis for the selection index model which, along with trait specific genetic parameters, was used to estimate the genetic progress attributed to each set of selection strategies. A summary of the simulation scenarios modelled in this study is in Table 7.

The GS Only and the Selective GS Only scenarios assumed that GS genotypes were the unique source of information for selection in multiplier candidates, i.e. not combined with phenotypes. This assumes that a relevant and effective training population for genomic selection is available outside of the multiplier itself, and thus, no investment in recording is required to maintain this training population.

Genetic lag

The genetic lag between tiers was calculated as the difference in average merit of progeny $(O^j_{y,T})$ in the nucleus and the lower tiers at a given point in time. Average genetic lag at year 20 in a higher $(T = H)$ and lower $(T = L)$ tier for trait j were calculated as:

$$Lag^j_{T=H,T'=L} = \frac{O^j_{y=20,T=H} - O^j_{y=20,T'=L}}{b^j_{T'=L}},$$

where $b^j_{T'=L}$ is the annual rate of genetic progress of trait j, in the lower tier, between years 20 and 30, a period when the rate of genetic progress had stabilized.

Genetic trend in the commercial flock

The model predicted specific genetic trends for different traits, after the implementation of the alternative scenarios. The breeding program was simulated over 40 years from the moment when performance recording and genotyping were implemented in the multiplier tier. The reference trend for comparison was based on the selection practices from the base scenario.

The annual sum of expressions (EBV^j_y) for trait j across the different age groups, i.e. seven ewe age groups, of the commercial flock in a given year y, was computed as:

$$EBV^j_y = \sum_{e=1}^{7} \left[\left(O^j_{T=M,y-e} + \Delta^j_{T=M,s=m,y-e} \right) \cdot \varepsilon^j_e \right].$$

where $O^j_{T=M,y-e}$ is the average genetic merit of offspring born in the multiplier tier in year y from the start of a new recording strategy, after year e following the first mating of a ram in that tier of the breeding structure, Δ is the selection differential of the trait j in the respective year in the given T tier, for the selected males $s = m$, and ε^j_e is the discounted expression of trait j in age group e in the commercial flock, expressed as a proportion of their total lifetime sum of expressions, as described in the equation that calculates ε^j_e as the sum of DGE expressed as a proportion of its total lifetime sum of expressions.

Table 7 Description of simulation scenarios modelled to the multiplier tier of a multi-tiered breeding scheme

Scenario	Policies				
	Performance recording	DNA parentage	Genomic selection (GS)	Genotyping strategy	Nucleus replacement policy
Base scenario[a]	–	No	No	–	Closed
Pheno + GS	Complete	No	Yes	All	Closed
Pheno + selective GS	Complete	No	Yes	Selected	Closed
Pheno + parentage	Complete	Yes	No	All	Closed
Phenotypic selection	Simple	No	No	–	Closed
Pheno + GS + open	Complete	No	Yes	All	Open
Pheno + parentage + open	Complete	Yes	No	All	Open
GS Only	–	No	Yes	All	Closed
Selective GS Only	–	No	Yes	Selected	Closed

[a] Refers to the base scenario in which no performance recording or genetic merit selection is undertaken in the multiplier flock

Economic evaluation

The economic impact of implementing recording efforts was calculated as the product of the direct trait expressions (EBV_y^j), their economic values (ev_j), and the number of animals affected (n^j, described in Table 8) within the breeding scheme. Additional revenue was summed across tiers for the different scenarios. The additional revenue (R_y) realised in the commercial flock in year y was calculated as:

$$R_{y,T=C} = \sum EBV_y^j \cdot ev_j \cdot n^j.$$

In the multiplier tier, the additional revenue realised from trait improvements after implementation of recording efforts was calculated as:

$$R_{y,T=C} = \sum \left(D_{y,T=M}^j \cdot ev_j \cdot nd_{T=M} \right) \cdot \rho_d^j$$
$$+ \sum \left(EBV_{y,T=M}^j \cdot ev_j \cdot no_{T=M} \right) \cdot \rho^j,$$

where D_y^j and EBV_y^j are the average genetic merit of dams and offspring respectively, born in tier T of the breeding structure for trait j in year y, while $nd_{T=M}$ and $no_{T=M}$ are the respective numbers of dams and offspring in the multiplier tier, ρ_d^j and ρ^j are the proportion of dams and offspring expressing the jth trait, after discounting the percentage of ewes mated to terminal sires and lambs sold as store (Table 1).

The net profit relative to the base scenario was calculated from the extra revenue minus the cost as:

$$\pi_y = \sum_y \left(R_{y,T=M} + R_{y,T=C} - C_y^{TOT} \right) \cdot \left(\frac{1}{1+r} \right)^y,$$

where r is a 7% annual discount rate and C^{TOT} is the total recording and selection related costs in year y calculated as:

$$C_y^{TOT} = C_y^{dna} + C_y^{eid} + C_y^{rec} + C_y^{ge},$$

where C_y^{dna} are the parentage and genomic selection costs, C_y^{eid} is the cost of electronic identification, C_y^{rec} are the estimated recording or direct measurement costs and C_y^{ge} is the genetic evaluation cost in scenarios for which parentage or genotypes were available.

These cost components were calculated as:

$$C_y^{dna} = \left(\$_{dna,y} \cdot n_{dna,y} \right) + \left(\$_{SNP,y} \cdot n_{SNP,y} \right),$$
$$C_y^{eid} = \left(\$_{eid,y} \cdot n_{eid,y} \right), C_y^{rec} = \left(\$_{rec,y} \cdot n_{rec,y} \right);$$
$$\text{and } C_y^{ge} = \left(\$_{ge,y} \cdot n_{ge,y} \right),$$

where $ represents the price of the different component costs, i.e. DNA parentage and SNP tests ($\$_{dna}$), electronic identification ($\$_{eid}$), phenotype recording practices ($\$_{rec}$) and genetic evaluation ($\$_{ge}$), which were assumed to remain constant over time, and n is the number of animals tested in year y. Table 8 presents these parameters for the different scenarios.

The discounted cumulative net present value of additional profit ($CNPV$) was calculated as the difference between the profit after implementation of recording efforts (π'), from year 0 to year y, for any given multiplier recording scenario expressed as a deviation from the profit obtained in the base scenario (π), computed as:

$$CNPV_y = \sum_{i=1}^{y} \pi_i' - \pi_i.$$

Results

Economic impact

Relative to the base scenario, the annual additional revenue that was generated by performance recording in the multiplier, grew steadily from 2 years after the introduction of new scenarios, and stabilized to a constant value after 11 years in all scenarios. The Pheno + GS + open scenario presented the largest benefits. This scenario reached constant annual revenues of $724 K from year 11 onwards. The next largest benefits were obtained in

Table 8 Prices of recording associated components and number of animals tested in the different scenarios

| Cost component | $/unit | Animals tested ($n$)[a] | | | | | | | |
| | | Scenarios[b] | | | | | | | |
		Pheno + GS	Pheno + selective GS	Pheno + par	Phenotypic selection	Pheno + GS + open	Pheno + par + open	GS only	Selective GS only
DNA Parentage Test	20.00	0	0	10,507	0	0	10,507	0	0
5K SNP Test	50.00	10,507	3921	0	0	10,507	0	10,507	3921
EID	1.50	10,507	10,507	10,507	10,507	10,507	10,507	10,507	10,507
Recording	2.00	17,507	17,507	17,507	17,507	17,507	17,507	0	0
Genetic evaluation	2.00	17,507	17,507	17,507	0	17,507	17,507	17,507	17,507

[a] Number of animals tested, identified, recorded and evaluated annually in different scenarios

[b] Scenario are described in Table 7

scenarios Pheno + GS, and Pheno + Selective GS. Both scenarios reached constant annual revenues of $713 K, also from year 11 onwards. The next best results arose from selection on phenotypes and parentage, represented in scenarios Pheno + parentage + open and Pheno + parentage, which produced annual increases in revenue of $576 K and $553 K, respectively. The annual revenues of genomic selection only scenarios, GS only and Selective GS only, stabilized at $530 K. Phenotypic selection alone had the smallest benefits, with stabilized marginal revenue of $13 K, also achieved after 11 years.

The cost of recording efforts in the multiplier tier of the breeding scheme had the biggest impact during establishment in year 1 because of the implementation costs when all breeding ewes were genotyped, or simply identified for phenotypic selection. In the following years, costs stabilized due to a fixed number of animals being tested, recorded and evaluated, and due to the assumption that relative prices remained constant throughout the simulation planning horizon.

The phenotypic selection strategy did not include parentage assignment or genomic selection, and therefore, had the smallest overall annual cost among all scenarios, at $108 K after completion of the establishment phase. Nevertheless, the very modest benefits that arose from this strategy did not offset the low costs. The annual cost of the Pheno + parentage scenario was $318 K. The annual cost of the Selective GS only was $290 K. The cost of the Pheno + selective GS was $329 K per year, while the annual cost for GS only was $594 K, and Pheno + GS was $634 K. Because fewer animals were tested in selective GS genotyping scenarios, these had lower costs when compared to the equivalent scenarios in which all lambs born were genotyped. There was no difference in costs between open and closed nucleus versions in Pheno + GS and Pheno + parentage scenarios, which were otherwise identical strategies.

In scenarios which involved both performance recording and parentage, the cost of parentage assignment through DNA was the largest cost component, comprising 70% of the total recording cost. Genomic testing was the most significant cost component in GS scenarios (from 64 to 88%), when compared to the cost of trait measurements (10 to 36%), electronic identification (9 to 27%) and genetic evaluation (10 to 15%).

Scenario comparison and cost-benefit analysis

Figure 1 presents the cumulative net present value (CNPV) that results from implementation of recording procedures in the multiplier tier over successive years, relative to the base scenario. Scenario Pheno + selective GS achieved breakeven the earliest, after 18 years. The profitability of this scenario in year 30 was about $903 K.

Pedigree selection represented by the Pheno + parentage scenarios, in both open and closed nucleus, reached breakeven in years 25 and 29, respectively, producing CNPV of $226 K and $58 K in year 30. Scenarios with complete phenotyping and GS genotyping of all lambs did not achieve breakeven within the simulated horizon. The CNPV reaches breakeven only in scenarios in which phenotypes were combined with parentage information or when only a selected subset of candidates was genotyped.

In general, the cost of parentage testing and genomic selection greatly influenced the CNPV of recording the multiplier tier. Under the current costs of genotyping, the most profitable selection strategies will take 18 to 29 years to achieve financial breakeven, thus resulting in long periods of large financial deficits. However, if genotyping costs decrease to $25 or less, then recording a sheep multiplier tier in the conditions included in the present study becomes more attractive. For instance, under lower GS genotyping costs of $25, as opposed to $50 per test, the long-term profit of the Pheno + GS scenario increased to $549 K in year 30, compared to a CNPV of −$2800 K. The breakeven point and CNPV of a range of scenarios under lower parentage test prices are in Fig. 2.

Selection differentials

Changes in selection differentials that result from the selection index model, which assumed performance recording implemented in year 1, were the main drivers of variation in benefits among scenarios. All selection differentials for the "base scenarios" are 0 for animals within the multiplier tier as there was no information to select on. The set of differentials for rams selected in the recorded multiplier tier to mate commercial breeding ewes are in Table 9. For clarity, trait selection differentials in the nucleus for ewes and rams contribute to overall genetic progress, and differentials in the multiplier and commercial tiers contribute to decreased genetic lag between higher and lower tiers.

Differentials for number of lambs born and weaning weight maternal were larger, and differentials for ewe mature weight were smaller, for GS scenarios when compared to selection based on phenotypes only, or on phenotypes and parentage. The phenotypic selection scenario produced the smallest selection differentials, which reflects the low index accuracy in commercial rams selected on phenotypes only, in the multiplier tier (Table 10). Selection differentials of ewes selected in the recorded multiplier differed between the open and the closed nucleus. The differences were limited to BCS, weaning weight maternal and Stay, which were bigger in the open nucleus scenarios when compared to closed nucleus scenarios. These moderate differences in replacement policy caused modest

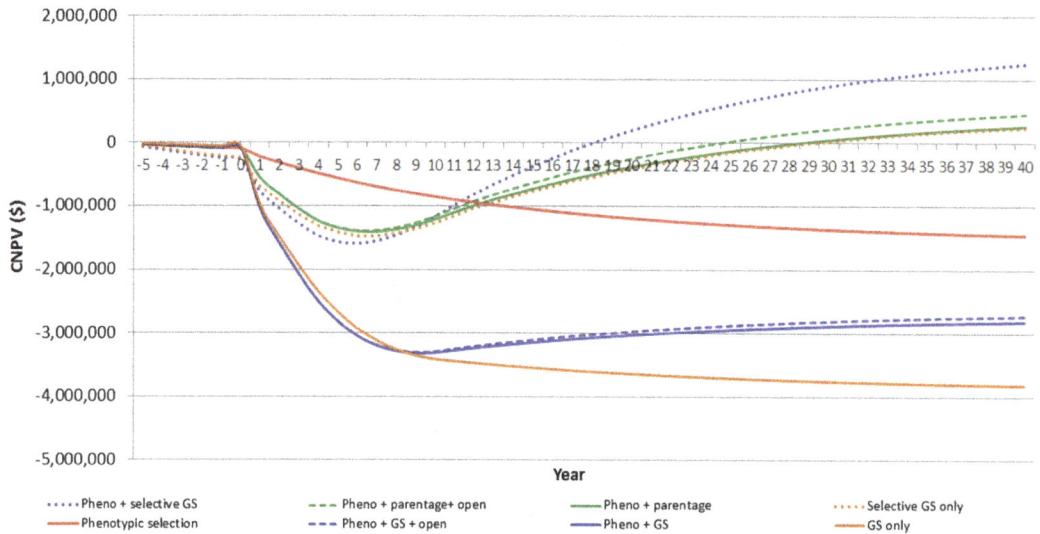

Fig. 1 Cumulative net present value (*CNPV*) of simulation scenarios, relative to the base scenario, in different selection strategies of the multiplier tier of the breeding scheme. Scenarios based on different selection strategies, described as: *Pheno + GS* phenotypic recording and genomic selection, *Pheno + parentage* phenotypic recording and parentage for pedigree selection, *Pheno + GS + open* phenotypic recording, genomic selection and the open nucleus, *Pheno + parentage + open* phenotypic recording, parentage for pedigree selection and the open nucleus, *Pheno + selective GS* phenotypic recording and alternative genotyping for genomic selection of physically sound candidates with potential to become replacement ewes and future only, *Selective GS only* genomic selection without performance recording in the multiplier by genotyping for genomic selection only physically sound candidates with potential to become replacement ewes and future rams. *Phenotypic selection* selection based on phenotypes only, *GS only* GS without performance recording in the multiplier

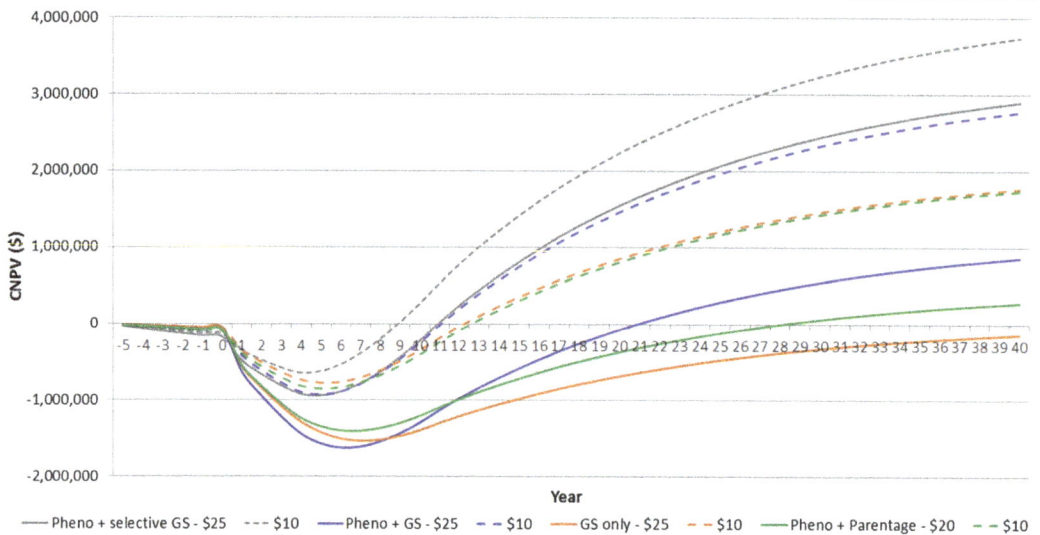

Fig. 2 Cumulative net present value (CNPV) of alternative selection strategies, assuming recording the multiplier tier under genotyping test prices at $10 and $25. See description of scenarios in Fig. 1 and details in Table 7

changes in differentials of rams selected in the multiplier tier to be mated in commercial flocks.

Table 10 presents the index accuracies of selection in the different tiers and scenarios. The accuracies were based on the number of phenotypic records of the different information sources, and underpin the genetic progress achieved through the range of selection scenarios.

Genetic contributions

The set of genetic expressions (ε), used in the calculation of gene flow from the multiplier tier through to the

Table 9 Selection differentials of recorded multiplier breeding rams for profit traits (units/year) in the different scenarios

Trait (abbreviation)	Unit	Pheno + GS[a]	Pheno + parentage	Phenotypic selection	Pheno + GS + open	Pheno + parentage + open	GS only[b]
Carcass weight (CWT)	kg	0.674	0.598	0.293	0.665	0.581	0.645
Weaning weight (WWT)	kg	0.868	0.759	0.334	0.857	0.739	0.835
Number of lambs born (NLB)	Lambs	0.053	0.020	0.016	0.054	0.023	0.056
Ewe mature weight (EWT)	kg	0.548	0.698	0.297	0.540	0.678	0.342
Ewe body condition score (BCS)	Score	0.058	0.042	0.000	0.057	0.041	0.032
Survival maternal (SURm)	Lambs	0.008	0.010	0.000	0.010	0.012	0.000
Weaning weight maternal (WWTm)	kg	0.090	0.040	0.000	0.093	0.047	0.087
Stayability (Stay)	Binary	0.012	0.018	0.000	0.010	0.018	0.001
Lamb survival (SUR)	Lambs	0.003	0.004	0.000	0.003	0.004	0.000

Scenario description in Fig. 1 and details in Table 7

[a] Selection differentials used in Pheno + GS and Pheno + selective GS scenarios

[b] Selection differentials used in GS only and Selective GS only scenarios

Table 10 Index accuracies of selection for breeding ewes and rams mated in different tiers in alternative scenarios

Scenario	Nucleus		Multiplier		Multiplier to commercial
	Ewes	Rams	Ewes	Rams	Rams
Pheno + GS	0.43	0.49	0.43	0.49	0.49
Pheno + selective GS	0.43	0.49	0.43	0.49	0.49
Pheno + parentage	0.36	0.38	0.32	0.38	0.37
Phenotypic selection	0.36	0.38	0.18	0.21	0.14
Pheno + GS + open	0.53	0.49	0.53	0.49	0.49
Pheno + parentage + open	0.46	0.38	0.45	0.38	0.38
GS only	0.44	0.49	0.32	0.33	0.33
Selective GS only	0.44	0.49	0.32	0.33	0.33

Scenario description in Fig. 1 and details in Table 7

commercial tier, are in Fig. 3. The results show that traits of the lamb at birth and at slaughter were expressed earlier, followed by ewe hogget traits and traits expressed annually by adult ewes, respectively. After the use of a ram, there will still be an impact of that selection expressed 10 years later through ewes that stay in the flock, and their female descendants. In addition, while slaughtered lamb traits and ewe annual trait expressions decrease slowly after peaking at 6 years, hogget trait expressions peak in the second year and drop fast after 5 years. It could be expected that the "open multiplier" presents a different timing and extent of expressions, given the difference in the age distribution of ewes. However, the replacement of nucleus ewes with older proven multiplier replacements did not affect the genetic expressions at the commercial level.

Ram and ewe selection on performance and genetic records in the multiplier tier resulted in a lift in genetic merit in both multiplier and commercial tiers. This superiority was expressed as the rate of genetic progress in units of the different breeding objective traits, and as the average genetic lag between the different tiers, expressed in years, for the different traits.

Genetic lags

Table 11 shows the average genetic lag between the nucleus and the commercial tier for the various traits in year 20, in the different scenarios. Trait lags in the commercial tier are represented for the base scenario (*Base*) and after recording was implemented (*With recording*) in the multiplier tier. A reduction from 1 to 4 years in the lag between the nucleus and the commercial tier was achieved with implementation of recording efforts in the multiplier tier. Phenotypic selection caused a reduction of less than a year, while DNA parentage and GS reduced the lag between the nucleus and the commercial tier in more than 3 years. Phenotypic selection alone did not reduce the genetic lag for BCS, SURm, WWTm and Stay. Consequently, phenotypic selection of these traits in the multiplier tier had modest economic impact when compared to the base scenario. The small reduction in genetic lag observed for lamb survival, especially in Phenotypic selection, GS only and Selective GS only scenarios, reflects its low rate of genetic progress and/or the small difference in merit between tiers.

The reduction in genetic lag for the different traits varied considerably across scenarios. This was the result of differences in rates of genetic progress between traits in different scenarios. The difference in estimated trait merit reflects the 8 to 40% higher rates of gain in commercial progeny with the recorded multiplier scenarios, compared to the base scenario. By year 20, scenarios assuming GS in the multiplier tier presented the largest

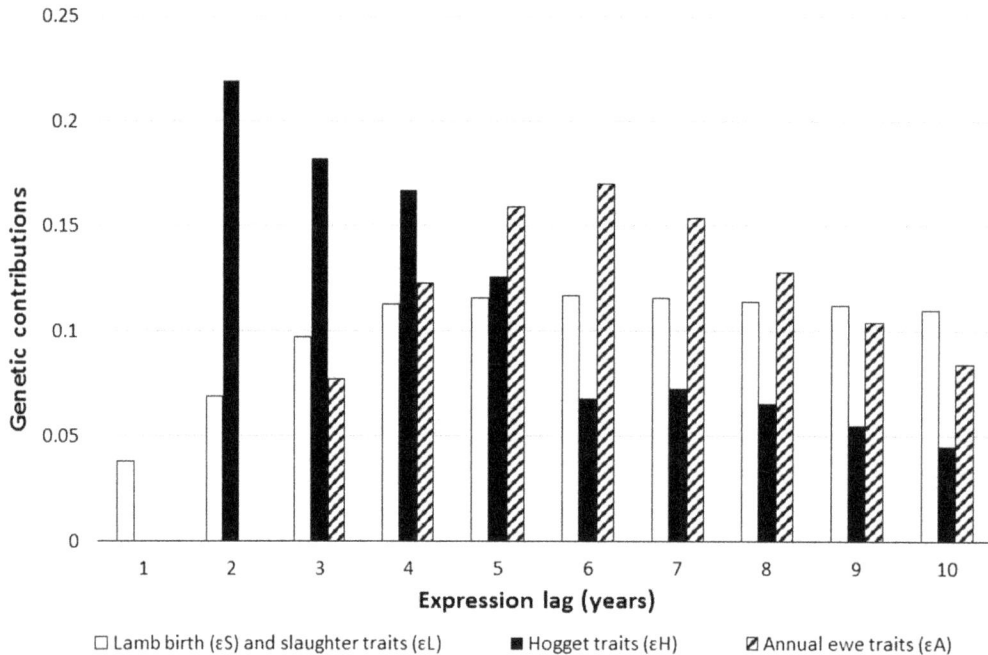

Fig. 3 Proportions of genetic contributions for different trait categories from ram selection in the multiplier flock as expressed in the self-replacing commercial ewe flock in years after recording and genotyping were introduced. Number of years from when a ram is first mated in the commercial tier. ε^L = lamb birth traits, ε^S = lamb slaughter traits, ε^H = hogget traits and ε^A = annual ewe traits

difference, i.e. more progress, in genetic merit of traits such as CWT, WWT, and SURm. The DNA parentage strategy resulted in more progress of traits such as NLB, BCS, WWTm and Stay in the commercial tier, when compared to the base scenario.

Discussion

The objective of this study was to assess the potential benefits that arise from introducing performance recording, parentage testing and genomic selection to the multiplier tier of a multi-tiered breeding scheme, which supplies improved rams to a large number of commercial ewes. The 2 to 7% gains in rate of genetic progress resulting from these recording efforts reflect the importance of trait expressions in the much larger number of animals at the base population of the breeding scheme.

The model considered the cost-benefit of information in commercially-managed sheep flocks. It demonstrated the effectiveness of performance records when combined with parentage information (gains of 5 to 7%), and when associated with genomic selection (gain of 6%), from the perspective of genetic progress. Genomic selection can have a dramatic effect on the reliability of breeding values, especially for sex-limited traits with high accuracy in some livestock industries [23]. Sise and Amer [24] predicted that the sheep meat industry in New Zealand could achieve a 5% increase in the annual rate of genetic

progress by adopting multi-trait selection indexes with genomic selection in the nucleus tier. In this study, we also predicted the largest annual benefits, i.e. additional revenue, in scenarios assuming DNA parentage and those assuming GS.

These results were likely due to the greater amount of information used in GS scenarios, which combine genome-based and pedigree-based relationships, as described by Dekkers [22] and Tusell et al. [25]. The results of this study were similar to those from Horton et al. [8] which estimated higher long-term net dollar gains from the use of genomic selection in a range of scenarios assuming various levels of prediction accuracy. The authors described how different tiers were able to benefit from the extra genetic gain derived from genomic testing, and how the degree of multiplication is important when calculating benefits which should offset the increased costs of genomic testing. Horton et al. [8] considered that recording the multiplier tier was not cost-effective because of the prohibitive testing prices. In this study, benefits for commercial flocks were conditional on assumed accuracies of prediction, which ultimately depend on the amount of data measured in the reference population, especially when animals with genotypes and records of their own were available.

According to Pickering et al. [9], the rate of genetic gain can be lifted when selecting young rams on a dual

Table 11 Genetic lag (years) between nucleus and the commercial tier of the breeding scheme by year 20 in different scenarios

Trait	Base scenario	Pheno + GS	Pheno + selective GS	Pheno + parentage	Phenotypic selection	Pheno + GS + open	Pheno + parentage + open	GS only	Selective GS only
	Base	With recording							
CWT	10.6	7.6	7.6	7.8	9.4	6.7	6.7	7.9	7.9
WWT	10.6	7.7	7.7	7.8	9.5	6.7	6.7	8.0	8.0
NLB	11.0	8.2	8.2	8.3	8.8	7.4	8.0	8.0	8.0
EWT	10.8	7.8	7.8	8.0	9.8	7.1	7.1	9.8	9.8
BCS	10.6	7.2	7.2	6.8	10.6	7.2	7.2	9.5	9.5
SURm	11.5	8.5	8.5	8.2	11.5	8.5	8.0	14.0	14.0
WWTm	11.5	8.3	8.3	8.3	11.5	7.5	8.2	8.4	8.4
Stay	9.2	4.8	4.8	4.8	9.2	5.2	4.8	11.0	11.0
SUR	10.9	7.0	7.0	7.1	10.6	6.1	6.0	13.0	13.0

See scenarios description of Fig. 1 and details in Table 7

purpose index (a New Zealand sheep national selection index [20]) with genomic information, from which the majority of the benefit comes from the increased accuracy of breeding value for sex-limited traits and those recorded later in life. However, the study of Pickering et al. [9] was based on direct benefits from recording only in the breeder flocks, and did not consider benefits from recording the multiplier tier.

This study found large differences in benefits and costs across scenarios, and in the timing at which cumulative benefits offset the cumulative costs. Under current genotyping prices, the implementation of the GS policy with higher revenue was not cost-effective and both the time to breakeven and the long-term *CNPV* were not economically attractive. High implementation costs will likely moderate the practical application of the selection in GS scenarios that were investigated in this simulation study, particularly the cost of parentage testing and genomic selection. The cost of genomic technology is dropping rapidly and the expectation is that prices will further decrease, with encouraging possibilities developed in the area of genotyping by sequencing (GBS) [26]. Reduction in genotyping costs and policies which restrict genotyping to animals for which the most benefit can be generated, will likely be a key to achieve higher adoption of recording efforts and genotyping among producers. According to van der Werf et al. [27], genotyping approximately 20% of a young sire crop, would result in close to maximum overall additional benefits of genomic selection. In this study, reduced genotyping costs from either lower prices and/or reduced proportion of genotyped selection candidates changed the *CNPV* and repositioned scenarios accordingly. We demonstrated this when cost-effectiveness was achieved through the selective genotyping policy (Fig. 1), and through the sensitivity analysis, in

which genotyping prices were reduced by 50% or more, with increased profitability (Fig. 2). The long-term nature of the benefits that are associated with informed selection of multiplier candidates demonstrates the critical importance of achieving lower costs of information. This is particularly important for large-scale commercial operations that need to adopt genomic technology on a wider scale, than nucleus operations.

The structure of the breeding scheme that was simulated in this study assumed a larger than necessary multiplier tier, relative to the size of nucleus and commercial tiers. This resulted in economic losses when genotyping all lambs, compared to pedigree selection, even with the larger selection differentials (Table 9). In a recorded multiplier scenario, which initially incurs higher recording and genotyping costs, these losses could be avoided by having a smaller multiplier flock while still allowing sufficient selection intensity so that genetic progress can be maintained. In a different breeding structure, with a smaller relative size of the multiplier population, it is likely that genomic selection policies, with a larger proportion of genotyped animals, would produce positive results.

The selection scenarios that resulted in the earliest and largest financial returns were parentage assignment assuming current DNA testing prices. Under lower genotyping costs, the scenarios that combined performance recording with genomic selection ranked as the most profitable. Information on genotypes or parentage becomes critical to achieving significant rates of genetic progress, because the low heritability of maternal traits (i.e. traits measured late in life) limits the effectiveness of phenotypic selection. Phenotypic information is only useful for genetic evaluation if records can be combined with parentage and/or genomic relationship information;

otherwise it is very difficult to estimate genetic merit with sufficient accuracy to be useful.

The selection policy simulated in the GS only scenario was not profitable under the accuracy of genomic predictions assumed in this study. In practice, it is difficult to achieve a high GS accuracy for traits with low heritability, such as maternal traits, which depend on more phenotypic data to derive prediction equations of high accuracy [6]. Higher prediction accuracies can be achieved through GS if an appropriate reference population exists to predict the phenotype of individuals from their genotype. Information collected in the multiplier tier could be useful to increase the reference population. This is because multi-tiered breeding schemes normally have large numbers of related animals with potential to generate performance data and genomic information in different tiers. The accuracy of genomic prediction can be increased if the reference population includes information with the highest average genetic relationship to the reference set [28]. Further investigation into the additional contribution of genomic selection to the accuracy of prediction in the nucleus might provide additional support for recording and genotyping in the multiplier tier.

In addition to overall profitability and time to break-even, higher chances of adoption of profitable selection strategies might depend on the existence of infrastructure to facilitate performance recording. Infrastructure to collect and store records requires investment in tracking individual animal production and development of uniform procedures for measurement, sampling, testing, data analysis and selection of candidates. Identifying and keeping the most productive commercial ewes for further lambing opportunities decreases the number of replacement ewes required and could potentially increase the flock overall lifetime performance [29]. In this study, the open nucleus resulted in only modest changes in *CNPV* when compared to the closed nucleus. There was not enough additional genetic gain to offset recording costs, or to compensate for eventual increases in generation interval. According to Garrick et al. [12], if the additional information from performance recording results in delays and increased generation interval, this would erode benefits from higher accuracy that are created by the introduction of older multiplier ewes as nucleus flock replacements.

An open nucleus policy might be more beneficial in the presence of genotype by environment interaction (G×E) where the nucleus tier is not fully representative of the commercial environment, and or if inbreeding management is an issue in the breeding program. In the presence of G×E, goal traits expressed in the commercial tier may differ from the reference trait expressed in the nucleus tier. Phenotypes and genotypes recorded in the multiplier flock, which is frequently maintained in a commercial environment, could be used to predict genetic merit in multi-tier breeding populations. This could inform the extent to which genomic accuracy of prediction in the nucleus tier could impact trait expressions in the commercial tier, or vice versa.

In summary, recording in the multiplier tier reduced the lag between the nucleus and the commercial tier by 2 to 3 years. This benefit has an important economic impact for commercial lamb production. Similar results were suggested by Hill [15] in pig populations using multiplier tiers, and by Horton et al. [8] in sheep. According to Bichard [11], genetic lags can be considerable and their actual size is determined by the annual rate of progress in the nucleus tier, flock age structure in the lower tiers, and origin and degree of selection intensity. Bichard [11] reported a lag of 11 years between nucleus and commercial tiers in three-tiered sheep breeding schemes. According to Blair and Garrick [13], the typical genetic lag in the New Zealand sheep industry is between 5 and 8 years, based on a two-tier breeding structure, and assuming that rams transferred from the nucleus to the commercial tier are the average of those born in the same year in the nucleus.

Breakeven and overall profitability depended highly on key pieces of information that enabled higher accuracy of prediction and consequently the highest level of genetic progress. Because net benefits were relatively modest when compared to the scale of investment required, adoption of recording efforts and genotyping by producers is likely to be low at the base levels of DNA test cost modelled here. Reductions in DNA technology costs, or recording policies that either mitigate problems with G×E interactions or facilitate accurate selection in nucleus candidates via DNA parentage (eliminating pedigree errors) and genomic prediction, may change the likelihood of the adoption of recording in multiplier tiers in the future. Nevertheless, scientific modelling of breeding schemes can assist in quantifying the genetic and economic impacts of selection alternatives. An additional selection scenario, which would be worthwhile to investigate and has not been included in the current study, is the presence of G×E and its impact on the prediction accuracy of genomic selection.

Conclusions

Our findings demonstrate that performance recording and parentage assignment in the multiplier tier can generate long-term economic benefits in multi-tiered breeding programs. Implementing recording in the multiplier tier reduces the long genetic lag between the nucleus and commercial tiers. Such recording is justified if the breeding scheme captures the benefits through more profit generated in the commercial tier. The investment in

Genetic and economic benefits of selection based on performance recording and genotyping in lower...

47

phenotyping is only worthwhile if parentage or genomic information is also available. Genomic selection also has the potential to significantly increase the benefits of recording, especially under reduced genotyping costs or when a subset of candidates is tested, as opposed to all lambs born. Finally, genomic selection policies in multiplier tiers might be feasible in the near future, given the expected reduction in genotyping costs, which are critical drivers of the magnitude of benefits, and of the time required to break-even investments in recording.

Authors' contributions
BFSS, JPG, JVDW and PRA conceptualized and designed the study. BFSS, PRA and TJB developed the simulation model. BFSS performed the simulations and wrote the first draft manuscript. All authors reviewed and collaborated to complete the draft of the manuscript. All authors read and approved the final manuscript.

Author details
[1] AbacusBio Limited, PO Box 5585, Dunedin 9058, New Zealand. [2] School of Environmental and Rural Science, University of New England, Armidale, NSW, Australia. [3] Cooperative Research Centre for Sheep Industry Innovation, Armidale, NSW, Australia.

Acknowledgements
The authors would like to acknowledge Beef + Lamb New Zealand Genetics consortium (Grant No. BLG Project Code 10447) for funding and supporting this research project. The authors are grateful for the support given by University of New England—School of Environmental and Rural Science. The authors are thankful for the valid contributions of the reviewers of this paper, and the contributions from John C McEwan.

Competing interests
The authors declare that they have no competing interests.

References
1. Holst PJ. Recording and on-farm evaluations and monitoring: breeding and selection. Small Rumin Res. 1999;34:197–202.
2. Caja G, Carné S, Salama AAK, Ait-Saidi A, Rojas-Olivares MA, Rovai M, et al. State-of-the-art of electronic identification techniques and applications in goats. Small Rumin Res. 2014;121:42–50.
3. Lewis RM, Simm G. Selection strategies in sire referencing schemes in sheep. Livest Prod Sci. 2000;67:129–41.
4. Meuwissen THE, Hayes BJ, Goddard ME. Prediction of total genetic value using genome-wide dense marker maps. Genetics. 2001;157:1819–29.
5. Schaeffer LR. Strategy for applying genome-wide selection in dairy cattle. J Anim Breed Genet. 2006;123:218–23.
6. Van der Werf JHJ. Potential benefit of genomic selection in sheep. Proc Adv Anim Breed Genet. 2009;18:38–41.
7. Daetwyler HD, Hickey JM, Henshall JM, Dominik S, Gredler B, Van Der Werf JHJ. Accuracy of estimated genomic breeding values for wool and meat traits in a multi-breed sheep population. Anim Prod Sci. 2010;50:1004–10.
8. Horton BJ, Banks RG, Van der Werf JHJ. Industry benefits from using genomic information in two- and three-tier sheep breeding systems. Anim Prod Sci. 2015;55:437–46.
9. Pickering NK, Dodds KG, Auvray B, Mcewan JC. The impact of genomic selection on genetic gain in the New Zealand sheep dual purpose selection index. Proc Adv Anim Breed Genet. 2013;20:175–8.
10. Sise JA, Auvray B, Dodds KG, Amer PR. SNiP and cut: quantifying the potential benefits of genomic selection tools for genetic fault elimination in sheep. Proc N Z Soc Anim Prod. 2008;68:33–6.
11. Bichard M. Dissemination of genetic improvement through a livestock industry. Anim Prod. 1971;13:401–11.
12. Garrick DJ, Blair HT, Clarke JN. Sheep industry structure and genetic improvement. Proc N Z Soc Anim Prod. 2000;60:175–9.
13. Blair HT, Garrick DJ. Application of new technologies in sheep breeding. N Z J Agric Res. 2007;50:89–102.
14. Van der Werf JHJ. Multi-trait selection index MTINDEX20T model. 1999. http://www-personal.une.edu.au/~jvanderw/software.htm. Accessed 25 Oct 2016.
15. Hill WG. Prediction and evaluation of response to selection with overlapping generations. Anim Prod. 1974;18:117–39.
16. McClintock AE, Cunningham EP. Selection in dual purpose cattle populations: defining the breeding objective. Anim Prod. 1974;18:237–47.
17. Amer PR. Economic accounting of numbers of expressions and delays in sheep genetic improvement. N Z J Agric Res. 1999;42:325–36.
18. Berry DP, Madalena FE, Cromie AR, Amer PR. Cumulative discounted expressions of dairy and beef traits in cattle production systems. Livest Sci. 2006;99:159–74.
19. Hazel LN, Dickerson GE, Freeman AE. The selection index—then, now, and for the future. J Dairy Sci. 1994;77:3236–51.
20. Byrne TJ, Amer PR, Fennessy PF, Hansen P, Wickham BW. A preference-based approach to deriving breeding objectives: applied to sheep breeding. Animal. 2012;6:778–88.
21. Auvray B, McEwan JC, Newman SA, Lee M, Dodds KG. Genomic prediction of breeding values in the New Zealand sheep industry using a 50K SNP chip. J Anim Sci. 2014;92:4375–89.
22. Dekkers JCM. Prediction of response to marker-assisted and genomic selection using selection index theory. J Anim Breed Genet. 2007;124:331–41.
23. Berry DP, Wall E, Pryce JE. Genetics and genomics of reproductive performance in dairy and beef cattle. Animal. 2014;8:105–21.
24. Sise JA, Amer PR. SNP predictors to accelerate the rate of genetic progress in sheep. Proc Adv Anim Breed Genet. 2009;18:220–3.
25. Tusell L, Gilbert H, Riquet J, Mercat MJ, Legarra A, Larzul C. Pedigree and genomic evaluation of pigs using a terminal-cross model. Genet Sel Evol. 2016;48:32.
26. Dodds KG, McEwan JC, Brauning R, Anderson RM, Van TC, Kristjánsson T, Clarke SM. Construction of relatedness matrices using genotyping-by-sequencing data. BMC Genomics. 2015;16:1047.
27. Van der Werf JHJ, Banks RG, Clark SA, Lee SJ, Daetwyler HD, Hayes BJ, et al. Genomic selection in sheep breeding programs. In: Proceedings of the 10th world congress of genetics applied to livestock production, 17–22 August 2014, Vancouver. 2014. https://asas.org/docs/default-source/wcgalp-proceedings-oral/351_paper_10381_manuscript_1648_0.pdf?sfvrsn=2.
28. Hayes BJ, Bowman PJ, Daetwyler HD, Kijas JW, Van der Werf JHJ. Accuracy of genotype imputation in sheep breeds. Anim Genet. 2012;43:72–80.
29. Turner HN, Brown G, Ford G. The influence of age structure on total productivity in breeding flocks of Merino sheep. I. Flocks with a fixed number of breeding ewes, producing their own replacements. Aust J Agric Res. 1968;19:443–75.

4

A genome-wide association study in a large F2-cross of laying hens reveals novel genomic regions associated with feather pecking and aggressive pecking behavior

Vanessa Lutz[1]*, Patrick Stratz[1], Siegfried Preuß[1], Jens Tetens[2], Michael A. Grashorn[1], Werner Bessei[1] and Jörn Bennewitz[1]

Abstract

Background: Feather pecking and aggressive pecking in laying hens are serious economic and welfare issues. In spite of extensive research on feather pecking during the last decades, the motivation for this behavior is still not clear. A small to moderate heritability has frequently been reported for these traits. Recently, we identified several single-nucleotide polymorphisms (SNPs) associated with feather pecking by mapping selection signatures in two divergent feather pecking lines. Here, we performed a genome-wide association analysis (GWAS) for feather pecking and aggressive pecking behavior, then combined the results with those from the recent selection signature experiment, and linked them to those obtained from a differential gene expression study.

Methods: A large F2 cross of 960 F2 hens was generated using the divergent lines as founders. Hens were phenotyped for feather pecks delivered (FPD), aggressive pecks delivered (APD), and aggressive pecks received (APR). Individuals were genotyped with the Illumina 60K chicken Infinium iSelect chip. After data filtering, 29,376 SNPs remained for analyses. Single-marker GWAS was performed using a Poisson model. The results were combined with those from the selection signature experiment using Fisher's combined probability test.

Results: Numerous significant SNPs were identified for all traits but with low false discovery rates. Nearly all significant SNPs were located in clusters that spanned a maximum of 3 Mb and included at least two significant SNPs. For FPD, four clusters were identified, which increased to 13 based on the meta-analysis (FPD_{meta}). Seven clusters were identified for APD and three for APR. Eight genes (of the 750 investigated genes located in the FPD_{meta} clusters) were significantly differentially-expressed in the brain of hens from both lines. One gene, *SLC12A9*, and the positional candidate gene for APD, *GNG2*, may be linked to the monomanine signaling pathway, which is involved in feather pecking and aggressive behavior.

Conclusions: Combining the results from the GWAS with those of the selection signature experiment substantially increased the statistical power. The behavioral traits were controlled by many genes with small effects and no single SNP had effects large enough to justify its use in marker-assisted selection.

*Correspondence: vanessa.lutz@uni-hohenheim.de
[1] Institute of Animal Science, University of Hohenheim, 70599 Stuttgart, Germany
Full list of author information is available at the end of the article

Background

Feather pecking in laying hens is a serious economic and welfare issue that can be observed in commercial and non-commercial chicken flocks. In spite of extensive research on feather pecking during the last decades, the motivation for this behavior is still unclear. The most widespread theory on the origin of feather pecking is that it is a redirected feeding and foraging behavior [1]. Some authors reported that feather pecking is related to dust-bathing [2]. Environmental factors such as light intensity [3], stocking density [4], and food form [5] can influence feather pecking. Feather pecking behavior has also been associated with fear [6–9]. Other studies suggested that the underlying motivation for feather pecking is feather eating [10–13] or that it is the consequence of a general hyperactivity disorder [14]. Feather pecking is often confounded with aggressive pecking but these two behaviors are clearly distinguishable, both in terms of form and motivation; aggressive pecks are delivered in an upright body posture, are mainly directed to the head of the other birds and aim at establishing and maintaining social hierarchy [15], while feather pecking is performed in a non-aggressive posture. Reports on the relationship between aggressive pecking and feather pecking show no consistent trend. While some authors found no correlation between the two behaviors, positive genetic and phenotypic correlations have been reported in lines selected for high and low feather pecking and their F2-crosses [16, 17]. Depending on the definition of the trait, study design, age of hens, statistical model applied, and data collection period, heritability estimates for feather pecking are low to moderate and range from 0.1 to 0.4, while heritability estimates for aggressive pecking range from 0.04 and 0.14 [17–20].

In a previous study, we analyzed two divergent lines that were selected for 11 generations for high (HFP) and low (LFP) feather pecking, respectively [20, 21]. We estimated genetic parameters within the lines and the phenotypic trend across generations. From the first round of selection onwards, the two lines differed in their means for feather pecking bouts. The highest selection response on the phenotypic scale was obtained during the first rounds of selection and thereafter, no clear trend was observed in the HFP line. The LFP line showed a constant low level of pecking behavior across the 11 generations of selection. Heritabilities of feather pecking estimated based on linear mixed models were equal to 0.15 and 0.01 in the HFP and LFP lines, respectively. The distribution of feather pecking bouts within each line and for each round of selection are discussed in detail in [21].

In addition, we performed a genome scan to map selection signatures in these two divergent HFP and LFP lines using an F_{ST}-based approach [20]. The analysis provided 17 genome-wide significant single-nucleotide polymorphisms (SNPs), most of which were located in clusters, which supports the presence of selection signatures.

These HFP and LFP lines formed the base population of the F2-population used in the current study, in which a genome-wide association analysis (GWAS) for feather pecking and aggressive pecking behavior was performed. The results obtained were combined with those from the previous selection experiment [20] in a meta-analysis, and then linked to those obtained from a differential gene expression study.

Methods
Experimental population

Chickens from a White Leghorn line were divergently selected for low and high feather pecking for 11 generations, resulting in a LFP and a HFP line. Selection took place for five generations at the Danish Institute of Animal Science [18] and then for five additional generations at the Institute of Animal Science, University Hohenheim, Germany [20]. From these two lines, a large F2 cross was established. Five sires and ten dams from generation 11 of each line were used to generate 10 F_1 families. Each HFP sire was crossed with two LFP full-sib dams and vice versa. Then, 10 F_1 sires were used to generate the F_2 families. Each sire was mated with eight F_1 hens four times by artificial insemination. A total of 960 F2 offspring were produced in four hatches, with an interval of 3 weeks between hatches.

Phenotypes

At 27 weeks of age, feather pecks delivered (FPD) and aggressive behavior [aggressive pecks delivered (APD) and aggressive pecks received (APR)] were recorded in groups of 36 to 42 hens. The applied ethogram was according to Savory [22] and Bessei et al. [16] and was as follows. Feather pecking was defined as a non-aggressive behavior and included forceful pecks, sometimes with feathers being pulled out and the recipient hen either tolerating this action or moving away. Aggressive pecking was defined as fast pecks towards the head and body of conspecifics. Usually, the hen that was attacked moved away but may have incurred tissue damage. For the behavioral observations, the hens were marked with numbered plastic batches on their backs. Seven observers visually recorded feather pecking and aggressive pecking within each pen during 20-min sessions for three consecutive days during daytime. Hatches 3 and 4 were observed twice for three consecutive days. The total number of observers varied between five and seven persons per observation day. The numbers of FPD, APD, and APR were summed over the entire observation period and standardized to an observation period of 420 min.

Heritabilities of FPD (APD, and APR), estimated with a linear mixed model in this F2 cross, were equal to 0.12 (0.27, and 0.27) [23]. Genetic and phenotypic correlations of 0.2 and 0.09, respectively, were obtained between FPD and APD [9]. Correlations of estimated breeding values between FPD and APR and between APD and APR were 0.18 and −0.23, respectively [17].

Genotypes

A total of 817 F2 hens were genotyped with the Illumina 60K chicken Infinium iSelect chip. For the remaining hens no samples were collected. A total of 57,636 SNPs were genotyped and after data filtering, 29,376 SNPs remained in the dataset. Based on positional information according to the chicken genome assembly galGal2.1, SNPs that were located on the sex chromosomes W or Z or in the linkage groups LGE22C19W28_E50C23 or LGE64, and SNPs that were not allocated to a specific chromosome or linkage group were excluded. In addition, SNPs with minor allele frequencies (MAF) lower than 0.03 and SNPs with a call frequency lower than 0.95 were filtered out.

Statistical analysis

In order to investigate the mapping resolution of the design, the linkage disequilibrium (LD) structure was investigated for the first nine chromosomes i.e. GGA1 to GGA9 (GGA for *Gallus gallus* chromosome). The Beagle Genetic Software Analysis [24, 25], which is included in the synbreed R package [26], was used to phase haplotypes and then the common LD measure r^2 was estimated using PLINK [27] for pairs of SNPs that were <5 Mb apart across the autosomes.

GWAS are frequently conducted using mixed linear models (e.g., [28]). In its simplest form, such models include a general mean, a fixed SNP effect and a random family effect. The latter is important to capture population stratification effects and, hence, to prevent inflation of type I errors (e.g., [29]). Previous studies showed that FPD, APD and APR are not normally distributed and that Poisson models should be used for the statistical analyses [17, 20]. Poisson models with fixed and random effects belong to a class of generalized linear mixed models (GLMM). Due to the lack of a closed form of expression of the likelihood for these models, approximate likelihood techniques are often used, as for example in the software ASReml [30]. However, for hypothesis testing, the behavior of these techniques has not been sufficiently well investigated, and Collins [31] recommended that GLMM should not be used for this purpose. Therefore, we used the following generalized linear model based on the Poisson distribution and no random effects for single-marker association analysis:

$$\eta_{ijm} = H_j + S_i + D_i + b_m x_{im}, \tag{1}$$

where η_{ijm} is the linear predictor for hen i and SNP m, H_j is the fixed hatch effect, S_i and D_i are the fixed sire and dam effects, respectively, x_{im} denotes the number of copies of the minor allele of SNP m ($x = 0$, 1, or 2), and b_m is the regression coefficient for SNP m. Thus, instead of fitting a random family effect, we included fixed sire and dam effects in the model to account for population stratification effects.

In a previous study, we detected substantial permanent environmental effects for FPD, APD and APR [17], which could also be caused by dominant gene effects. Because dominance and additive gene effects tend to be correlated such that larger dominance deviations are observed for genes with larger additive effects [32], we tested only genome-wide significant SNPs from Model (1) or from the meta-analysis (described below) for dominance effects using the following Poisson model:

$$\eta_{ijm} = H_j + S_i + D_i + b_m x_{im} + \tilde{b}_m z_{im} \tag{2}$$

where z_{im} is an indicator variable, which is 1(0) if the individual is heterozygous (homozygous) at SNP m and $\tilde{b}_m$ is a fixed regression coefficient, which is a dominance estimate. The other terms are defined as in Model (1).

To correct for multiple-testing, we applied a Bonferroni-type correction as:

$$p_{genome-wide} = 1 - (1 - p)^{\#SNP},$$

where the number (#) of SNPs was equal to 29,376. The genome-wide significance level was set at $p_{genome-wide} \leq 0.05$. Because Bonferroni's correction is very conservative, we considered an additional nominal significant level; i.e. $p < 5 \times 10^{-5}$. To estimate the number of false positives among the significant SNPs, we calculated a false discovery rate (FDR) q value for each association test by using the software QVALUE [33]. The FDR q value of the significant SNP with the largest p value provided an estimate of the proportion of false positives among the significant SNPs.

A meta-analysis was performed using the data from the selection experiment and the F2-cross experiment. We combined the p values from both studies using the inverse Chi square method of Fisher [34], known as Fisher's combined probability test, as follows:

$$\chi^2_{2k} \sim -2 \sum_{i=1}^{k} \ln{(p_i)},$$

where p_i is the p value for the ith hypothesis test and k is the number of studies being combined (i.e., $k = 2$ in our study). The significance levels were used for the p value obtained from the meta-analysis were the same as those for the GWAS (Model 1).

Cluster identification

We assumed that a causative mutation is in LD with several SNPs, and thus built clusters of SNPs, which provided strong evidence for trait-associated chromosomal regions compared to single significant SNPs, although of course it cannot be guaranteed that the mutation is within these clusters. A cluster contained at least two significant SNPs ($\leq 5 \times 10^{-5}$), with a maximum distance of 3 Mb between them. The bounds of each cluster were identified using the LD structure as well as the p values of SNPs with lower statistical support, as follows. Starting from the midpoint of the cluster of significant SNPs ($p \leq 5 \times 10^{-5}$) and moving in both directions up to 1.5 Mb on each side, we searched for weakly significant SNPs. The weakly significant SNPs ($p \leq 5 \times 10^{-4}$) at a maximum distance of 1.5 Mb from the cluster midpoint in both directions were used as the cluster bounds.

Differential gene expression analysis

Within each FDP_{meta} cluster, genes were investigated for differential expression. Expression data were generated in an earlier study [35]. In brief, the brains of nine hens each from the HFP and LFP line were collected after slaughter. RNA was extracted from the whole brain, reverse-transcribed into cDNA and then converted into labeled cRNA by *in vitro* transcription. Following this procedure, 1.65 μg of each single cRNA sample was hybridized on the Chicken Gene Expression Microarray (4 × 44 K format, Agilent Technologies) and fluorescent signal intensities were detected. The quantile-normalized and log2-transformed data were averaged across the hens within each line. A total of 1083 transcripts included in the microarray gene expression chip were located within the FDP_{meta} clusters. The average expression levels of these genes only were compared between the two lines using a standard Welch t test. Correction for multiple-testing was performed using Bonferroni's test, assuming 1083 independent tests. Sequences of probes with no assigned gene or only a LOC number were subjected to BLAST analysis against the most recent genome database galGal 5.0 (assembly GCA_000002315.3) to identify the corresponding gene. Results of the expression analysis were subsequently compared to the candidate genes that were identified within the associated clusters. Clusters that contained differentially-expressed transcripts were checked for potential enrichment of those transcripts, because this indicates the presence of cis-acting QTL. The corrected p values obtained in the original study [35] were used to separate transcripts into three categories of

significance i.e. $p \leq 0.1$, $p \leq 0.05$, and $p \leq 0.01$, respectively. For each of these categories, the proportions of significantly differentially-expressed genes within clusters were compared to genome- and chromosome-wide proportions.

Results

Results of the LD analysis are in Fig. 1 and illustrated as a plot of the LD against the physical distance of the loci up to 5 Mb. Figure 1 shows that for small distances, the level of LD was high and decreased as distance increases, especially for distances larger than 1.5 Mb. This holds true for all nine investigated chromosomes.

The GWAS (Model 1) revealed 45 (20, 19, and 58) significant SNPs at $p < 5 \times 10^{-5}$ for ADP (APR, FDP, and FDP_{meta}). The FDR for the significant SNPs associated with ADP, APR, FDP, and FDP_{meta} were <0.025, 0.07, 0.05, and 0.01, respectively. Lists of these significant SNPs are in Additional file 1: Table S1 and Additional file 2: Table S2. Plots of the test statistics for the GWAS (i.e., $-\log10 p$ values) are in Fig. 2. For APD, four genome-wide significant SNPs were identified; i.e., two on GGA1 and two on GGA5 (Table 1). The latter two SNPs also showed a significant dominance effect ($p = 0.01$, results from Model 2, not shown). For FDP_{meta}, nine genome-wide significant SNPs were identified (Table 1) with none showing a significant dominance effect.

Results from the cluster analyses are in Tables 2 and 3. For FDP, four clusters were identified, and for FDP_{meta} 13 clusters were identified. Only the cluster on GGA8 overlapped between the two traits. Seven of the nine genome-wide significant FDP_{meta} SNPs were located within clusters on GGA8 and 9. For APD, seven clusters were identified and the four genome-wide significant SNPs were located within two clusters on GGA1 and 5. For APR, three clusters were identified on GGA7 and almost all the significant SNPs were located in clusters (see Additional file 1: Table S1 and Additional file 2: Table S2).

Results from the gene expression analysis are in Table 4. Nine of the 26 probe sets that showed significant results (nominal p value ≤ 0.0001) were assigned to a LOC number or were not assigned to any gene. BLAST analysis identified the corresponding gene for only one of these. The 26 probes represented 22 different genes (Table 4). Sixteen of the 1083 probes showed a significant differential expression level, among which seven had a fold difference >2, and one a fold difference of 7.8. Six of the Bonferroni's test-corrected significant transcripts were located within the same cluster, i.e. cluster number 9. The largest number of differentially-expressed transcripts was observed on GGA9, among which eight were experiment-wide significant and four were significant probes that mapped to clusters 9 and 10.

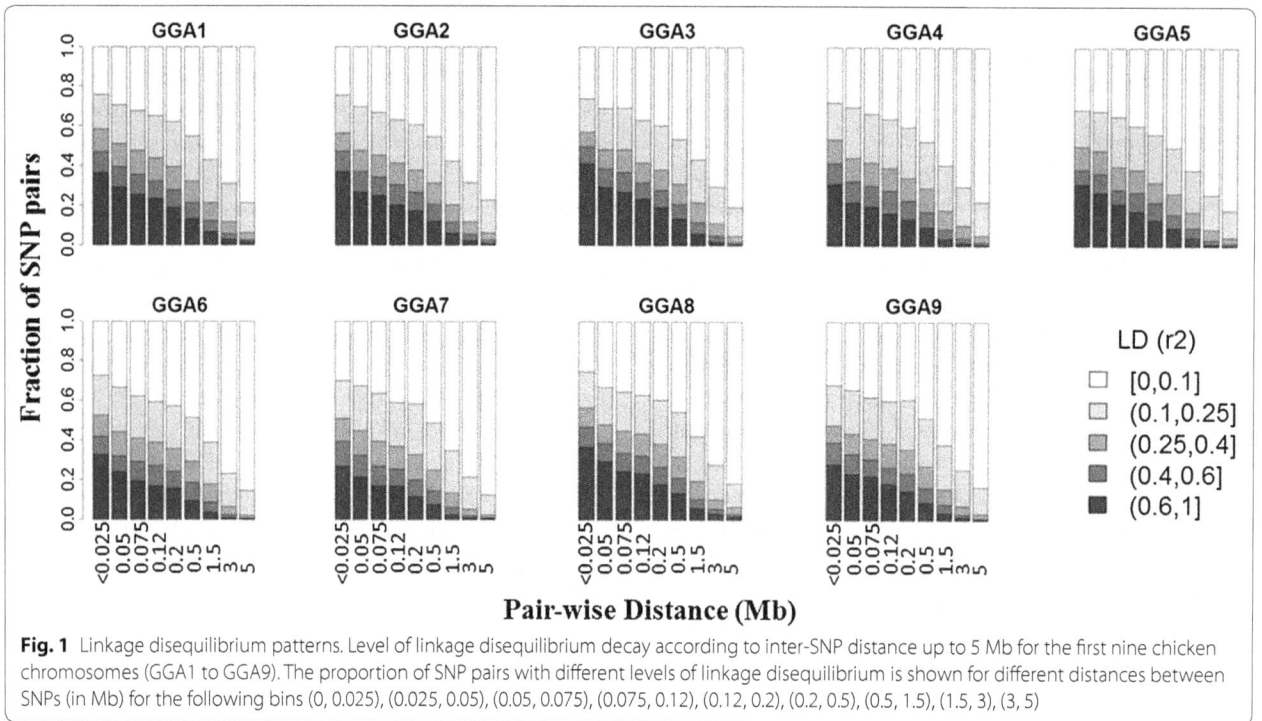

Fig. 1 Linkage disequilibrium patterns. Level of linkage disequilibrium decay according to inter-SNP distance up to 5 Mb for the first nine chicken chromosomes (GGA1 to GGA9). The proportion of SNP pairs with different levels of linkage disequilibrium is shown for different distances between SNPs (in Mb) for the following bins (0, 0.025), (0.025, 0.05), (0.05, 0.075), (0.075, 0.12), (0.12, 0.2), (0.2, 0.5), (0.5, 1.5), (1.5, 3), (3, 5)

In the previous expression study [35], 16.5, 9.7, and 2.3% of the annotated probe sets were significantly differentially-expressed with corrected p values <0.1, 0.05 and 0.01, respectively. For the individual chromosomes tested in this study, marked deviations from these fractions were found for GGA8 and GGA19 (Fig. 3). Among the seven FDP_{meta} clusters that harbored differentially-expressed transcripts, substantial enrichment was found for FPD_{meta} cluster 4 and a moderate enrichment for FPD_{meta} cluster 9 (Fig. 3). FDP_{meta} cluster 10 showed a slight enrichment only for p values <0.01 (Fig. 3).

Discussion

Experimental design and statistical analysis

We used an experimental F2-design, which has frequently been analyzed using classical linkage analyses. However, we applied single-marker GWAS, which was justified by the high level of LD between adjacent SNPs (Fig. 1). In addition, the decay of LD for SNPs separated by more than 1.5 Mb shows that the mapping resolution for these distances was generally high. Intuitively, this might be surprising, because it is usually assumed that an F2-design results in very long range LD. However, a recent simulation study showed that this holds true only if the founder lines of the F2 cross are 'distantly' related. If they are 'closely' related, the mapping resolution is high (and sometimes even higher than in the founder lines) [36]. In the current study, the founder lines were

separated by 11 generations, and thus they can be considered to be between closely and distantly related, which resulted in the high mapping resolution for distances >1.5 Mb.

Several significant trait-associated SNPs were identified for the traits included in this study and the FDR of these significant SNPs was low. In addition, nearly all significant SNPs were located within clusters. The power to map significant FDP-associated SNPs was substantially increased by combining the results from the association mapping study in the F2 cross and the selection signature results obtained in the earlier study, as can be deduced from the roughly three-fold larger number of significant SNPs for FDP_{meta} compared to FPD. This shift in power was also observed in an experiment on bovine data [37]. Intermediate gene frequencies and high F_{ST} values (only for FDP_{meta}) were obtained in the earlier selection signature experiment [20] for the genome-wide significant SNPs (see Table 1). This earlier study pointed to divergent gene frequencies in the HFP and LFP lines. Such a gene frequency pattern was expected for these genome-wide significant SNPs, because the variance contributed by an additive gene is maximized at these values. The assumption of the Fisher's combined probability test is that the p values to be combined are independent. In our study, individuals from the same population were used; i.e., a sample of individuals from the HFP and LFP lines for selection signature mapping [20] and F2 individuals

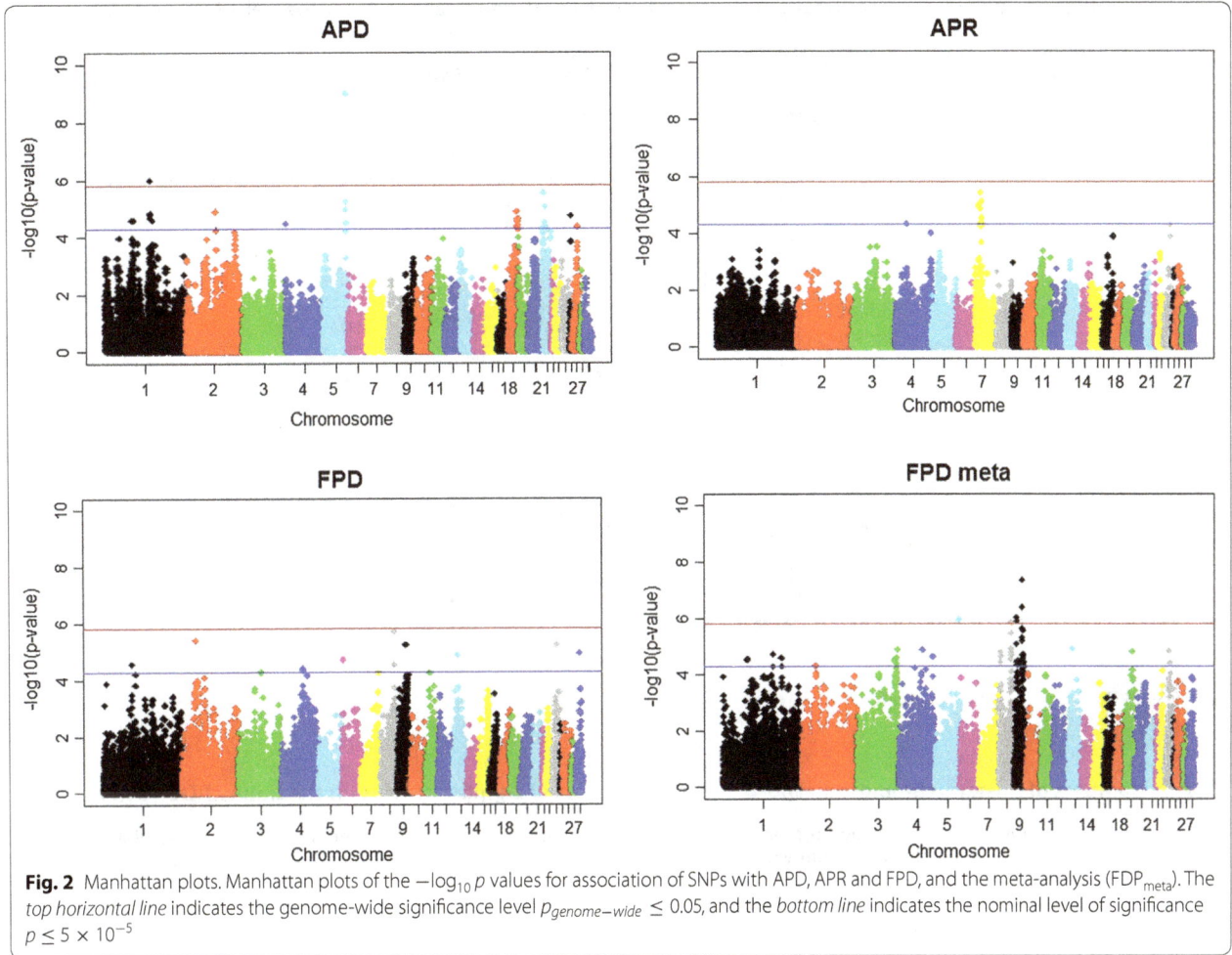

Fig. 2 Manhattan plots. Manhattan plots of the $-\log_{10} p$ values for association of SNPs with APD, APR and FPD, and the meta-analysis (FPD_{meta}). The *top horizontal line* indicates the genome-wide significance level $p_{genome-wide} \leq 0.05$, and the *bottom line* indicates the nominal level of significance $p \leq 5 \times 10^{-5}$

Table 1 List of genome-wide significant SNPs for the traits APD and FDP_{meta}

Trait	SNP	Chr	Position	$-\log_{10}(p)$	Gene frequency	F_{ST}
APD	Gga_rs14552589	5	57353834	6.8	0.13	–
	GGaluGA290503	5	57401911	6.4	0.13	–
	Gga_rs13923655	1	116041775	6.0	0.44	–
	Gga_rs15388609	1	116062599	5.8	0.44	–
FDP_{meta}	GGaluGA341482	9	17128657	7.4	0.45	0.76
	Gga_rs14676055	9	16629471	6.4	0.44	0.80
	GGaluGA341217	9	16764865	6.4	0.44	0.80
	Gga_rs13766455	9	5961337	6.0	0.46	0.82
	Gga_rs16519883	5	59368007	5.9	0.44	0.91
	Gga_rs14667686	9	6739756	5.9	0.48	0.92
	Gga_rs14652254	8	23911149	5.8	0.48	0.97
	Gga_rs15930799	8	23892743	5.8	0.48	0.97
	Gga_rs14652966	8	24679820	5.8	0.41	0.84

Chr chromosome number

Position in bp

Gene frequency in the F2-design

P value obtained from Model (1)

FST-value obtained from the previously conducted selection signature experiment

Table 2 Numbers of clusters, chromosomal positions, and numbers of significant SNPs for the traits FDP and FDP$_{meta}$

Trait	Cluster number	Chr	Start/end position in bp 3 Mbp interval	Length in Mb	Number of SNPs $p \leq 5 \times 10^{-5}$	Number of SNPs $p_{genome-wide} \leq 0.05$
FPD	1	3	58,834,628–59,725,450	0.89	3	0
	2	4	53,335,653–53,945,398	0.61	6	0
	3	6	3059,760–3075 330	0.02	2	0
	4	8	25,309,634–25,399,547	0.09	2	0
FPD$_{meta}$	1	1	58,412,953–58,831,069	0.42	3	0
	2	1	149,753,999–150,465,791	0.71	2	0
	3	2	37,372,218–39,828,657	2.46	2	0
	4	3	102,969,523–105,470,402	2.50	2	0
	5	3	107,262,448–109,945,836	2.68	3	0
	6	4	87,030,671–87,082,448	0.05	2	0
	7	8	3612,454–5410,229	1.80	3	0
	8	8	23,799,410–26,002,938	2.20	9	3
	9	9	5650,341–7645,421	2.00	5	2
	10	9	16,342,044–18,770,002	2.43	13	3
	11	9	18,726,350–20,815,056	2.09	4	0
	12	19	6883,105–8064,270	1.18	2	0
	13	24	2480,724–3900,089	1.42	3	0

Chr Chromosome

Significance level $p \leq 5 \times 10^{-5}$ and $p_{genome-wide} \leq 0.05$

Table 3 Numbers of clusters, chromosomal positions, and numbers of significant SNPs for the traits APD and APR

Trait	Cluster number	Chr	Start/end position in bp 3 Mbp interval	Length in Mb	Number of SNPs $p \leq 5 \times 10^{-5}$	Number of SNPs $p_{genome-wide} \leq 0.05$
APD	1	1	64,103,417–67,037,983	2.93	3	0
	2	1	116,041,775–117,435,846	1.39	6	2
	3	2	83,445,347–86,114,050	2.67	2	0
	4	4	33,821–552,165	0.52	7	0
	5	5	56,835,282–58,214,037	1.38	6	2
	6	18	8135,718–101,911,44	2.06	11	0
	7	21	504,778–3009,557	2.50	7	0
APR	1	7	6241,588–6327,771	0.09	3	0
	2	7	9746,560–12,631,641	2.89	10	0
	3	7	13,378,513–14,679,901	1.30	5	0

Chr Chromosome

Significance level $p \leq 5 \times 10^{-5}$ and $p_{genome-wide} \leq 0.05$

obtained from these lines for association mapping. However, a different type of information was used in each experiment, i.e. in the selection signature experiment differences in gene frequencies between the two lines were used, whereas in the association analysis SNP genotypes and trait phenotypes were used. A correlation of nearly 0 was found between the p values obtained in the selection signature and those in the association studies ($r = -0.003$), which provided further evidence for the independence of these studies.

Comparison of results with literature reports

Buitenhuis et al. [38] conducted a microsatellite-based linkage study to map QTL for feather pecking and identified QTL on GGA1 and 2. We also found significant clusters on these chromosomes, but a detailed comparison of the results was hampered by the wide confidence intervals obtained in the QTL linkage study. Recently, Recoquillay et al. [39] conducted a QTL linkage study for several behavior and production traits in Japanese quail. They did not detect a QTL for feather pecking but

Table 4 Genes located in one of the FPD$_{meta}$ clusters (Table 2) that were significantly differentially-expressed (nominal p value $\leq$ 0.0001) in the HFP and LFP lines

ProbeSetID[a]	Chr[b]	Position (Mb)[b]	FPD$_{meta}$ cluster	−log10 p	Gene symbol	Gene name	Nfold	Reg
A_87_P022983	3	104.30	4	4.58	_WDR35_	WD repeat domain 35	7.80	Up
A_87_P021624	3	104.33	4	5.85	_LAPTM4A_	lysosomal protein transmembrane 4 alpha	1.28	Down
A_87_P018137	3	104.80	4	5.23	_HS1BP3_	HCLS1 binding protein 3	2.53	Up
A_87_P254443	3	104.84	4	4.28	_LDAH_	lipid droplet associated hydrolase	1.34	Up
A_87_P176188	3	105.39	4	4.16	_LOC769627_	Unknown[c]	1.94	Down
A_87_P304288	8	3.73	7	5.64	_LOC101751271_	1-phosphatidylinositol phosphodiesterase-like	1.88	Down
A_87_P052241	8	4.03	7	4.01	_MTA1_	metastasis associated 1	1.19	Down
A_87_P079496	8	25.66	8	4.58	_GLIS1_	GLIS family zinc finger 1	2.02	Down
A_87_P016336	8	26.00	8	4.10	_TTC4_	tetratricopeptide repeat domain 4	1.37	Down
A_87_P022335	8	26.00	8	4.35	_PARS2_	prolyl-tRNA synthetase 2, mitochondrial (putative)	1.36	Up
A_87_P139413	9	5.67	9	4.17	_AQP12_	aquaporin 12	1.78	Up
A_87_P012759	9	5.67	9	6.92	_AQP12_	aquaporin 12	1.67	Up
A_87_P077026	9	5.68	9	4.09	_PAK2_	p21(RAC1)activated kinase 2	1.96	Up
A_87_P280878	9	5.69	9	7.95	_PAK2_	p21(RAC1)activated kinase 2	1.81	Up
A_87_P285338	9	5.76	9	5.38	_RNF168_	ring finger protein 168	1.27	Down
A_87_P017768	9	5.98	9	4.12	_PPP1R7_	protein phosphatase 1, regulatory (inhibitor) subunit 7	1.21	Down
A_87_P223178	9	5.98	9	4.05	_PPP1R7_	protein phosphatase 1, regulatory (inhibitor) subunit 7	1.28	Down
A_87_P023784	9	6.18	9	6.00	_ETV5_	ets variant 5	1.40	Down
A_87_P077621	9	16.69	10	4.17	_SLC12A9_	solute carrier family 12 (potassium/chloride transporters), member 9	1.51	Down
A_87_P005339	9	16.78	10	6.92	_CYP2J6L1_	cytochrome P450 2J6-like 1	2.24	Up
A_87_P177293	9	16.78	10	4.09	_CYP2J6L1_	cytochrome P450 2J6-like 1	1.97	Up
A_87_P077646	9	16.79	10	7.95	_CYP2J2L5_	cytochrome P450 2J2-like 5	2.25	Up
A_87_P181713	19	6.94	12	4.07	_FAM101B_	family with sequence similarity 101 member B	2.17	Down
A_87_P017169	19	7.26	12	4.86	_PTRH2_	peptidyl-tRNA hydrolase 2	1.16	Down
A_87_P011731	19	8.05	12	7.95	_CA4_	carbonic anhydrase IV	1.74	Down
A_87_P018194	24	25.84	13	4.19	_VPS26B_	VPS26 retromer complex component B	3.85	Up

The experiment-wide significant genes (Bonferroni corrected, $p \leq 0.05$) are written in underline

Italic: Gene symbols

[a] Unique Agilent ID for the 60mer probe on the Agilent Chicken Gene Expression Microarrays

[b] Chromosomal assignment and position according to genome release galGal2.1

[c] Recording was discontinued and the probe set could not be assigned to any gene

reported QTL for aggressive pecking on chromosomes 1 and 2. The corresponding position of the QTL on quail chromosome 1 on the chicken genome [39] was close to cluster number 1 for APD (Table 3), but the QTL on quail chromosome 2 could not be confirmed. Flisikowski et al. [40] suggested the genes _dopamine receptor D4_ (_DRD4_) and _DEAF1 transcription factor_ (_DEAF1_) as candidates for feather pecking and found significant trait associations in brain samples from the HFP and LFP lines. These lines were the same as used in Grams et al. [20] and in our study to create the F2 cross. _DRD4_ and _DEAF1_ are located on GGA5. We did identify one cluster for FDP$_{meta}$ on GGA5, but it was not in the vicinity of these candidate genes. No single SNP in the chromosomal region that included these genes showed a nominal significant p value. In addition, although two probes were located in _DRD4_ and three in _DEAF1_, none of these showed significant differential expression in the HFP and LFP lines. Thus, based on results from the current study, the candidate status of these genes was not supported.

Comparison of our study with reports from the literature revealed few congruent results, which can be due to several reasons. First, it is very likely that different ethograms were used in these studies, resulting in different

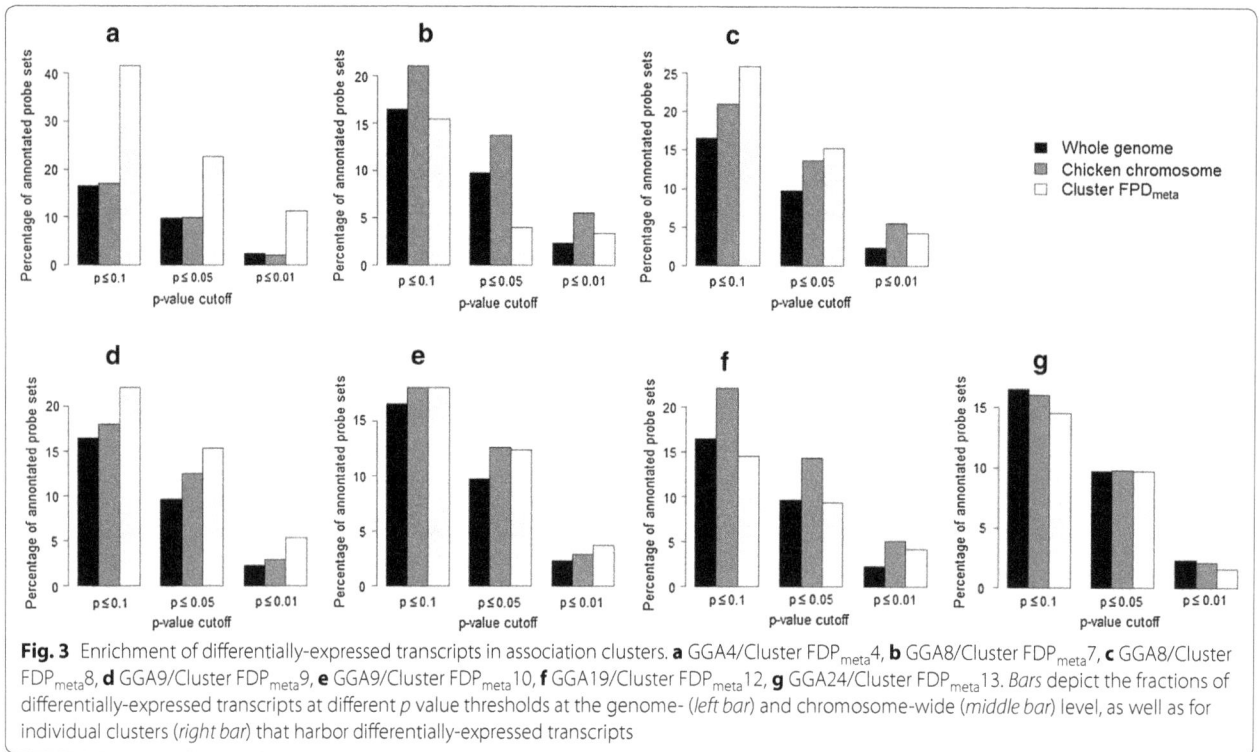

Fig. 3 Enrichment of differentially-expressed transcripts in association clusters. **a** GGA4/Cluster FDP$_{meta}$4, **b** GGA8/Cluster FDP$_{meta}$7, **c** GGA8/Cluster FDP$_{meta}$8, **d** GGA9/Cluster FDP$_{meta}$9, **e** GGA9/Cluster FDP$_{meta}$10, **f** GGA19/Cluster FDP$_{meta}$12, **g** GGA24/Cluster FDP$_{meta}$13. *Bars* depict the fractions of differentially-expressed transcripts at different *p* value thresholds at the genome- (*left bar*) and chromosome-wide (*middle bar*) level, as well as for individual clusters (*right bar*) that harbor differentially-expressed transcripts

definitions of the traits. Second, in addition to differences in mapping procedures and in the genetic maps used, the size of the experimental populations also differed substantially between studies, with the largest size being used in the current study. Finally, it is also possible that significant associations were not confirmed simply because they do not segregate in other populations.

Candidate gene identification

The association clusters spanned more than 20 Mbp for all analyzed traits, i.e. a region comprising hundreds of genes, which makes the identification of candidate genes very speculative. However, inclusion of gene expression data can be used to classify positional candidate genes on a functional basis, as was done in the current study, which was based on genome-wide expression data that were restricted to association clusters to reduce the multiple-testing burden. Differentially-expressed genes that are located within QTL regions can indicate the presence of a cis-acting regulatory mutation. However, hundreds of differentially-expressed transcripts were located within the association clusters, which made such an assumption very speculative. However, enrichment of such transcripts within clusters compared to the whole genome or individual chromosomes supports the hypothesis that differential expression can, at least partly, be explained by cis-acting regulatory mechanisms. In that case, it is

expected that enrichment is stronger for more stringent *p* value cutoffs. The most substantial enrichment in the current study was obtained for FDP$_{meta}$ cluster 4 (Fig. 3). However, no functionally plausible candidate gene was identified within this region.

Positional candidate gene *SLC12A9* in FDP$_{meta}$ cluster 10 on GGA9 exhibited experiment-wide significant differential expression between the HFP and LFP lines. However, for this cluster only a slight enrichment was observed for the most stringent *p* value cutoff. Nevertheless, *SLC12A9* remains a functionally very plausible candidate gene for this QTL. It belongs to a family of nine genes that code for electroneutral cation–chloride-cotransporters [41]. Although the function of this gene is unclear, other *SLC12* transporters are known to be crucial in the control of the electrochemical chloride gradient that is required for hyperpolarizing the postsynaptic inhibition that is mediated by GABA$_A$ and glycine receptors [42]. This is remarkable, because reduction of postsynaptic GABA$_A$ receptor currents is also an effect of serotonin mediated by 5-HT$_2$ receptors [43]. There is a growing body of evidence that brain monoamines, such as serotonin and dopamine, are involved in the occurrence of feather pecking and aggressive pecking in hens [44–48] and in aggressive behavior in humans [49]. Kops et al. [47] showed that differences in dopamine turnover between a low mortality and a control hen line were largest, in particular, in the arcopallium region of the brain. Another purely positional candidate gene for feather pecking was located in FDP$_{meta}$ cluster 9, i.e. *CLSTN2* (*calsyntenin 2*), which is also involved in post-synaptic signaling related to excitatory synaptic transmission [50].

For APD, the *GNG2* (*G protein subunit gamma 2*) gene was identified as a positional candidate gene in FDP$_{meta}$ cluster 5 on GGA5 (Table 3). This gene is also involved in monoamine signaling, particularly in postsynaptic signaling at serotonergic (KEGG pathway ko04726) and dopaminergic (KEGG pathway ko04728) synapses.

Shared environment and associated effects

Behavior traits involve interactions between individuals. Statistical models that include interaction or associated effects have developed, as reviewed by Bijma [51] and Ellen et al. [52], which have shown that these effects can substantially contribute to the heritable variation in survival of hens related to feather pecking and cannibalism [52]. Indeed, these interactions might also be another possible explanation for the low genetic trend in later generations in our selection experiment [20]. In a recent study, we chose the simplest form to capture shared environment effects and associated effects by fitting a random pen effect to the model [17]. Since pen variances were very small, they were not included in the current study. Moreover, the size of the pens used here was rather large for social interaction models.

Conclusions

Several significant trait-associated clusters of SNPs were identified, especially for the trait FPD$_{meta}$ but also for aggressive pecking. However, behavioral traits, appeared to be controlled by many genes with small effects and no single SNP was promising for selection purposes. However, understanding the motivation for feather pecking is of interest in its own right. In-depth sequence-based association analyses of the clusters identified in this study and subsequent identification of candidate genes from a small list of putative positional genes will help to formulate and validate hypotheses for the expression of this abnormal behavior pattern. Clearly, for this purpose additional data need to be collected.

Authors' contributions
MG, WB, and VG conducted the field experiment; SP performed the genotyping and cleaned up the genotypic data; PS performed the LD structure analysis; VG performed the remaining statistical analysis; VG, SP, JT, and JB interpreted the results and wrote the paper; WB and JB initiated and oversaw the project. All authors read and approved the final manuscript.

Author details
[1] Institute of Animal Science, University of Hohenheim, 70599 Stuttgart, Germany. [2] Division of Functional Breeding, Department of Animal Sciences, Georg-August-University Göttingen, 37077 Göttingen, Germany.

Acknowledgements
This study was partially supported by a grant from the German Research Foundation (Deutsche Forschungsgemeinschaft, DFG).

Competing interests
The authors declare that they have no competing interests.

References
1. Blokhuis HJ. Feather-pecking in poultry: its relation with ground-pecking. Appl Anim Behav Sci. 1986;16:63–7.
2. Vestergaard KS, Lisborg L. A model of feather pecking development which relates to dustbathing in the fowl. Behaviour. 1993;126:291–308.
3. Kjaer JB, Vestergaard K. Development of feather pecking in relation to light intensity. Appl Anim Behav Sci. 1999;62:243–54.
4. Savory CJ, Mann JS, MacLeod MG. Incidence of pecking damage in growing bantams in relation to food form, group size, stocking density, dietary tryptophan concentration and dietary protein source. Br Poult Sci. 1999;40:579–84.
5. Aerni V, El-Lethey H, Wechsler B. Effect of foraging material and food form on feather pecking in laying hens. Br Poult Sci. 2000;41:16–21.
6. Vestergaard KS, Kruijt JP, Hogan JA. Feather pecking and chronic fear in groups of red junglefowl: their realtions to dustbathing, rearing environment and social status. Anim Behav. 1993;45:1127–40.
7. Jones RB. Fear and adaptability in poultry: insights, implications and imperatives. World Poult Sci J. 1996;52:131–74.
8. Jensen P, Keeling L, Schütz K, Andersson L, Mormède P, Brändström H, et al. Feather pecking in chickens is genetically related to behavioural and developmental traits. Physiol Behav. 2005;86:52–60.
9. Grams V, Bögelein S, Grashorn MA, Bessei W, Bennewitz J. Quantitative genetic analysis of traits related to fear and feather pecking in laying hens. Behav Genet. 2015;45:228–35.
10. McKeegan DEF, Savory CJ. Feather eating in layer pullets and its possible role in the aetiology of feather pecking damage. Appl Anim Behav Sci. 1999;65:73–85.
11. McKeegan DEF, Savory CJ. Feather eating in individually caged hens which differ in their propensity to feather peck. Appl Anim Behav Sci. 2001;73:131–40.
12. Harlander-Matauschek A, Bessei W. Feather eating and crop filling in laying hens. Arch Geflügelkd. 2005;69:241–4.
13. Lutz V, Kjaer JB, Iffland H, Rodehutscord M, Bessei W, Bennewitz J. Quantitative genetic analysis of causal relationships between feather pecking, feather eating and general locomotor activity in laying hens using structural equation models. Poult Sci. 2016;95:1757–63.
14. Kjaer JB. Feather pecking in domestic fowl is genetically related to locomotor activity levels: implications for a hyperactivity disorder model of feather pecking. Behav Genet. 2009;39:564–70.
15. Bilcík B, Keeling LJ. Changes in feather condition in relation to feather pecking and aggressive behaviour in laying hens. Br Poult Sci. 1999;40:444–51.
16. Bessei W, Bauhaus H, Bögelein S. The effect of selection for high and low feather pecking on aggression—related behaviours of laying hens. Arch Geflügelkd. 2013;77:10–4.
17. Bennewitz J, Bögelein S, Stratz P, Rodehutscord M, Piepho HP, Kjaer JB, et al. Genetic parameters for feather pecking and aggressive behavior in a large F2-cross of laying hens using generalized linear mixed models. Poult Sci. 2014;93:810–7.
18. Kjaer JB, Sørensen P, Su G. Divergent selection on feather pecking behaviour in laying hens (*Gallus gallus domesticus*). Appl Anim Behav Sci. 2001;71:229–39.

19. Rodenburg TB, Buitenhuis AJ, Ask B, Uitdehaag KA, Koene P, van der Poel JJ, et al. Heritability of feather pecking and open-field response of laying hens at two different ages. Poult Sci. 2003;82:861–7.

20. Grams V, Wellmann R, Preuß S, Grashorn MA, Bessei W, Bennewitz J. Genetic parameters and signatures of selection in two divergent laying hen lines selected for feather pecking behaviour. Genet Sel Evol. 2015;47:77.

21. Piepho HP, Lutz V, Kjaer JB, Grashorn MA, Bennewitz J, Bessei W. The presence of extreme feather peckers in groups of laying hens. Animal. 2016. doi:10.1017/S1751731116001579.

22. Savory CJ. Feather pecking and cannibalism. World Poult Sci J. 1995;51:215–9.

23. Grams V, Bessei W, Piepho HP, Bennewitz J. Genetic parameters for feather pecking and aggressive behavior in laying hens using Poisson and linear models. In: Proceedings of the 10th world congress on genetics applied to livestock production: 17–22 August 2014; Vancouver. 2014.

24. Browning SR, Browning BL. Rapid and accurate haplotype phasing and missing-data inference for whole-genome association studies by use of localized haplotype clustering. Am J Hum Genet. 2007;81:1084–97.

25. Browning BL, Browning SR. A unified approach to genotype imputation and haplotype-phase inference for large data sets of trios and unrelated individuals. Am J Hum Genet. 2008;84:210–23.

26. Wimmer V, Albrecht T, Auinger HJ, Schön CC. Synbreed: a framework for the analysis of genomic prediction data using R. Bioinformatics. 2012;28:2086–7.

27. Purcell S, Neale B, Todd-Brown K, Thomas L, Ferreira MAR, Bender D, et al. PLINK: a tool set for whole-genome association and population-based linkage analyses. Am J Hum Genet. 2007;81:559–75.

28. Yang J, Zaitlen NA, Goddard ME, Visscher PM, Price AL. Advantages and pitfalls in the application of mixed-model association methods. Nat Genet. 2014;46:100–6.

29. Pausch H, Flisikowski K, Jung S, Emmerling R, Edel C, Go K. Genome-wide association study identifies two major loci affecting calving ease and growth-related traits in cattle. Genetics. 2011;187:289–97.

30. Gilmour AR, Gogel BJ, Cullis BR, Thompson R. ASReml user guide release 3.0. Hemel Hempstead: VSN International Ltd; 2009.

31. Collins D. Generalized linear mixed models. In: Gilmour AR, Gogel BJ, Cullis BR, Thompson R, editors. ASReml user guide release 3.0. Hemel Hempstead: VSN International Ltd; 2009.

32. Wellmann R, Bennewitz J. The contribution of dominance to the understanding of quantitative genetic variation. Genet Res (Camb). 2011;93:139–54.

33. Storey JD, Tibshirani R. Statistical significance for genomewide studies. Proc Natl Acad Sci USA. 2003;100:9440–5.

34. Fisher RA. Statistical methods for research workers. 4th ed. London: Oliver & Boyd; 1932.

35. Wysocki M, Preuss S, Stratz P, Bennewitz J. Investigating gene expression differences in two chicken groups with variable propensity to feather pecking. Anim Genet. 2013;44:773–7.

36. Bennewitz J, Wellmann R. Mapping resolution in simulated porcine F2 populations using dense marker panels. In: Proceedings of the 10th World Congress on Genetics Applied to Livestock Production: 17–22 August 2014; Vancouver.

37. Schwarzenbacher H, Dolezal M, Flisikowski K, Seefried F, Wurmser C, Schlötterer C, et al. Combining evidence of selection with association analysis increases power to detect regions influencing complex traits in dairy cattle. BMC Genomics. 2012;13:48.

38. Buitenhuis AJ, Rodenburg TB, Van Hierden YM, Siwek M, Cornelissen SJB, Nieuwland MGB, et al. Mapping quantitative trait loci affecting feather pecking behavior and stress response in laying hens. Poult Sci. 2003;82:1215–22.

39. Recoquillay J, Pitel F, Arnould C, Leroux S, Dehais P, Moreno C, et al. A medium density genetic map and QTL for behavioral and production traits in Japanese quail. BMC Genomics. 2015;16:10.

40. Flisikowski K, Schwarzenbacher H, Wysocki M, Weigend S, Preisinger R, Kjaer JB, et al. Variation in neighbouring genes of the dopaminergic and serotonergic systems affects feather pecking behaviour of laying hens. Anim Genet. 2009;40:192–9.

41. Gagnon KB, Delpire E. Physiology of SLC12 transporters: lessons from inherited human genetic mutations and genetically engineered mouse knockouts. Am J Physiol Cell Physiol. 2013;304:C693–714.

42. Blaesse P, Airaksinen MS, Rivera C, Kaila K. Cation-chloride cotransporters and neuronal function. Neuron. 2009;61:820–38.

43. Feng J, Cai X, Zhao J, Yan Z. Serotonin receptors modulate GABA$_A$ receptor channels through activation of anchored protein kinase C in prefrontal cortical neurons. J Neurosci. 2001;21:6502–11.

44. van Hierden YM, Korte SM, Ruesink EW, van Reenen CG, Engel B, Korte-Bouws GAH, et al. Adrenocortical reactivity and central serotonin and dopamine turnover in young chicks from a high and low feather-pecking line of laying hens. Physiol Behav. 2002;75:653–9.

45. van Hierden YM, Koolhaas JM, Kost'ál L, Výboh P, Sedlacková M, Rajman M, et al. Chicks from a high and low feather pecking line of laying hens differ in apomorphine sensitivity. Physiol Behav. 2005;84:471–7.

46. Kops MS, de Haas EN, Rodenburg TB, Ellen ED, Korte-Bouws GAH, Olivier B, et al. Effects of feather pecking phenotype (severe feather peckers, victims and non-peckers) on serotonergic and dopaminergic activity in four brain areas of laying hens (Gallus gallus domesticus). Physiol Behav. 2013;120:77–82.

47. Kops MS, de Haas EN, Rodenburg TB, Ellen ED, Korte-Bouws GAH, Olivier B, et al. Selection for low mortality in laying hens affects catecholamine levels in the arcopallium, a brain area involved in fear and motor regulation. Behav Brain Res. 2013;257:54–61.

48. Kops MS, Kjaer JB, Güntürkün O, Westphal KGC, Korte-Bouws GAH, Olivier B, et al. Serotonin release in the caudal nidopallium of adult laying hens genetically selected for high and low feather pecking behavior: an in vivo microdialysis study. Behav Brain Res. 2014;268:81–7.

49. Fernandez-Castillo N, Cormand B. Aggressive behavior in humans: genes and pathways identified through association studies. Am J Med Genet B Neuropsychiatr Genet. 2016;171:676–96.

50. Hintsch G, Zurlinden A, Meskenaite V, Steuble M, Fink-Widmer K, Kinter J, et al. The calsyntenins—a family of postsynaptic membrane proteins with distinct neuronal expression patterns. Mol Cell Neurosci. 2002;21:393–409.

51. Bijma P. The quantitative genetics of indirect genetic effects: a selective review of modeling issues. Heredity (Edinb). 2014;112:61–9.

52. Ellen ED, Rodenburg TB, Albers GAA, Bolhuis JE, Camerlink I, Duijvesteijn N, et al. The prospects of selection for social genetic effects to improve welfare and productivity in livestock. Front Genet. 2014;5:337.

5

A predictive assessment of genetic correlations between traits in chickens using markers

Mehdi Momen[1], Ahmad Ayatollahi Mehrgardi[1*], Ayoub Sheikhy[2], Ali Esmailizadeh[1,3], Masood Asadi Fozi[1], Andreas Kranis[4], Bruno D. Valente[5], Guilherme J. M. Rosa[5,6] and Daniel Gianola[5,6,7]

Abstract

Background: Genomic selection has been successfully implemented in plant and animal breeding programs to shorten generation intervals and accelerate genetic progress per unit of time. In practice, genomic selection can be used to improve several correlated traits simultaneously via multiple-trait prediction, which exploits correlations between traits. However, few studies have explored multiple-trait genomic selection. Our aim was to infer genetic correlations between three traits measured in broiler chickens by exploring kinship matrices based on a linear combination of measures of pedigree and marker-based relatedness. A predictive assessment was used to gauge genetic correlations.

Methods: A multivariate genomic best linear unbiased prediction model was designed to combine information from pedigree and genome-wide markers in order to assess genetic correlations between three complex traits in chickens, i.e. body weight at 35 days of age (BW), ultrasound area of breast meat (BM) and hen-house egg production (HHP). A dataset with 1351 birds that were genotyped with the 600 K Affymetrix platform was used. A kinship kernel ($\mathbf{K}$) was constructed as $\mathbf{K} = \lambda \mathbf{G} + (1 - \lambda)\mathbf{A}$, where $\mathbf{A}$ is the numerator relationship matrix, measuring pedigree-based relatedness, and $\mathbf{G}$ is a genomic relationship matrix. The weight (λ) assigned to each source of information varied over the grid $\lambda = (0, 0.2, 0.4, 0.6, 0.8, 1)$. Maximum likelihood estimates of heritability and genetic correlations were obtained at each λ, and the "optimum" λ was determined using cross-validation.

Results: Estimates of genetic correlations were affected by the weight placed on the source of information used to build $\mathbf{K}$. For example, the genetic correlation between BW–HHP and BM–HHP changed markedly when λ varied from 0 (only $\mathbf{A}$ used for measuring relatedness) to 1 (only genomic information used). As λ increased, predictive correlations (correlation between observed phenotypes and predicted breeding values) increased and mean-squared predictive error decreased. However, the improvement in predictive ability was not monotonic, with an optimum found at some $0 < \lambda < 1$, i.e., when both sources of information were used together.

Conclusions: Our findings indicate that multiple-trait prediction may benefit from combining pedigree and marker information. Also, it appeared that expected correlated responses to selection computed from standard theory may differ from realized responses. The predictive assessment provided a metric for performance evaluation as well as a means for expressing uncertainty of outcomes of multiple-trait selection.

*Correspondence: mehrgardi@uk.ac.ir
[1] Department of Animal Science, Faculty of Agriculture, Shahid Bahonar University of Kerman (SBUK), Kerman, Iran
Full list of author information is available at the end of the article

Background

The increasing availability of genome-wide dense molecular markers [e.g., single nucleotide polymorphisms (SNPs)] has opened new avenues for obtaining additional genetic gain in breeding of elite animals and plants by exploiting "genomic selection" methods. These techniques have become important tools in modern breeding programs [1, 2]. Many statistical methods with parametric or non-parametric formulations have been proposed to predict either genomic estimated breeding values (GEBV) of animals or yet-to-be observed phenotypes [1, 3–6].

Most prediction studies have been based on single-trait (uni-variate) statistical models. However, in practice, animals and plants often must be evaluated for several economically important traits. Multiple-trait model predictions have been typically regarded as better than uni-variate predictions [7]. For example, milk yield and composition in dairy cattle or grain yield and resistance to disease in plants are often analyzed with multiple-trait methods [8, 9]. Multi-trait models based on pedigree information represent the typical modeling strategy used to capitalize on genetic evaluation of several correlated traits before genomic selection methods became popular [10]. A multiple-trait analysis requires knowledge of phenotypic and genetic correlations among characters [7]. These correlations indicate the extent to which measurements on one trait inform about other traits [11], and predictions based on single-trait models do not exploit the extra information provided by other traits.

Multiple-trait genomic selection models (MT-GS) have been explored and tested in research only to a limited extent [12]. A genome-based multiple-trait analysis may also offer insight into mechanisms that create trait associations, such as pleiotropy and linkage disequilibrium (LD) between quantitative trait loci (QTL) and markers [13]. One hypothesis is that correlation parameters that are inferred using whole-genome dense molecular markers may give a novel picture of the genetic correlation between traits [14, 15]. However, the sources of genetic and genomic correlations may be distinct [13, 16]. Genomic correlations depend in part on linkage disequilibrium (LD) relationships between markers and QTL, which are unknown, while genetic correlations are in part a function of LD between QTL. Multivariate genome-based models may produce "missing": situation in which the genetic correlation is undetected by the markers, "excessive": LD between markers increase the magnitude of the pleiotropy effects of the QTL, or even "spurious": there is no pleiotropy but LD between markers and/or pairs of QTL may produce pseudo pleiotropy (abbreviated as MES) genetic correlations and, as a consequence, distort expectations about outcomes of multiple-trait selection.

The objective of this study was to infer genetic and genomic correlations between three traits measured in broilers by exploring linear combinations of pedigree-based (genealogical) and marker-based relationship matrices. As advocated by [17], a predictive approach was used to gauge parameter estimates and to provide an empirical test of the extent of genetic association between traits.

Methods

Data

The data consisted of records on 1351 birds from a commercial broiler chicken line that had undergone several generations of selection using the traditional multiple-trait genetic evaluations at the Aviagen Ltd Company (Aviagen Ltd, Newbridge, UK). The traits considered were body weight at 35 days of age (BW), ultrasound area of breast meat (BM), and hen-house production (HHP, total number of eggs laid between weeks 28 and 54). Some features of the dataset and pedigree information are in Table 1. All birds had phenotype records and a known sire and dam, and there were 326 and 274 paternal half-sib and full-sib groups in the sample, respectively. This dataset has also been used in other studies by Abdollahi-Arpanahi et al. [18] and Morota et al. [19].

Phenotype correction

Prior to implementing the genome-enabled trivariate prediction model, we pre-corrected phenotypes to eliminate all known nuisance non-genetic sources of variation. This correction was based on uni-variate mixed effects models; BW and BM were corrected for a combined effect of sex, hatch week, contemporary group of parents, and pen in the growing farm. HHP was corrected for random hatch effects, with a general mean as the sole fixed effect. Figure 1 shows a scatter plot of pre-corrected phenotypes for these traits. A positive association between BW and BM is suggested, whereas the scatter plots for the pairs BM-HHP and BW-HHP do not indicate concomitant variation.

Table 1 Pedigree information and features of the chicken data used

Total birds in the pedigree	1675
Number of sires	326
Number of dams	592
Number of full-sib groups	274
Number of progeny with records and known sire and dam	1351
Number of inbreds (pedigree-based inbreeding >0)	159
Inbreeding coefficient range all birds in the pedigree (%)	0.4 to 10.9

Genotyping

The 1351 birds were genotyped using an 600 K Affymetrix SNP chip. SNPs with a minor allelic frequency (MAF) lower than 1% and a call frequency lower than 0.95 were filtered out. Missing genotypes were imputed locus by locus using the Beagle software version 3.3.2 [20]. After quality control, 354,364 SNPs remained for statistical analyses.

Whole-genome prediction models

Tri-variate linear models were used for estimating (co)variance components and for predicting genomic breeding values. Such models were an extension of a typical single-trait model with random pedigree or genome-based effects, which can be represented as:

$$\mathbf{y}_t = \mathbf{1}\mu_t + \mathbf{Z}\mathbf{g}_t + \boldsymbol{\epsilon}_t; \quad t = 1, 2, 3, \quad (1)$$

where, $\mathbf{y}_t$ is a vector of $m \times 1$ pre-corrected phenotypes for trait t ($m = 1351$); μ_t is a general constant and 1 is a vector of ones; $\mathbf{Z}$ is an incidence matrix (an identity matrix in all cases) that allocates records to breeding values; $\mathbf{g}_t$ is a vector of additive genetic effects or of direct genomic breeding values, and $\boldsymbol{\epsilon}_t$ is a vector of residuals for trait t. It was assumed that $\mathbf{g}_t \sim \left(0, \mathbf{K}\sigma_{\mathbf{g}_t}^2\right)$ where $\sigma_{\mathbf{g}_t}^2$ is the additive genetic or genomic variance of trait t, and $\mathbf{K}$ ($m \times m$) reflects a covariance structure that results from the combined use of pedigree and marker information, as described later. Random residuals were assumed to follow a normal distribution $\boldsymbol{\epsilon}_t \sim N\left(0, \mathbf{I}_t\sigma_{\epsilon_t}^2\right)$, where $\mathbf{I}_t$ is an $m \times m$ identity matrix and $\sigma_{\epsilon_t}^2$ is the residual variance for trait t; this term represents variation of pre-corrected phenotypes that is not explained by additive genomic effects. The vectors $\mathbf{g}_t$ and $\boldsymbol{\epsilon}_t$ were assumed to be independent. The multi-variate model was:

$$\begin{pmatrix} \mathbf{y}_1 \\ \mathbf{y}_2 \\ \mathbf{y}_3 \end{pmatrix} = \begin{bmatrix} \mathbf{I}_1 & 0 & 0 \\ 0 & \mathbf{I}_2 & 0 \\ 0 & 0 & \mathbf{I}_3 \end{bmatrix} \begin{bmatrix} \mu_1 \\ \mu_2 \\ \mu_3 \end{bmatrix} + \begin{bmatrix} \mathbf{Z}_1 & 0 & 0 \\ 0 & \mathbf{Z}_2 & 0 \\ 0 & 0 & \mathbf{Z}_3 \end{bmatrix} \begin{bmatrix} \mathbf{g}_1 \\ \mathbf{g}_2 \\ \mathbf{g}_3 \end{bmatrix} + \begin{bmatrix} \boldsymbol{\epsilon}_1 \\ \boldsymbol{\epsilon}_2 \\ \boldsymbol{\epsilon}_3 \end{bmatrix},$$

$$(2)$$

where, $\mathbf{y}_t$, μ_t, $\mathbf{Z}_t$, $\mathbf{g}_t$ and $\boldsymbol{\epsilon}_t$ are as before. The vector of multi-trait additive genetic or genomic breeding values was distributed as $\begin{bmatrix} \mathbf{g}_1 \\ \mathbf{g}_2 \\ \mathbf{g}_3 \end{bmatrix} \sim N(0, \mathbf{K} \otimes \mathbf{Q})$, where $\mathbf{K}$ is a kinship or kernel matrix (described later) and $\mathbf{Q}$ is the (3×3) matrix of pedigree- or marker-based covariances among traits. The multivariate residual distribution was assumed to be $\begin{bmatrix} \boldsymbol{\epsilon}_1 \\ \boldsymbol{\epsilon}_2 \\ \boldsymbol{\epsilon}_3 \end{bmatrix} \sim N(0, \mathbf{I} \otimes \mathbf{R})$, where $\mathbf{R}$ is the (3×3) residual covariance matrix among traits. The Kronecker product ($\otimes$) notation applies to the residual covariance since all traits were measured on all birds.

Pedigree-based and whole-genome relationship matrices

In a genomic best linear unbiased prediction model (GBLUP), a genomic relationship matrix ($\mathbf{G}$) computed from marker data replaces the pedigree-based matrix ($\mathbf{A}$) of standard BLUP applications. The genomic relationship matrix intends to measure the realized fraction of alleles shared, rather than the expected fraction, as is the case for $\mathbf{A}$ [21, 22]. Genomic relationship matrices can be calculated in different ways (e.g., [23]) but here we used two known alternatives, as described next. First, VanRaden [22] proposed the $m \times m$ matrix $\mathbf{G}_V = \frac{\mathbf{WW}'}{2\sum p_i q_i}$, which renders $\mathbf{G}$ analogous to the numerator relationship matrix $\mathbf{A}$ due to the denominator, $2\Sigma p_i q_i$. Here, $\mathbf{W}$ is a $m \times p$ centered matrix of SNP genotype codes 0, 1 and 2 ($p = 354,364$) and p_i is the MAF at locus i. Second, Forni

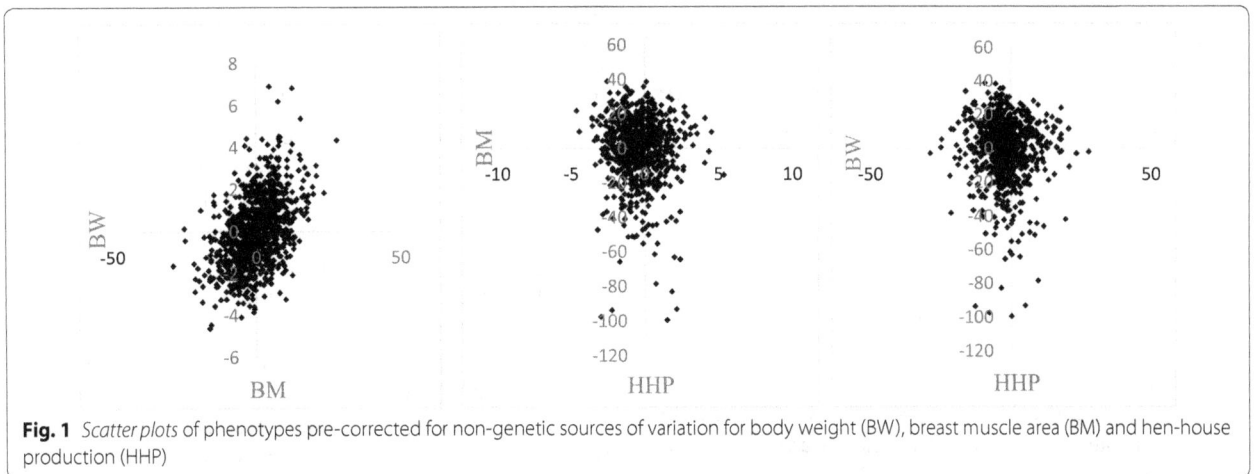

Fig. 1 *Scatter plots* of phenotypes pre-corrected for non-genetic sources of variation for body weight (BW), breast muscle area (BM) and hen-house production (HHP)

et al. [21] suggested a modification of the denominator, $\mathbf{G_F} = \frac{\mathbf{WW'}}{\{\text{trace}[\mathbf{WW'}]\}/m}$, which attempts to attain compatibility of the genomic relationship matrix with $\mathbf{A}$ when either the average level inbreeding is low or when the number of generations back to the base population is small.

An alternative to using any given $\mathbf{G}$ is to combine genomic and pedigree information into a single kinship "kernel" matrix (in the sense of [24]). A "kernel" matrix that exploits genealogy information together with marker-based information could potentially capture parts of the genetic covariance among traits that is not accounted for by either $\mathbf{A}$ or $\mathbf{G}$ alone. We followed multiple-kernel ideas [25] and used the kernel $\mathbf{K} = \lambda\mathbf{G} + (1 - \lambda)\mathbf{A}$, where λ is a parameter (weight) bounded between 0 and 1, and $\mathbf{G} = \mathbf{G_V}$ or $\mathbf{G_F}$. For example, if $\lambda = 0$, pedigree information "dominates" in the model, which retrieves a traditional pedigree-based BLUP. Our expectation was that a specific combination of $\mathbf{A}$ and $\mathbf{G}$ matrices would provide the "best" estimates of parameters, as gauged by prediction of outcomes, as opposed to using either $\mathbf{A}$ or $\mathbf{G}$ alone or both, with (co)variance components estimated in training samples. To assess the best value of λ, we applied the grid $\lambda = (0, 0.2, 0.4, 0.6, 0.8, 1)$ and evaluated the ensuing predictive abilities over such a grid.

When using marker- and pedigree-based relationship matrices together, scaling of genomic relationship matrices is needed for interpretation of parameters in the context of theory, e.g., in relation to a base population [26]. Estimates of parameters may be distorted if a genomic relationship matrix is not on the same scale as the pedigree-based relationship matrix. A reasonable rescaling may be achieved by using genomic relationship matrices with elements that range between 0 and 2, which are the minimum and maximum values of $\mathbf{A}$, respectively. To render $\mathbf{G}$ on the same scale as $\mathbf{A}$, we used the *map min-max-function* that is widely used in machine learning, e.g. [27], as follows:

$$Gs_{ij} = \frac{(Gs_{max} - Gs_{min}) \times (G_{ij} - G_{min})}{G_{max} - G_{min}} + Gs_{min}. \quad (3)$$

Here, Gs_{ij} is a scaled element of the $\mathbf{G_V}$ or $\mathbf{G_F}$ matrix and G_{ij} is typical element of $\mathbf{G_V}$ or $\mathbf{G_F}$; $Gs_{max} = 2$ and $Gs_{min} = 0$ are the minimum and maximum values elements that the scaled matrix is allowed to take, respectively, and G_{min} and G_{max} are the maximum and minimum entries of the $\mathbf{G_V}$ or $\mathbf{G_F}$ matrix, respectively. While $\mathbf{G_V}$ and $\mathbf{G_F}$ may contain negative off-diagonals, this is not the case for the scaled matrices used here.

Model fitting and validation
Variance and covariance components were estimated with multiple-trait restricted maximum likelihood

(REML) via an average information algorithm (AI) implemented in the WOMBAT program [28]. The software provides point estimates of (co)variance components and their asymptotic standard errors. Matrix $\mathbf{K} = \lambda\mathbf{G} + (1 - \lambda)\mathbf{A}$ was used as kinship matrix, where $\mathbf{G}$ was either the unscaled or scaled versions $\mathbf{G_V}$ or $\mathbf{G_F}$.

We used a cross-validation scheme with 20 randomly constructed training and testing sets to assess predictive ability over the grid of λ values. We randomly partitioned the whole data into training (60%) and testing (40%) sets in each of the 20 repetitions. After a model was fitted to the training set data, we compared its predictions against realized values in the test set. Predictive ability was measured by mean squared error (MSE) and by the correlation between predicted and observed phenotypes in the testing set.

Realized versus expected genetic regressions between traits
We also evaluated predictive relationships between pairs of traits, i.e., BW-BM, BW-HHP and BM-HHP, according to the cross-validation scheme described earlier. Over the predefined grid $\lambda = (0, 0.2, 0.4, 0.6, 0.8, 1)$, we computed least-squares estimates of the regression of the phenotype for trait x on DGV for trait y, and vice versa, for each pair of traits for each of the 20 validation sets. These realized regressions were compared to expected genetic regressions deduced from REML (co)variance component estimates as:

$$b_{(x,y)}(\lambda) = r_{G(x,y)}(\lambda) \times \sqrt{\sigma^2_{G(x)}(\lambda)}/\sqrt{\sigma^2_{G(y)}(\lambda)},$$

where, $r_{G(x,y)}(\lambda)$ is the estimated genetic correlation between traits x and y, and $\sigma^2_{G(x)}(\lambda)$ and $\sigma^2_{G(y)}(\lambda)$ are the genetic variances estimated by REML over the predefined grid of λ.

Results
Heritability
Table 2 shows the heritability estimates obtained for each λ value, both for unscaled and scaled genomic relationship matrices. A low to moderate heritability was found for BW, BM and HHP. When using pedigree-based information only, heritability estimates (standard errors in parenthesis) were 0.187 (0.049), 0.244 (0.052) and 0.315 (0.074), respectively (Table 2). These estimated heritabilities changed to 0.165 (0.039), 0.255 (0.042) and 0.196 (0.052) with an unscaled $\mathbf{G_F}$, and to 0.156 (0.06), 0.243 (0.04) and 0.174 (0.04) with an unscaled $\mathbf{G_V}$. Scaling the genomic relationship matrices increased heritability estimates relative to those obtained from unscaled matrices. Estimated heritabilities in the present study were lower than in [29, 30] using the same population from which

Table 2 Estimates of heritability for body weight (BW), ultrasound area of breast meat (BM) and hen-house egg production (HHP) obtained by placing varying weights (λ) on the pedigree-based relationship matrix (A) and on Forni's (G_F) or VanRaden's (G_V) relationship matrix

Regularization parameter (λ)	G_F						G_V					
	Unscaled			Scaled			Unscaled			Scaled		
	h^2_{BW}	h^2_{BM}	h^2_{HHP}	h^2_{BW}	h^2_{BM}	h^2_{HHP}	h^2_{BW}	h^2_{BM}	h^2_{HHP}	h^2_{BW}	h^2_{BM}	h^2_{HHP}
A (λ = 0)	0.187	0.244	*0.315*	0.187	0.244	0.315	0.187	0.244	*0.315*	0.187	0.244	*0.315*
λ = 0.20	0.226	0.291	0.309	0.234	0.297	0.340	0.223	0.291	0.299	0.230	0.295	0.311
λ = 0.40	*0.232*	*0.303*	0.285	0.278	0.348	0.360	*0.227*	*0.299*	0.270	*0.247*	0.318	0.295
λ = 0.60	0.219	0.297	0.258	0.315	0.395	0.374	0.213	0.290	0.239	0.245	*0.323*	0.272
λ = 0.80	0.197	0.281	0.230	*0.335*	0.431	*0.377*	0.189	0.272	0.209	0.227	0.316	0.245
G (λ = 1)	0.165	0.255	0.196	0.313	*0.442*	0.361	0.156	0.243	0.174	0.193	0.293	0.214

The largest estimates are italics

our dataset was drawn but with a larger sample size from four generations of three commercial lines, at varying intensities of selection in the Aviagen UK breeding program. For example, in [29, 30] estimates for BW ranged from 0.326 (0.011) to 0.399 (0.015), whereas in our study they ranged from 0.156 (0.06) to 0.187 (0.049). Our result is based on a small subset of birds taken from the overall population; therefore it is expected that estimated heritabilities h^2 would differ from those obtained using all available data, which would account for past selection.

Here, we used scaled kinship matrices to obtain "genetic parameters" which do not necessarily correspond to only those from standard pedigree-based additive genetic relationships or realized genomic pairwise similarities. Following VanRaden [22], if the expectation of **G** is **A**, then; $E(K) = E(\lambda G + (1 - \lambda)A) = A$. However, if one uses a scaled G_V, it follows from the scaling formula that $E(Gs_{ij}) = 2E(G_{ij} \times G_{min}/(G_{max} - G_{min}))$. The latter expectation cannot be written in a closed form, because this requires knowledge of the distributions of G_{min} and G_{max}.

Our multiple-trait GBLUP analysis indicated that the highest heritability estimates were not obtained at the extremes (0 or 1) of the λ grid. For example, the highest genomic heritability for BW was obtained at λ = 0.4(0.8), for unscaled (scaled) G_F, and at λ = 0.4 for the two versions of G_V. Scaling G_F and G_V always increased heritability estimates. For BM, higher heritabilities were obtained when scaling was applied. More specifically, the highest estimates were obtained at λ = 1 (G_F) and λ = 0.6 (G_V). With unscaled matrices, the highest heritabilities were obtained at λ = 0.4 and λ = 0.2. For HHP, the highest heritabilities were found at the extreme values of λ: λ = 0 for unscaled G_F and G_V (scaled or unscaled) and λ = 1 for scaled G_F. Adding genomic information had little impact on heritability estimates of HHP, except with scaled G_F.

Our findings illustrate a fairly obvious point made by Legarra et al. [26]: genomic heritability and its estimates are not invariant with respect to how **G** is constructed. Hence, inferences and comparisons between results from different studies must be done with care. In short, our results with a multiple-trait model indicated that a pedigree-marker based kernel (**K**) had an impact on heritability estimates and that scaling of the genomic relationship matrix led to higher "heritability" estimates, especially for G_F.

Genetic correlations
Estimates of correlations are in Table 3 and Figs. 2 and 3. Estimates of residual and phenotypic correlations were less sensitive to λ than genetic correlations, so our discussion concentrates on the latter. All parameters were estimated for each λ and for each of the two genomic relationship matrices. When using a pedigree-marker based kinship matrix (**K**), estimates of genetic correlations for BW-HHP and BM-HHP changed gradually when λ increased from 0 to 1. Results are shown graphically in Figs. 2 and 3 for G_F and G_V, respectively. Changes were more pronounced for the genetic correlation between BW and HHP, which decreased in absolute value from −0.192 (λ = 0) to −0.02 (G_F, unscaled), −0.019 (G_F, scaled), and 0.033 (G_V scaled or unscaled) with λ = 1. Estimates of the genetic correlation between BM and HHP were always negative and tended to decrease in absolute value as λ increased. They decreased from about −0.206 when only pedigree-based information (λ = 0) was used to −0.154 when only genomic information (λ = 1) was used to construct **K** from scaled or unscaled versions of G_F. BW and BM presented large positive genetic correlation estimates that ranged from 0.484 with the pedigree-based model to 0.497 (0.525) when only G_F (G_V) was used. It was insensitive to scaling

Table 3 Phenotypic (r_P) and environmental (r_e) correlations between body weight (BW), ultrasound area of breast meat (BM) and hen-house egg production (HHP) from a tri-variate analysis with varying weights (λ) on the pedigree-based relationship matrix (A) and on Forni's (G_F) or VanRaden's (G_V) relationship matrix

Regularization parameter (λ)	$r_{e(BW,BM)}$	$r_{e(BW,HHP)}$	$r_{e(BM,HHP)}$	$r_{P(BW,BM)}$	$r_{P(BW,HHP)}$	$r_{P(BM,HHP)}$
G_F						
Unscaled						
A ($\lambda = 0$)	0.480 (0.034)	−0.026 (0.058)	−0.010 (0.063)	0.480 (0.023)	−0.066 (0.036)	−0.065 (0.039)
$\lambda = 0.20$	0.481 (0.036)	−0.023 (0.059)	−0.003 (0.064)	0.479 (0.023)	−0.067 (0.037)	−0.065 (0.039)
$\lambda = 0.40$	0.482 (0.035)	−0.034 (0.057)	−0.011 (0.061)	0.479 (0.023)	−0.065 (0.037)	−0.063 (0.039)
$\lambda = 0.60$	0.482 (0.033)	−0.047 (0.053)	−0.020 (0.057)	0.479 (0.023)	−0.062 (0.036)	−0.061 (0.039)
$\lambda = 0.80$	0.481 (0.031)	−0.058 (0.049)	−0.028 (0.053)	0.479 (0.023)	−0.059 (0.036)	−0.059 (0.038)
G ($\lambda = 1$)	0.479 (0.029)	−0.068 (0.045)	−0.035 (0.049)	0.479 (0.023)	−0.059 (0.036)	−0.062 (0.038)
Scaled						
$\lambda = 0.20$	0.484 (0.035)	−0.022 (0.059)	−0.006 (0.064)	0.479 (0.024)	−0.074 (0.039)	−0.073 (0.041)
$\lambda = 0.40$	0.484 (0.036)	−0.024 (0.059)	−0.006 (0.063)	0.478 (0.027)	−0.079 (0.042)	−0.080 (0.045)
$\lambda = 0.60$	0.485 (0.035)	−0.034 (0.057)	−0.012 (0.061)	0.477 (0.031)	−0.080 (0.047)	−0.084 (0.049)
$\lambda = 0.80$	0.484 (0.033)	−0.050 (0.052)	−0.062 (0.078)	0.477 (0.035)	−0.072 (0.052)	−0.084 (0.055)
G ($\lambda = 1$)	0.479 (0.029)	−0.068 (0.045)	−0.035 (0.049)	0.481 (0.038)	−0.052 (0.056)	−0.082 (0.059)
G_V						
Unscaled						
$\lambda = 0.20$	0.480 (0.036)	−0.033 (0.058)	−0.004 (0.063)	0.480 (0.023)	−0.066 (0.037)	−0.066 (0.039)
$\lambda = 0.40$	0.480 (0.035)	−0.046 (0.055)	−0.011 (0.059)	0.480 (0.023)	−0.064 (0.036)	−0.064 (0.039)
$\lambda = 0.60$	0.479 (0.033)	−0.060 (0.051)	−0.018 (0.055)	0.480 (0.023)	−0.061 (0.036)	−0.063 (0.039)
$\lambda = 0.80$	0.476 (0.030)	−0.070 (0.047)	−0.024 (0.051)	0.481 (0.023)	−0.059 (0.036)	−0.062 (0.038)
G ($\lambda = 1$)	0.474 (0.028)	−0.079 (0.044)	−0.030 (0.047)	0.481 (0.023)	−0.060 (0.036)	−0.065 (0.038)
Scaled						
$\lambda = 0.20$	0.484 (0.036)	−0.030 (0.059)	−0.006 (0.063)	0.479 (0.024)	−0.070 (0.037)	−0.071 (0.040)
$\lambda = 0.40$	0.483 (0.035)	−0.043 (0.056)	−0.010 (0.060)	0.479 (0.024)	−0.069 (0.038)	−0.072 (0.041)
$\lambda = 0.60$	0.482 (0.033)	−0.057 (0.053)	−0.018 (0.056)	0.479 (0.025)	−0.066 (0.039)	−0.072 (0.041)
$\lambda = 0.80$	0.478 (0.031)	−0.070 (0.048)	−0.025 (0.051)	0.481 (0.025)	−0.061 (0.039)	−0.072 (0.042)
G ($\lambda = 1$)	0.474 (0.028)	−0.078 (0.044)	−0.030 (0.047)	0.483 (0.025)	−0.056 (0.039)	−0.073 (0.042)

A: numerator relationship matix, **G$_F$**: Forni's relationship matrix, **G$_V$**: VanRaden's relationship matrix

of the **G** matrix. Standard errors of estimates for BM-BW (results not shown) tended to decrease when λ increased. There were no clear tendencies for the standard errors of estimates of the genetic correlation of BW with HHP and BM with HHP. In short, classical genetic correlations (based on **A**) and genomic correlations (based on **G**) were distinct, depending on the pairs of traits considered. However, varying λ from 0 to 1 produced very minor changes in estimates of the genetic correlation between BM and HHP, but large changes in estimates of the genetic correlation between BW and HHP. Estimates of the genetic correlations between BW and BM were insensitive to λ.

The differences that were observed in estimates of genetic correlations depended on the type of information used. From theory, standard pedigree-based linear models capture expected genetic covariation, whereas marker-based models capture genetic covariation that is

marked by SNPs. Our results are important from the perspective of multiple-trait genomic analysis because they indicate that estimates of genetic correlations between some traits may depend on the type of information used. This was clearly the case for the genetic correlation between BW and HHP.

Multiple-trait pedigree or marker-based prediction was designed to exploit genetic correlations between target characters and indicator traits [16], especially when a lowly heritable target trait is genetically correlated with an indicator that has a higher heritability. Our results indicate that estimates of genomic correlation between characters may reaffirm or disagree with expectations that are developed from a pedigree-based analysis. For example, on the one hand, the genetic and genomic correlations between BW and BM were insensitive to λ values, i.e., estimates of the genomic correlation and of the genetic correlation derived from the infinitesimal model were the

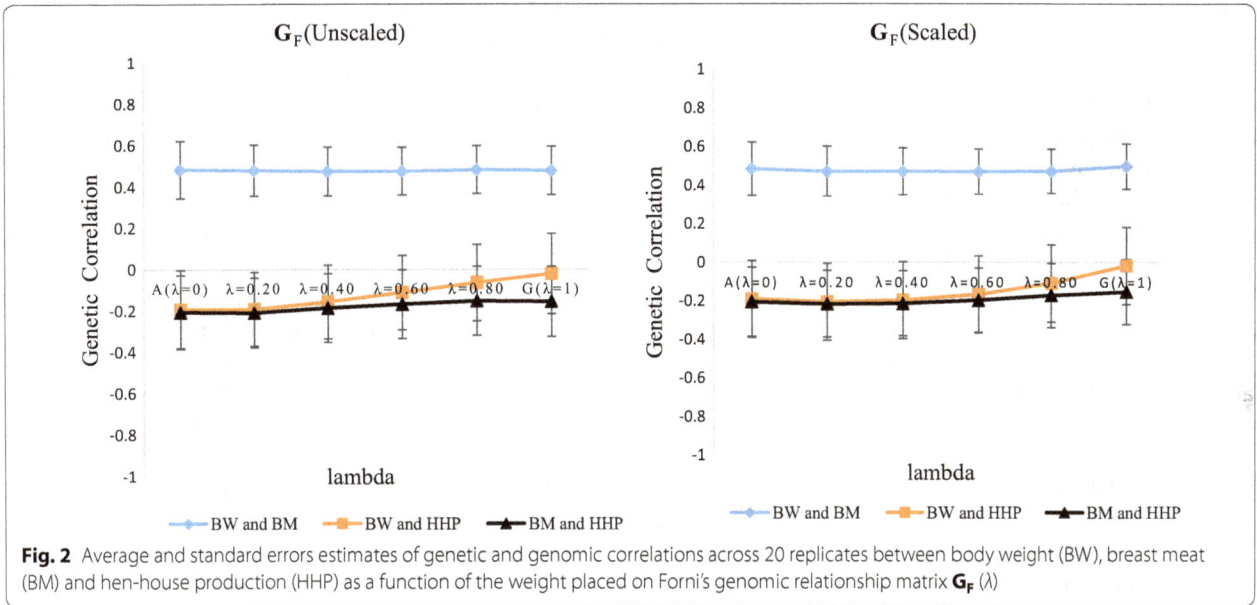

Fig. 2 Average and standard errors estimates of genetic and genomic correlations across 20 replicates between body weight (BW), breast meat (BM) and hen-house production (HHP) as a function of the weight placed on Forni's genomic relationship matrix $\mathbf{G_F}$ (λ)

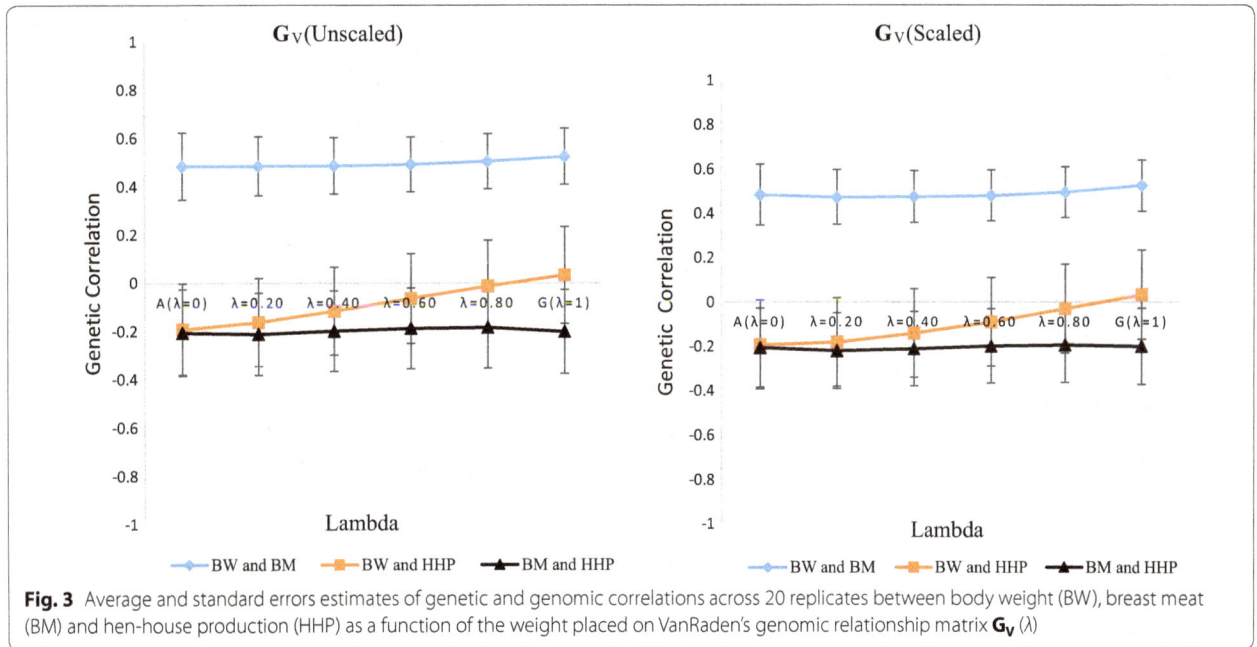

Fig. 3 Average and standard errors estimates of genetic and genomic correlations across 20 replicates between body weight (BW), breast meat (BM) and hen-house production (HHP) as a function of the weight placed on VanRaden's genomic relationship matrix $\mathbf{G_V}$ (λ)

same. On the other hand, when considering BW and HHP, the estimate of the pedigree-based genetic correlation was equal to 0.2, whereas the estimate of the genomic correlation was close to 0. This illustrates a situation where part of the covariance between a pair of traits was not detected by SNPs ("missing correlation"). Sources of genetic correlation may be lost in a multiple-trait marker-based analysis. In the case of BM and HHP, the classical genetic correlation was estimated at −0.20 and the genomic

correlation at −0.15. The pedigree-based analysis suggested a stronger genetic correlation.

Care should be exercised when interpreting and using genetic parameters that are assessed via molecular markers, as predictions for complex traits based on pedigree data may differ significantly from those based on SNP data. For this reason, we explored whether the two sources of information could be combined in some "optimal" manner.

Predictive ability

The question of how to arrive at a "best" estimate of a genetic correlation (i.e., for which the greatest advantage of predicting ability is obtained) was examined and, to accomplish this objective, we used the predictive approach advocated by Lo et al. [17]. Figure 4 shows boxplots with the distributions of predictive correlations and mean squared errors for the cross-validation with 20 random repetitions. Some of the plots (e.g., BW) show a mild advantage of using a linear combination of $\mathbf{G}$ and $\mathbf{A}$ as kinship kernel. For BW, the largest correlation and lowest MSE were obtained with unscaled $\mathbf{G_F}$ and $\mathbf{G_V}$. In terms of the predictive correlation for BW, the largest values were obtained with scaled $\mathbf{G_V}$ and unscaled $\mathbf{G_F}$, both at $\lambda = 0.8$. For BM, the largest predictive correlation was achieved with unscaled $\mathbf{G_V}$ and scaled $\mathbf{G_F}$, at $\lambda = 0.4$ and 0.8, respectively. For HHP, both scaled $\mathbf{G_V}$ and $\mathbf{G_F}$ resulted in better performance, and the largest predictive correlations were obtained at $\lambda = 0.2$.

The lowest MSE for BW was achieved for unscaled $\mathbf{G_F}$ and scaled $\mathbf{G_V}$ at $\lambda = 0.6$. For BM, the lowest MSE was obtained with λ close to 1 using scaled $\mathbf{G_V}$ and $\mathbf{G_F}$. In addition, the scaled $\mathbf{G_V}$ and $\mathbf{G_F}$ produced the lowest MSE for HHP, with a slight superiority for values of λ close to 1. In terms of MSE, except for BW with $\mathbf{G_F}$, scaling of genomic relationship matrices yielded better results. Our findings are in agreement with Rodríguez-Ramilo et al. [31], who reported that when a larger weight was assigned to the numerator relationship matrix ($\mathbf{A}$), the predictive correlation was lower than when assigning more weight to the genomic relationship matrix ($\mathbf{G}$); a similar behavior was found for MSE. Rodríguez-Ramilo et al. [31] estimated λ by using Bayesian methods and reported that the posterior mean of λ depended on training sample size and the trait.

Our results indicate that multiple-trait genome-enabled predictions may be improved in some cases by combining $\mathbf{A}$ and $\mathbf{G}$ to quantify kinship. This result may also hold when prediction involves multiple selection lines or crossbred animals. Combining kernels can be viewed as a form of model averaging [25], with markers and pedigree playing complementary roles in prediction, e.g., markers may exploit similarity in state and LD, with $\mathbf{A}$ informing about similarity by descent.

Our results using dense SNPs (600 K Affymetrix platform) indicate that GBLUP with scaled or unscaled relationship matrices typically performed better than pedigree-based BLUP. However, in most cases, the largest correlation and lowest MSE were achieved using a linear combination of $\mathbf{A}$ and $\mathbf{G}$.

Regression coefficients

Figures 5, 6 and 7 show scatter plots and average (red dotted line) genetic regression coefficients of the three traits on the estimated direct genomic values (DGV) of other traits calculated from REML estimates of (co)variance components. The realized regression coefficients were computed at each λ value for the 20 cross-validation random samples and their medians are depicted as dark blue dotted lines on each plot. The REML regressions express the expected change in genetic value of trait i if the direct genomic value for trait j changes by one unit.

For BW and BM (Fig. 5), the expected and realized regressions were larger than 0 for all values of λ. In general, there was reasonable agreement between expected and realized regressions. However, for BW and HHP (Fig. 6), the expected genetic regressions were negative and moved toward 0 as λ increased, but the realized regressions (blue dotted lines) varied around 0 for all λ values. There was some apparent inconsistency between the expectations based on REML estimates and the cross-validation regression.

Figure 7 indicated a disagreement between expected genetic regressions and cross-validation regressions of BM on HHP when λ was close to 0. The expected regressions based on pedigree information were negative, while the cross-validation regressions were positive. The expected regressions of HHP phenotypes on DGV of BM and variances tended towards 0 as λ tended to 1, i.e., when more weight was placed on SNPs. The cross-validation regressions were much more affected by the value of λ than the expected regressions based on REML estimates.

Discussion

In genome-enabled prediction, there are different ways of incorporating molecular marker information into parametric and non-parametric models [24, 32]. Research with simulated and real data has consistently shown that single-trait GBLUP displays slightly better prediction accuracy when a trait is affected by a large number of QTL with small effects and as well as other genomic prediction methods for most traits [33, 34]. However, few studies on multiple-trait genomic prediction have been carried out with GBLUP, or have assessed estimates of genetic correlations when genomic or pedigree data were used. Similar to traditional pedigree-based genetic evaluations, the use of multiple-trait GBLUP is expected to increase the accuracy of predictions via "borrowing" of information from genetically correlated traits [35].

In order to explore a multiple-trait GBLUP model that also makes use of pedigree information, we constructed a pedigree-marker based kinship matrix ($\mathbf{K}$) as a linear

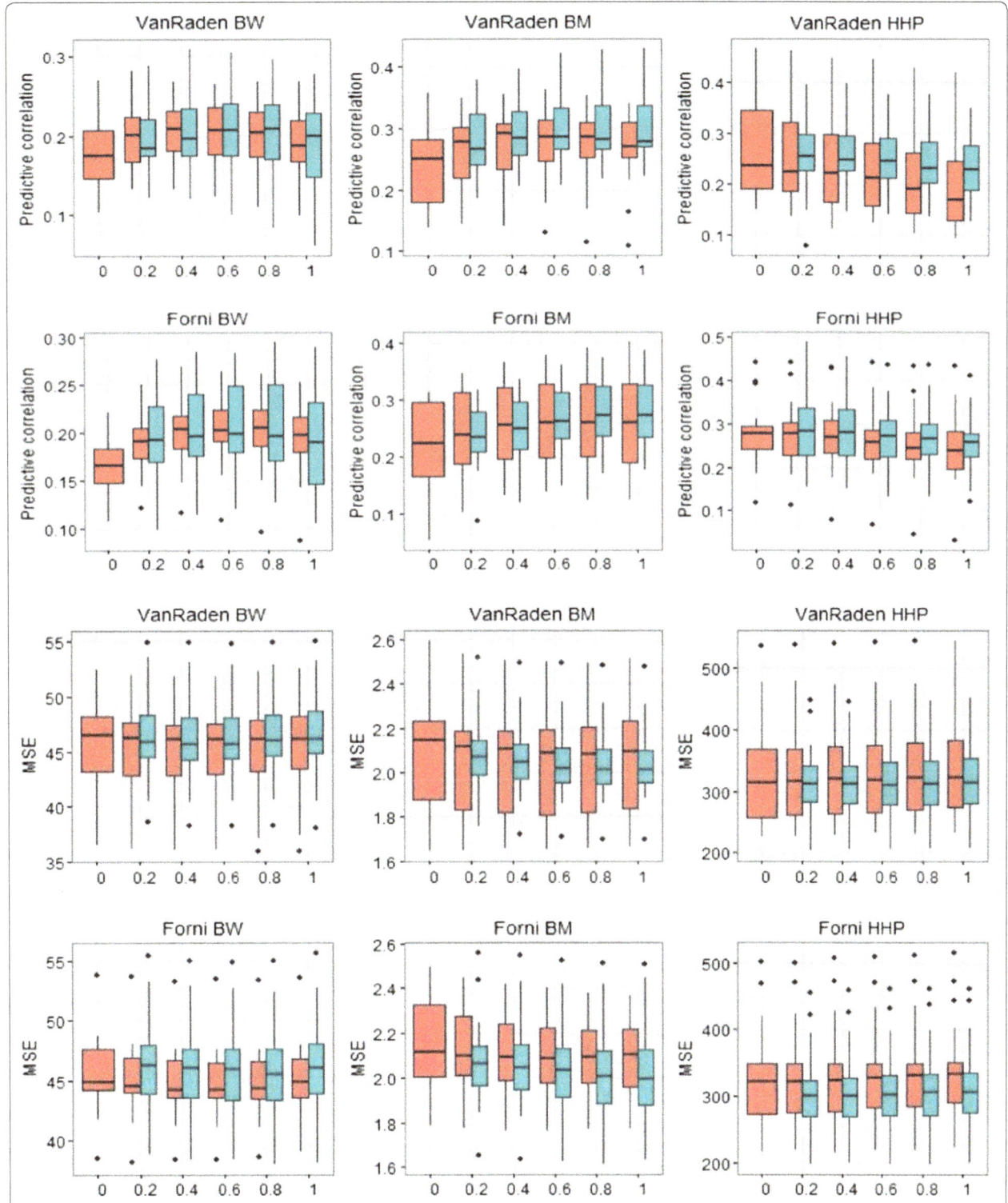

Fig. 4 *Boxplot* of predictive correlations across 20 replicates between phenotypes and predicted breeding values (*upper two rows*), and of mean squared errors (MSE) (*bottom two rows*) in testing sets. *Red* and *light blue colors* denote values for unscaled and scaled relationship matrices of Forni or VanRaden, respectively. Outliers are denoted as *black dots*, and the x-axis label denotes λ = 0, 0.2, 0.4, 0.6, 0.8, 1

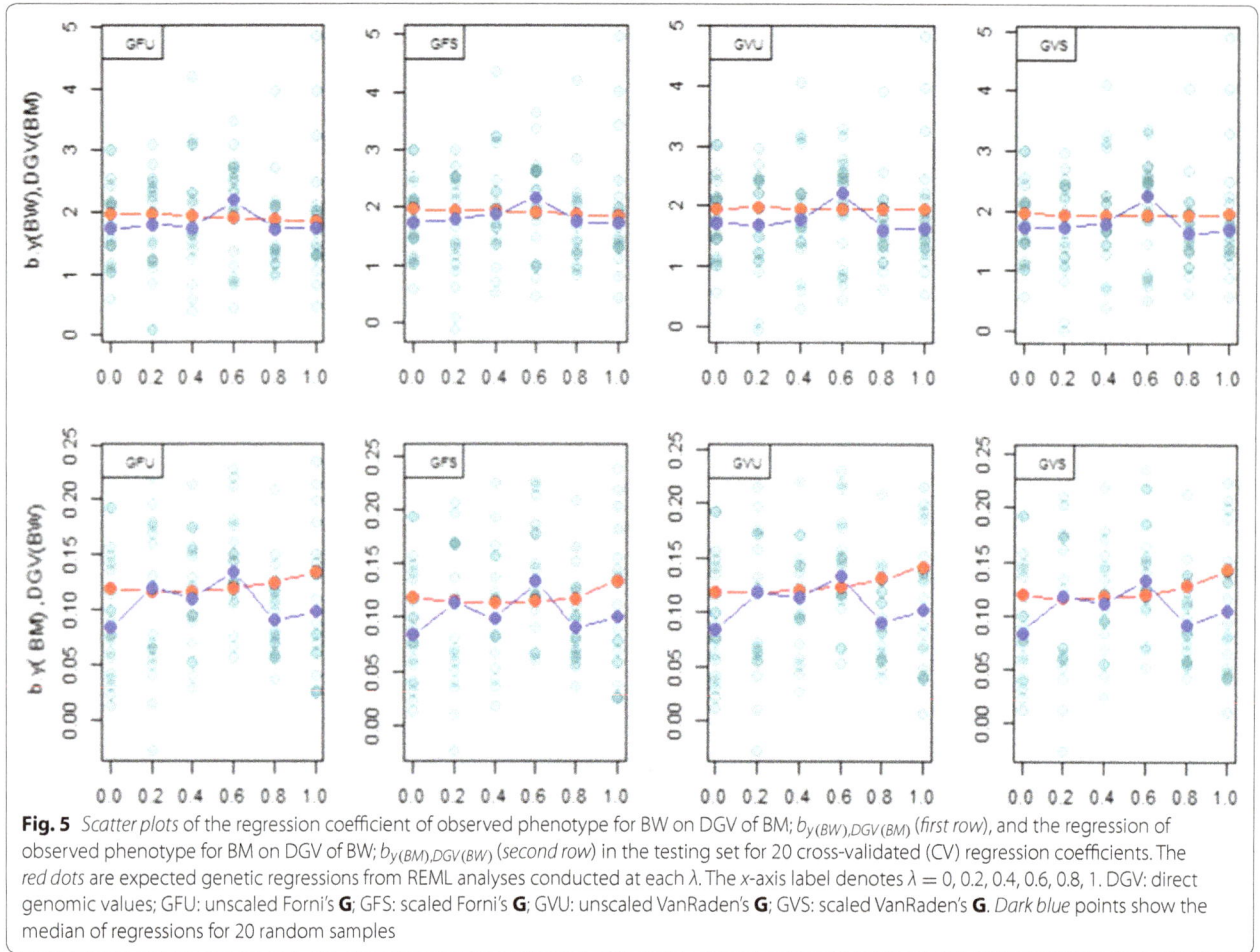

Fig. 5 *Scatter plots* of the regression coefficient of observed phenotype for BW on DGV of BM; $b_{y(BW),DGV(BM)}$ (*first row*), and the regression of observed phenotype for BM on DGV of BW; $b_{y(BM),DGV(BW)}$ (*second row*) in the testing set for 20 cross-validated (CV) regression coefficients. The *red dots* are expected genetic regressions from REML analyses conducted at each λ. The x-axis label denotes $\lambda = 0, 0.2, 0.4, 0.6, 0.8, 1$. DGV: direct genomic values; GFU: unscaled Forni's **G**; GFS: scaled Forni's **G**; GVU: unscaled VanRaden's **G**; GVS: scaled VanRaden's **G**. *Dark blue* points show the median of regressions for 20 random samples

combination of pedigree and marker-based relationships between animals, defined as $\mathbf{K} = \lambda \mathbf{G} + (1 - \lambda) \mathbf{A}$. Predictive ability of the model and parameter estimates were obtained over a grid of values of λ that varied between 0 and 1, e.g., $\lambda = 0$ implied that all weight was assigned to pedigree, and none to SNPs.

One important factor to take into account when combining marker- and pedigree-based relationship matrices is that such matrices are on the same scale. The elements of the additive relationship matrix are the numerators of Wright's correlation coefficients that represent the relative genetic variances and covariances among individuals. Consequently, the diagonals of **A** can be as large as 2, and relationships between two individuals can be greater than 1.

Traditionally, to quantify coefficients of relationship with respect to a base (reference) population, as discussed in [36, 37], the probability that alleles are identical by descent (IBD) was derived from pedigree information and from a base population consisting of founders. However, for relationships estimated from genetic markers

there is no obvious base population, and they estimate the proportion of the genome that is identical by state (IBS). In our data, genomic relationships measured by unscaled $\mathbf{G}_V$ and $\mathbf{G}_F$ can take negative values, whereas pedigree relationships are non-negative. In our data, no negative values were observed for full-sib genomic relationships but negative genomic additive relationships with small values near 0 were observed for unrelated individuals based on the pedigree (i.e., pedigree based relationship = 0). It remains to be seen whether genomic relationship measures can detect true 'negative genomic correlations' (if such correlations exist), which may be detectable using deep pedigree information and a definition of a base population. The genomic relationship matrices in our analyses were based on (IBS information and on frequencies of alleles to build the GRM.

Our results suggest that multiple-trait genetic predictions depended on the weight assigned to genomic data. Better predictions were often obtained when pedigree and SNP information were used simultaneously. Earlier studies using simulated or real data have explored

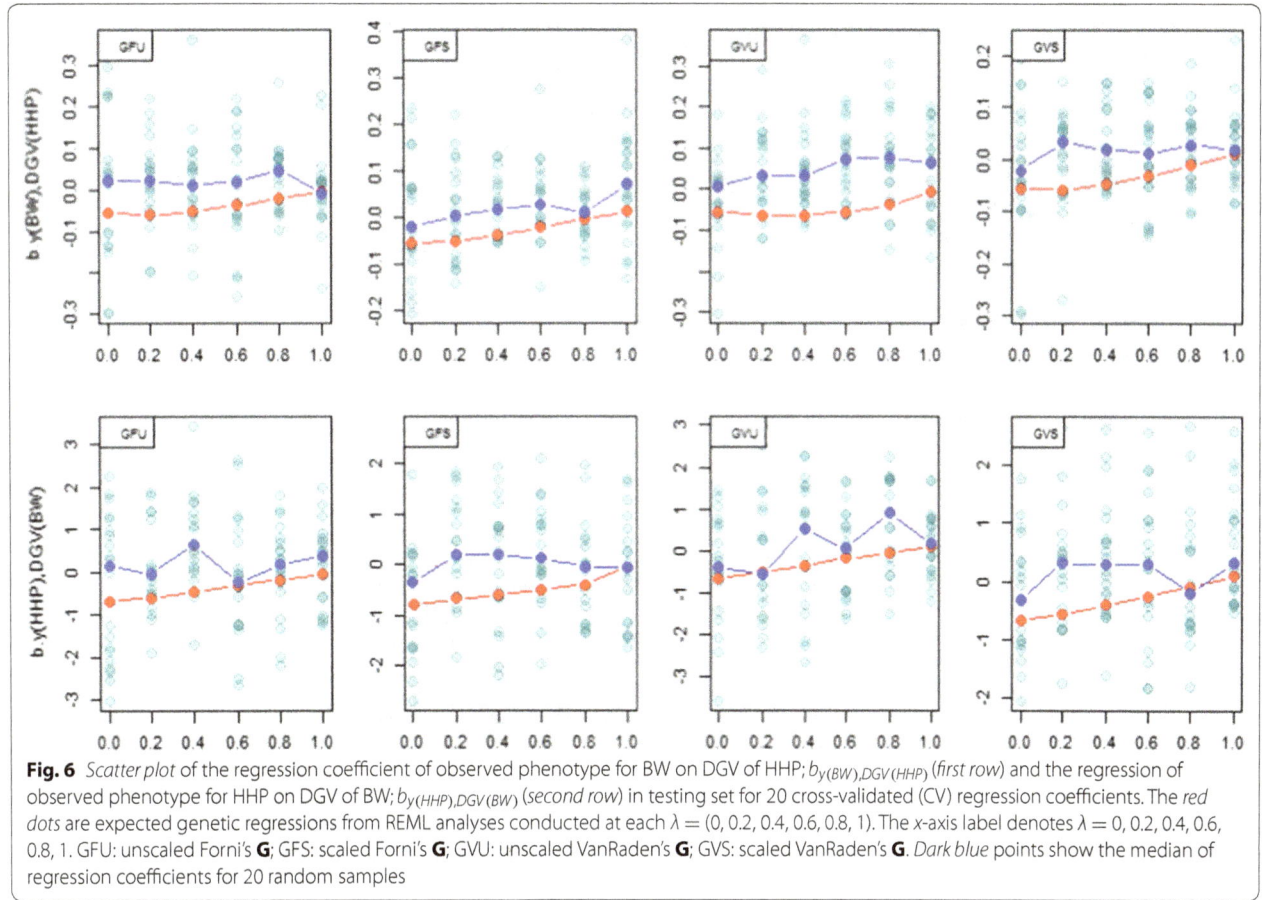

Fig. 6 *Scatter plot* of the regression coefficient of observed phenotype for BW on DGV of HHP; $b_{y(BW),DGV(HHP)}$ (*first row*) and the regression of observed phenotype for HHP on DGV of BW; $b_{y(HHP),DGV(BW)}$ (*second row*) in testing set for 20 cross-validated (CV) regression coefficients. The *red dots* are expected genetic regressions from REML analyses conducted at each $\lambda = (0, 0.2, 0.4, 0.6, 0.8, 1)$. The x-axis label denotes $\lambda = 0, 0.2, 0.4, 0.6, 0.8, 1$. GFU: unscaled Forni's **G**; GFS: scaled Forni's **G**; GVU: unscaled VanRaden's **G**; GVS: scaled VanRaden's **G**. *Dark blue* points show the median of regression coefficients for 20 random samples

the potential superiority of multiple-trait over single-trait genomic prediction with a focus on the relationship between traits in terms of differences in heritability, genetic correlations and number of indicator traits (e.g., [35, 38, 39]). De Los Campos et al. [40] indicated that potential problems may emerge when trying to infer genetic parameters using molecular markers that are imperfectly associated with genotypes at causal loci. Gianola et al. [13] showed that correlation parameters that are inferred from markers (i.e., genomic correlations) can give a distorted picture of the genetic correlation between traits. The sources of genetic correlation are pleiotropy (i.e., the same QTL affects more than one trait) and LD between QTL. When markers are used, marker-QTL LD and LD relationships among markers intervene in the genomic correlation.

Here, we examined the impact of combining **A** and **G** on estimates of the genomic correlation between three chicken traits and evaluated outcomes using a predictive framework. Some studies [12, 41] have shown superiority of multiple-trait prediction over single-trait prediction, and combining pedigree with marker information was

found to be better than when using either **A** and **G** alone [32].

Our estimates of genetic correlations depended on the choice of λ. For example, on the one hand for BW and HHP, when using pedigree as the only measure of similarity ($\lambda = 0$), the genetic correlation was -0.20, but it shifted to near 0 or was even positive ($\mathbf{G_V}$) when only marker information was used. On the other hand, the estimate of the genetic correlation between BW and BM was stable with respect to λ, while the estimate of the genetic correlation between BM and HHP was only slightly affected, i.e. changing from -0.21 to -0.15 for $\mathbf{G_F}$, and from -0.22 to -0.20 for $\mathbf{G_V}$ for $\lambda = 0$ and $\lambda = 1$, respectively. Clearly, genomic data provide a distinct measure of similarity between individuals, and this translates into differential capturing of genetic signals. For instance, most off-diagonal entries of **A** were zero but all entries of **G** were non-null.

In order to increase the accuracy of predictions by using pedigree and genomic information jointly, Legarra et al. [42] proposed a single-step procedure that enhances relationship information for non-genotyped animals,

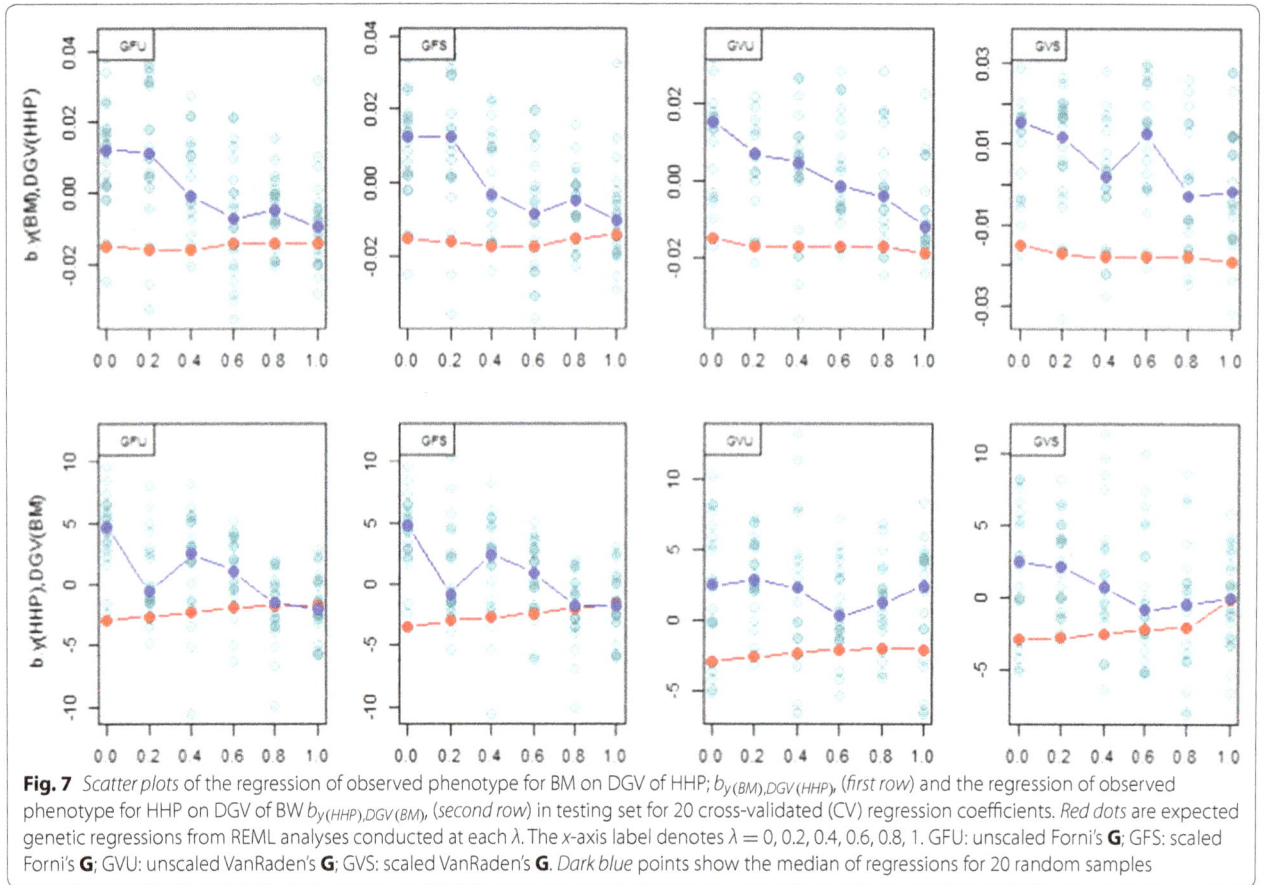

Fig. 7 *Scatter plots* of the regression of observed phenotype for BM on DGV of HHP; $b_{y(BM),DGV(HHP)}$, (*first row*) and the regression of observed phenotype for HHP on DGV of BW $b_{y(HHP),DGV(BM)}$, (*second row*) in testing set for 20 cross-validated (CV) regression coefficients. *Red dots* are expected genetic regressions from REML analyses conducted at each λ. The *x*-axis label denotes $\lambda = 0, 0.2, 0.4, 0.6, 0.8, 1$. GFU: unscaled Forni's **G**; GFS: scaled Forni's **G**; GVU: unscaled VanRaden's **G**; GVS: scaled VanRaden's **G**. *Dark blue* points show the median of regressions for 20 random samples

without requiring major changes in the implementation of a standard BLUP model. In the study of Aguilar et al. [43], a three-fold increase in accuracy of GEBV was found for traits related to conception rate in Holstein dairy cows, with low heritability, when using a genomic-based relationship combined with a pedigree-based relationship in a multiple trait model [43]. Using 18 quantitative traits in Holstein dairy cattle, Tsuruta et al. [44] reported that prediction accuracies increased when a multiple-trait genomic prediction model was used compared to a single-trait model, but the increase depended on the trait being predicted. However, Bao et al. [45] did not observe clear benefits when four traits were included in a multiple-trait genomic prediction model for soybeans compared to a single-trait model. However, these studies did not examine the impact of using combined genetic and genomic relationships. This shows that the effect of genetic correlations on multi-trait genomic prediction depends on the information type being used to construct **K**, with an impact on accuracy of prediction.

In our study, predictive ability was measured using the correlation between predicted genetic values and observed phenotypic values, and mean squared error

of these predictions. As shown in Fig. 4, the optimum weight placed on genomic relationships was trait-dependent. We took the view point that a "best" estimate of the genetic correlation would correspond to the linear combination of **A** and **G** (with a specific weight on each one) that delivered the best predictive ability, which was found by searching the weight (λ) placed on genomic versus pedigree relationships. This type of predictive approach has been advocated in the statistical literature [17, 46, 47].

In general, combinations of **A** and **G** kernels yielded better predictions than when only **G** was used. In a GBLUP model, the entries of **G** reflect the actual extent of IBS relationships between individuals, but without making a clear reference to a base population [48]. This implies that genomic (co)variance parameters do not necessarily have the same meaning as standard classical multiple-trait models genetic parameters, such as the infinitesimal genetic correlation. According to [49], pedigree information, co-segregation and population-wide LD are three sources of genetic information that contribute to the predictive ability of genomic selection models. Co-segregation information can be captured by

IBD or IBS relationships and, when a pedigree is not deep enough, relatedness among individuals that is inferred from markers may improve prediction. However, how does one decide if an estimate of genetic correlation derived from genomic data is better than a pedigree-based estimate?

In classical quantitative genetics, a genetic correlation between traits arises due to either genes that have an effect on both traits (pleiotropy), or due to LD between genes that affect different traits [50]. When investigating the basis of a genetic correlation, an important question is to determine the extent to which these two forces act on the genetic parameters [13, 51]. Multiple-trait QTL mapping methods may help distinguishing pleiotropy from linkage [52], but any such dissection in the absence of knowledge on QTL would be speculative.

Estimates of genetic correlations obtained from pedigree or from markers may differ either due to chance or other reasons, such as extent of LD between markers and the unknown QTL. One possible way of testing if such differences are systematic, is to examine pairs of estimates of pedigree- and marker-based correlations in re-samples from the dataset and constructing a paired comparison, by using either a parametric or a non-parametric approach. For example, the estimates of correlations could be z-transformed and a paired t test conducted.

In summary, combining pedigree- and marker-based information had an impact on predictive performance of multiple-trait models. Discerning the optimum weight placed on genomic and genealogical information is an important issue, and a grid-search scheme was used for that purpose. We found that estimates of genetic correlation obtained with **A** and **G** matrices were different, but depended on the trait. This indicates that multiple-trait marker-based prediction may be enhanced by the combined use of genealogy and marker information in the models.

Conclusions

To our knowledge, this is the first study with animal breeding data that explores how the weight placed on pedigree and marker information affects multiple-trait predictions. We designed a tri-variate genomic prediction model that exploited pedigree and marker information simultaneously. Use of a kinship matrix that is formed as a linear combination of pedigree- and marker-based relationships may enhance genome-enabled prediction, but the optimal weight placed on the two sources of information will differ between traits. Genetic correlation estimates from pedigree-based models may differ from those obtained from marker-based models, at least in some cases. Cross-validation was useful for gauging the genetic correlation in multiple-trait models.

Authors' contributions
MM carried out the study and wrote the first draft of the manuscript. DG and GJMR designed the experiment, supervised the study and critically contributed to the final version of manuscript. BDV and AAM participated in discussion and reviewed the manuscript. AS, AE and MAF contributed materials and revised the manuscript. All authors read and approved the final manuscript.

Author details
[1] Department of Animal Science, Faculty of Agriculture, Shahid Bahonar University of Kerman (SBUK), Kerman, Iran. [2] Department of Statistical, Faculty of Mathematic and Computer Science, Shahid Bahonar University of Kerman (SBUK), Kerman, Iran. [3] State Key Laboratory of Genetic Resources and Evolution, Yunnan Laboratory of Molecular Biology of Domestic Animals, Kunming Institute of Zoology, Chinese Academy of Sciences, Kunming 650223, China. [4] Roslin Institute, University of Edinburgh, Midlothian, UK. [5] Department of Animal Sciences, University of Wisconsin, Madison, WI, USA. [6] Department of Biostatistics and Medical Informatics, University of Wisconsin, Madison, WI, USA. [7] Department of Dairy Science, University of Wisconsin, Madison, WI, USA.

Acknowledgements
The first author wishes to acknowledge Aviagen (Midlothian, United Kingdom) for providing the data, and the Ministry of Science, Research and Technology of Iran for financially supporting his visit to the University of Wisconsin-Madison. Work was partially supported by the Wisconsin Agriculture Experiment Station under hatch Grant 142-PRJ63CV to DG.

Competing interests
The authors declare that they have no competing interests.

References
1. Meuwissen THE, Hayes BJ, Goddard ME. Prediction of total genetic value using genome-wide dense marker maps. Genetics. 2001;157:1819–29.
2. Hayes BJ, Bowman PJ, Chamberlain AJ, Goddard ME. Invited review: genomic selection in dairy cattle: progress and challenges. J Dairy Sci. 2009;92:433–43.
3. de Los Campos G, Hickey JM, Pong-Wong R, Daetwyler HD, Calus MP. Whole-genome regression and prediction methods applied to plant and animal breeding. Genetics. 2013;193:327–45.
4. Gianola D, Fernando RL, Stella A. Genomic-assisted prediction of genetic value with semiparametric procedures. Genetics. 2006;173:1761–76.
5. Habier D, Fernando RL, Dekkers JC. The impact of genetic relationship information on genome-assisted breeding values. Genetics. 2007;177:2389–97.
6. González-Recio O, Rosa GJM, Gianola D. Machine learning methods and predictive ability metrics for genome-wide prediction of complex traits. Livest Sci. 2014;166:217–31.
7. Henderson CR, Quaas RL. Multiple trait evaluation using relatives records. J Anim Sci. 1976;43:1188–97.
8. Jannink JL, Lorenz AJ, Iwata H. Genomic selection in plant breeding: from theory to practice. Brief Funct Genomics. 2010;9:166–77.
9. Kadarmideen HN. Genomics to systems biology in animal and veterinary sciences: progress, lessons and opportunities. Livest Sci. 2014;166:232–48.
10. Gianola D, Rosa GJM. One hundred years of statistical developments in animal breeding. Annu Rev Anim Biosci. 2015;3:19–56.
11. Thompson R, Meyer K. A review of theoretical aspects in the estimation of breeding values for multi-trait selection. Livest Prod Sci. 1986;15:299–313.
12. Jia Y, Jannink JL. Multiple-trait genomic selection methods increase genetic value prediction accuracy. Genetics. 2012;192:1513–22.
13. Gianola D, de los Campos G, Toro MA, Naya H, Schön CC, Sorensen D. Do molecular markers inform about pleiotropy? Genetics. 2015;201:23–9.
14. Korte A, Vilhjálmsson BJ, Segura V, Platt A, Long Q, Nordborg M. A mixed-model approach for genome-wide association studies of correlated traits in structured populations. Nat Genet. 2012;44:1066–71.
15. Maier R, Moser G, Chen GB, Ripke S, Coryell W, Potash JB, et al. Joint analysis of psychiatric disorders increases accuracy of risk prediction for

schizophrenia, bipolar disorder, and major depressive disorder. Am J Hum Genet. 2015;96:283–94.

16. Calus MP, Veerkamp RF. Accuracy of multi-trait genomic selection using different methods. Genet Sel Evol. 2011;43:26.

17. Lo A, Chernoff H, Zheng T, Lo SH. Why significant variables aren't automatically good predictors. Proc Natl Acad Sci USA. 2015;112:13892–7.

18. Abdollahi-Arpanahi R, Morota G, Valente BD, Kranis A, Rosa GJM, Gianola D. Differential contribution of genomic regions to marked genetic variation and prediction of quantitative traits in broiler chickens. Genet Sel Evol. 2016;48:10.

19. Morota G, Abdollahi-Arpanahi R, Kranis A, Gianola D. Genome-enabled prediction of quantitative traits in chickens using genomic annotation. BMC Genomics. 2014;15:109.

20. Browning SR, Browning BL. Rapid and accurate haplotype phasing and missing-data inference for whole-genome association studies by use of localized haplotype clustering. Am J Hum Genet. 2007;81:1084–97.

21. Forni S, Aguilar I, Misztal I. Different genomic relationship matrices for single-step analysis using phenotypic, pedigree and genomic information. Genet Sel Evol. 2011;43:1.

22. VanRaden PM. Efficient methods to compute genomic predictions. J Dairy Sci. 2008;91:4414–23.

23. Astle W, Balding DJ. Population structure and cryptic relatedness in genetic association studies. Stat Sci. 2009;24(4):451–71.

24. Morota G, Gianola D. Kernel-based whole-genome prediction of complex traits: a review. Front Genet. 2014;5:363.

25. De Los Campos G, Gianola D, Rosa GJM, Weigel KA, Crossa J. Semi-parametric genomic-enabled prediction of genetic values using reproducing kernel Hilbert spaces methods. Genet Res (Camb). 2010;92:295–308.

26. Legarra A, Christensen OF, Vitezica ZG, Aguilar I, Misztal I. Ancestral relationships using metafounders: finite ancestral populations and across population relationships. Genetics. 2015;200:455–68.

27. Gianola D, Okut H, Weigel KA, Rosa GJ. Predicting complex quantitative traits with Bayesian neural networks: a case study with Jersey cows and wheat. BMC Genet. 2011;12:87.

28. Meyer K. WOMBAT: a tool for mixed model analyses in quantitative genetics by restricted maximum likelihood (REML). J Zhejiang Univ Sci B. 2007;8:815–21.

29. Kapell DN, Hill WG, Neeteson AM, McAdam J, Koerhuis AN, Avendaño S. Genetic parameters of foot-pad dermatitis and body weight in purebred broiler lines in 2 contrasting environments. Poult Sci. 2012;91:565–74.

30. Kapell DN, Hill WG, Neeteson AM, McAdam J, Koerhuis AN, Avendaño S. Twenty-five years of selection for improved leg health in purebred broiler lines and underlying genetic parameters. Poult Sci. 2012;91:3032–43.

31. Rodríguez-Ramilo ST, García-Cortés LA, González-Recio Ó. Combining genomic and genealogical information in a reproducing kernel Hilbert spaces regression model for genome-enabled predictions in dairy cattle. PLoS One. 2014;9:e93424.

32. Crossa J, de Los Campos G, Pérez P, Gianola D, Burgueño J, Araus JL, et al. Prediction of genetic values of quantitative traits in plant breeding using pedigree and molecular markers. Genetics. 2010;186:713–24.

33. Wientjes YC, Veerkamp RF, Calus MP. The effect of linkage disequilibrium and family relationships on the reliability of genomic prediction. Genetics. 2013;193:621–31.

34. Wimmer V, Lehermeier C, Albrecht T, Auinger HJ, Wang Y, Schön CC. Genome-wide prediction of traits with different genetic architecture through efficient variable selection. Genetics. 2013;195:573–87.

35. Guo G, Zhao F, Wang Y, Zhang Y, Du L, Su G. Comparison of single-trait and multiple-trait genomic prediction models. BMC Genet. 2014;15:30.

36. Powell JE, Visscher PM, Goddard ME. Reconciling the analysis of IBD and IBS in complex trait studies. Nat Rev Genet. 2010;11:800–5.

37. Forneris NS, Steibel JP, Legarra A, Vitezica ZG, Bates RO, Ernst CW, et al. A comparison of methods to estimate genomic relationships using pedigree and markers in livestock populations. J Anim Breed Genet. 2016;133:452–62.

38. Hayashi T, Iwata H. A Bayesian method and its variational approximation for prediction of genomic breeding values in multiple traits. BMC Bioinformatics. 2013;14:34.

39. Schulthess AW, Wang Y, Miedaner T, Wilde P, Reif JC, Zhao Y. Multiple-trait- and selection indices-genomic predictions for grain yield and protein content in rye for feeding purposes. Theor Appl Genet. 2015;129:273–87.

40. de Los Campos G, Sorensen D, Gianola D. Genomic heritability: what is it? PLoS Genet. 2015;11:e1005048.

41. Clark SA, Hickey JM, Daetwyler HD, van der Werf JH. The importance of information on relatives for the prediction of genomic breeding values and the implications for the makeup of reference data sets in livestock breeding schemes. Genet Sel Evol. 2012;44:4.

42. Legarra A, Aguilar I, Misztal I. A relationship matrix including full pedigree and genomic information. J Dairy Sci. 2009;92:4656–63.

43. Aguilar I, Misztal I, Tsuruta S, Wiggans G, Lawlor T. Multiple trait genomic evaluation of conception rate in Holsteins. J Dairy Sci. 2011;94:2621–4.

44. Tsuruta S, Misztal I, Aguilar I, Lawlor T. Multiple-trait genomic evaluation of linear type traits using genomic and phenotypic data in US Holsteins. J Dairy Sci. 2011;94:4198–204.

45. Bao Y, Kurle JE, Anderson G, Young ND. Association mapping and genomic prediction for resistance to sudden death syndrome in early maturing soybean germplasm. Mol Breed. 2015;35:128.

46. Shmueli G. To explain or to predict? Stat Sci. 2010;25(3):289–310.

47. Geisser S. Predictive inference: an introduction. New York: Chapman & Hall; 1993.

48. Román-Ponce SI, Samoré AB, Dolezal MA, Bagnato A, Meuwissen TH. Estimates of missing heritability for complex traits in Brown Swiss cattle. Genet Sel Evol. 2014;46:36.

49. Habier D, Fernando RL, Garrick DJ. Genomic BLUP decoded: a look into the black box of genomic prediction. Genetics. 2013;194:597–607.

50. Falconer D, Mackay T. Introduction to quantitative genetics. Harlow: Longman Group Ltd.; 1995.

51. Vattikuti S, Guo J, Chow CC. Heritability and genetic correlations explained by common SNPs for metabolic syndrome traits. PLoS Genet. 2012;8:e1002637.

52. Stich B, Piepho HP, Schulz B, Melchinger A. Multi-trait association mapping in sugar beet (Beta vulgaris L.). Theor Appl Genet. 2008;117:947–54.

6

Whole-genome sequence-based genomic prediction in laying chickens with different genomic relationship matrices to account for genetic architecture

Guiyan Ni[1*], David Cavero[2], Anna Fangmann[1], Malena Erbe[1,3] and Henner Simianer[1]

Abstract

Background: With the availability of next-generation sequencing technologies, genomic prediction based on whole-genome sequencing (WGS) data is now feasible in animal breeding schemes and was expected to lead to higher predictive ability, since such data may contain all genomic variants including causal mutations. Our objective was to compare prediction ability with high-density (HD) array data and WGS data in a commercial brown layer line with genomic best linear unbiased prediction (GBLUP) models using various approaches to weight single nucleotide polymorphisms (SNPs).

Methods: A total of 892 chickens from a commercial brown layer line were genotyped with 336 K segregating SNPs (array data) that included 157 K genic SNPs (i.e. SNPs in or around a gene). For these individuals, genome-wide sequence information was imputed based on data from re-sequencing runs of 25 individuals, leading to 5.2 million (M) imputed SNPs (WGS data), including 2.6 M genic SNPs. De-regressed proofs (DRP) for eggshell strength, feed intake and laying rate were used as quasi-phenotypic data in genomic prediction analyses. Four weighting factors for building a trait-specific genomic relationship matrix were investigated: identical weights, $-(\log_{10}P)$ from genome-wide association study results, squares of SNP effects from random regression BLUP, and variable selection based weights (known as BLUP|GA). Predictive ability was measured as the correlation between DRP and direct genomic breeding values in five replications of a fivefold cross-validation.

Results: Averaged over the three traits, the highest predictive ability (0.366 ± 0.075) was obtained when only genic SNPs from WGS data were used. Predictive abilities with genic SNPs and all SNPs from HD array data were 0.361 ± 0.072 and 0.353 ± 0.074, respectively. Prediction with $-(\log_{10}P)$ or squares of SNP effects as weighting factors for building a genomic relationship matrix or BLUP|GA did not increase accuracy, compared to that with identical weights, regardless of the SNP set used.

Conclusions: Our results show that little or no benefit was gained when using all imputed WGS data to perform genomic prediction compared to using HD array data regardless of the weighting factors tested. However, using only genic SNPs from WGS data had a positive effect on prediction ability.

Background

Genomic prediction (GP) uses genomic information to obtain estimated breeding values, which are subsequently used to select candidate individuals [1]. GP has been widely implemented in livestock [2–4] and plant [5] breeding schemes. The availability of next-generation sequencing technologies has made it possible to apply GP with whole-genome sequencing (WGS) data. GP with WGS is expected to lead to higher predictive ability, since WGS data include a large number of genomic variants including most of the

*Correspondence: gyni.ni@agr.uni-goettingen.de
[1] Animal Breeding and Genetics Group, Georg-August-Universität, Göttingen, Germany
Full list of author information is available at the end of the article

causal mutations. Thus, prediction depends much less on linkage disequilibrium (LD) between single nucleotide polymorphisms (SNPs) and causal mutations. Furthermore, Georges [6] claimed that WGS data can measure segregation of SNPs properly, which is not the case of commercial chips, particularly for rare SNPs. Based on a simulation study, Pérez-Enciso et al. [7] stated that using WGS data did not increase prediction accuracy compared to high-density (HD) array data. In a first study using sequenced inbred lines of *Drosophila melanogaster*, prediction based on WGS data using ~2.5 million (M) SNPs did not increase accuracy compared to an approach using only ~5% of the segregating SNPs [8]. In cattle data, Hayes et al. [9] found that accuracy of GP was improved by only 2% with WGS data compared to the 800 K array data when using BayesRC and imputed 1000 Bull genomes data. In addition, Van Binsbergen et al. [10] reported that GP with imputed WGS data did not lead to a higher prediction accuracy, compared to the HD array data from more than 5000 Holstein–Friesian bulls. Brøndum et al. [11] showed that the reliability of GP could be improved by adding several significant quantitative trait loci (QTL), which were detected by genome-wide association studies (GWAS) of WGS data, to the regular 54 K bovine array data, especially for production traits. Thus, GP with WGS data could be attractive, although so far the expectations for higher accuracies have not been realized with real data on cattle.

In chicken, most previous studies regarding GP were based on commercial array data. For instance, Morota et al. [12] reported that GP accuracy was higher when using all available SNPs than when using only validated SNPs from a partial genome (e.g. coding regions), based on the 600 K SNP array data of 1351 commercial broiler chicken. Abdollahi-Arpanahi et al. [13] studied 1331 chicken which were genotyped with a 600 K Affymetrix platform and phenotyped for body weight; they reported that predictive ability increased by adding the top 20 SNPs with the largest effects that were detected in the GWAS as fixed effects in the genomic best linear unbiased prediction (GBLUP) model. So far, studies to evaluate the predictive ability with WGS data in chicken are rare. Heidaritabar et al. [14] studied imputed WGS data from 1244 white layer chickens, which were imputed from 60 K SNPs up to sequence level with 22 sequenced individuals as reference samples. They reported a small increase (~1%) in predictive ability for the trait 'number of eggs' by using WGS data compared to 60 K SNPs when using a GBLUP model, while there was no difference when using a BayesC model.

Regardless of the genotyping source (i.e. WGS data or array data) used, GBLUP has been widely used in GP studies. Besides GBLUP in its classical form, in which each SNP is assumed to have the same contribution to the genetic variance, several weighting factors for SNPs or parts of the SNP set were proposed to account for the genetic architecture [15–17]. De los Campos et al. [15] proposed a method using the $-(\log_{10}P)$ from GWAS as a weighting factor for each SNP to build a genomic relationship matrix (**G** matrix). They observed that prediction accuracy for human height was improved compared to the original GBLUP, based on ~6000 records that were drawn from a public human type-2 diabetes case–control dataset with a 500 K SNP platform. Zhou et al. [16] used LD phase consistency, or estimated SNP effects or both as weighting factors to build a weighted **G** matrix, and reported that GBLUP with those weighted **G** matrices did not lead to higher GP accuracy in a study based on 5215 Nordic Holstein bulls and 4361 Nordic Red bulls. Using a German Holstein dataset, Zhang et al. [17] reported that the performance of BLUP given genomic architecture (BLUP|GA), which puts an optimal weight on a subset of SNPs with the strongest effects from the training set was similar to that of GBLUP for somatic cell score (SCS), but that BLUP|GA outperformed GBLUP for fat percentage and milk yield. The advantages of BLUP|GA were larger when the datasets were relatively small.

The objective of this study was to compare results from genomic prediction analyses using both HD array data and WGS data that were performed with GBLUP models and a variety of weighting factors for specific SNPs in a purebred commercial brown layer chicken line.

Methods
Data
High-density array data
We used 892 female and male chickens from six generations from a purebred commercial brown layer line (see Additional file 1: Table S1 for the number of individuals in each generation). These chickens were genotyped with the Affymetrix Axiom® Chicken Genotyping Array (denoted as the HD array), which initially included 580 K SNPs. Genotype data were pruned by removing SNPs located on the sex chromosomes and in unmapped linkage groups, and SNPs with a minor allele frequency (MAF) lower than 0.5% or a genotyping call rate lower than 97%. Individuals with call rates lower than 95% were also discarded. After filtering, 336,224 SNPs that segregated for 892 individuals remained for analyses.

Imputed whole-genome sequence data
Data from re-sequencing that were obtained with the Illumina HiSeq2000 technology with a target coverage of 8× were available for 25 brown layer chickens of the same population (of which 18 were also genotyped with the HD array) and for another 25 white layer chickens. Chickens used for whole-genome sequencing were chosen from the older generations and with a maximum relationship with the chickens that were to be imputed [18, 19]. Data from

re-sequencing runs (brown and white layer chickens) were aligned to Build 4 of the chicken reference genome (galGal4) with BWA (version 0.7.9a-r786) [20] using default parameters for paired-end alignment and SNP variants were called using GATK (version 3.1-1-g07a4bf8, UnifiedGenotyper) [21]. Called variants (only for the 25 brown layers) were edited for depth of coverage (DP) and mapping quality (MQ) based on the following criteria: (1) for DP, outlier SNPs (at the top 0.5% of DP) were removed, then, mean and standard deviations of DP were calculated for the remaining SNPs and those that had a DP above and below 3 times the standard deviation from the mean were removed; and (2) for MQ, SNPs with a MQ lower than 30 (corresponding to a probability of 0.001 that their position on the genome was not correct) were removed. After filtering, within the set of 25 re-sequenced brown layers, 10,420,560 SNPs remained and were used as the reference dataset to impute HD array data up to sequence level. Imputation of all genotyped individuals was then performed using Minimac3 [22] which needs pre-phased data as input. The pre-phasing procedure was done with the BEAGLE 4 package [23]. Default numbers of iteration were used in pre-phasing and imputation. The imputation process did not use pedigree information. According to our previous study [24], phasing genotype data with BEAGLE 4 and further imputing with Minimac3 provided the highest imputation accuracy under different validation strategies. After imputation, post-imputation filtering criteria were applied per SNP, namely, SNPs with a MAF lower than 0.5% or SNPs with an imputation accuracy lower than 0.8 were removed. The imputation accuracy used here was the Rsq measurement from Minimac3, which was the estimated value of the squared correlation between true and imputed genotypes. After this step, 5,243,860 imputed SNPs were available for 892 individuals, which are hereafter denoted as WGS data.

In addition, SNPs, regardless of which dataset they were in, were classified into nine classes by gene-based annotation with the ANNOVAR software [25] by setting default parameters and using galGal4 as reference genome [26]. Our set of genic SNPs (SNP_genic) included all SNPs from the eight categories exon, splicing, ncRNA, UTR5′, UTR3′, intron, upstream, and downstream regions of the genome, whereas the ninth category included SNPs from intergenic regions. There were 2,593,054 SNPs characterized as genic SNPs from the WGS data (hereafter denoted as WGS_genic data) and 157,393 SNPs characterized as genic SNPs from the HD array data (hereafter denoted as HD_genic data).

Phenotypic observations

The quasi-phenotypic data were de-regressed proofs (DRP) for eggshell strength (ES), feed intake (FI), and arcsine transformed laying rate in the last third of the laying period (LR). The arcsine transformation of the latter trait was performed to achieve an approximate normalization. To obtain de-regressed proofs, a single trait BLUP animal model was performed for each trait using raw phenotypic and pedigree data, respectively. Estimated breeding values from these models were then de-regressed following Garrick et al. [27]. The de-regression process included removal of the parent average information.

Genomic prediction

Genomic prediction was performed using the following GBLUP model with different genomic relationship matrices that are described below:

$$\mathbf{y} = \mathbf{X}\mu + \mathbf{Z}\mathbf{g} + \mathbf{e},$$

where $\mathbf{y}$ is the vector of DRP of individuals in the training set for a specific trait; μ is the overall mean; $\mathbf{g}$ is the vector of additive genetic values (i.e. genomic breeding values) for all genotyped chickens; $\mathbf{e}$ is the vector of residual terms; $\mathbf{X}$ and $\mathbf{Z}$ are design matrices assigning DRP to the overall mean and additive genetic values, where the dimension of $\mathbf{Z}$ is the number of individuals in the training set times the number of all genotyped individuals.

A normal distribution of the residual term $\mathbf{e}$ is assumed $\mathbf{e} \sim N\left(0, \mathbf{R}\sigma_e^2\right)$, where $\mathbf{R}$ is a diagonal matrix, with diagonal element $R_{ii} = \left(1 - r_{DRPi}^2\right)/r_{DRPi}^2$ [28] for an individual i in the training set, where r_{DRPi}^2 is the reliability of DRP for individual i, and σ_e^2 is the residual variance. The distribution of the additive genetic values is assumed normal $\mathbf{g} \sim N\left(0, \mathbf{G}\sigma_g^2\right)$, where σ_g^2 is the additive genetic variance and $\mathbf{G}$ is a realized genomic relationship matrix including all genotyped individuals, which can be calculated with different approaches resulting in different GBLUP models.

The general approach to build a $\mathbf{G}$ matrix is:

$$\mathbf{G} = \frac{\mathbf{M}\mathbf{D}\mathbf{M}^{\mathrm{T}}}{2\sum_{i=1}^{m} p_i(1 - p_i)},$$

where $\mathbf{M}$ contains the corrected SNP genotypes with individuals in rows and SNPs in columns. The elements of column i of $\mathbf{M}$ are $0 - 2p_i$ (for homozygotes of the first allele), $1 - 2p_i$ (for heterozygotes), and $2 - 2p_i$ (for homozygotes of the second allele), where p_i is the frequency of the second allele at locus i from the current dataset. $\mathbf{D}$ is a diagonal matrix that contains the weight of each locus; these weights varied according to the scenario studied. An identity matrix was used ($\mathbf{D} = \mathbf{I}$) in the original GBLUP [29], which implies that all loci contribute equally to the variance–covariance structure. The resulting $\mathbf{G}$ matrix is denoted as $\mathbf{G}_{\mathbf{I}}$ in the following. De los Campos et al. [15] suggested using the corresponding $-(\log_{10}P)$ from a t test of a GWAS as weighting factors to consider the relative importance of different SNPs on a specific trait. The genomic relationship matrix including a $\mathbf{D}$ matrix based on this weighing factor will be denoted as $\mathbf{G_P}$. The corresponding P values were derived from different GWAS models, each being fitted for each trait of interest separately in the respective training set. In order to correct for population stratification and relationships between individuals, a principal component analysis (PCA) was performed on genomic data and significance among principal components (PC) was tested in advance with a Tracy Widom test as implemented in the program EIGENSTRAT [30]. Then, the PC with P values $\leq 10^{-100}$ (or ≤ 0.05) were used as fixed covariates in single-SNP GWAS runs. The resulting genomic relationship matrix was denoted as $\mathbf{G_{P100}}$ (or $\mathbf{G_{P005}}$). Genomic relationship matrices with weighting factors based on results from single-SNP GWAS may not adequately represent or may overweight regions because different SNPs can capture the effect from the same QTL due to long-range LD. However, a SNP effect is not corrected for any other SNP effect in a single-marker regression type GWAS. We also investigated the usefulness of weighting the $\mathbf{G}$ matrix with results from a random-regression BLUP (RRBLUP) in which random SNP effects are fitted simultaneously. Thus, for matrix $\mathbf{G_S}$, we used the squares of the estimated SNP effects of the respective trait as weighting factors to build matrix $\mathbf{D}$ (as was done in [28]). Finally, we also investigated BLUP|GA [17] in this study. To account for genetic architecture, the trait-specific genomic relationship matrix $\mathbf{G_z}$ was constructed as a weighted sum of a genetic architecture matrix $\mathbf{S}$ and a realized relationship matrix $\mathbf{G_I}$ (i.e. $\mathbf{G_z} = \omega\mathbf{S} + (1-\omega)\mathbf{G_I}$). The construction of the $\mathbf{S}$ matrix was similar to the construction of $\mathbf{G_S}$, but it was based only on selected SNPs according to the size of their absolute SNP effects (top%) from RRBLUP. The optimal choices for top% and ω were identified with a grid search strategy applied in the training population. The combinations for searching for optimal parameters were the same as in the original study of Zhang et al. [17] (top% within a range of [0.05, 10] and ω within the range [0.1, 0.99]). To make sure that the weighted $\mathbf{G}$ matrices were in the same scale as $\mathbf{G_I}$, all weighting factors were divided by their mean. To mimic the real situation in the best way and avoid over-fitting, all weighting factors in all models were derived exclusively from individuals in the respective training set. To assess whether focusing on functional information improves prediction accuracy, the original GBLUP was applied to the functional subset of the WGS data (HD array data) by building a genomic

relationship matrix $\mathbf{G_G}$ based on WGS_genic data (HD_genic data) with weights in $\mathbf{D}$ being 1.

Each approach mentioned above was investigated using fivefold random cross-validation (i.e. having 614 or 615 individuals in the training set and 178 or 179 individuals in the validation set) with five replications and was applied to both WGS and HD array data. Predictive ability was measured as the correlation between the obtained direct genomic values (DGV) and DRP for each trait of interest. DGV and corresponding variance components were estimated using ASReml 3.0 [31].

In layer chicken breeding, genomic breeding values are especially interesting for selecting the best individuals from full-sib families. Thus, we performed the Spearman's rank correlation to evaluate the ranking of full-sibs according to DRP and DGV in a randomly chosen full-sib family with 12 individuals. Results presented here were from the validation sets of the first replicate of a fivefold cross-validation.

Results and discussion
Data summary
Numbers of SNPs in different MAF bins for different datasets are shown in Fig. 1. The difference in the distribution of SNPs between HD array data and data from re-sequencing runs is illustrated in the top panel. The last bin ($0.48 < \text{MAF} \leq 0.5$) contains only half the number of SNPs since, in this bin, only one allele frequency class (25 out of 50 alleles) is represented, while in all other bins two frequency classes (e.g. 24 and 26 out of 50 alleles in the adjacent class) are reflected. The MAF distribution based on WGS data was significantly different from that based on HD data (tested with a χ^2-test, $P < 0.001$). For data from re-sequencing runs of the 25 sequenced chickens, the number of SNPs per bin decreased with increasing MAF. SNPs with a very small MAF are not so extremely overrepresented in the re-sequenced set as in other studies with sequenced data [32, 33], which could be due to two reasons. First, the size of the reference dataset was relatively small (25 chickens) and thus, some of the rare variants may not be captured. Second, the commercial layers have been subject to intensive within-line selection, which might have reduced the genetic diversity dramatically, and further resulted in a lack of rare SNPs [34]. Presumably, this problem can only be overcome with a larger sequenced reference set, which would allow higher imputation accuracies for rare SNPs. Numbers of SNPs in different MAF bins in the WGS data set before and after post-imputation filtering are in the bottom panel of Fig. 1. Unlike Van Binsbergen et al. [10], in which 429 sequenced individuals from several cattle breeds were used as a reference set for imputation process, we did not observe a clear

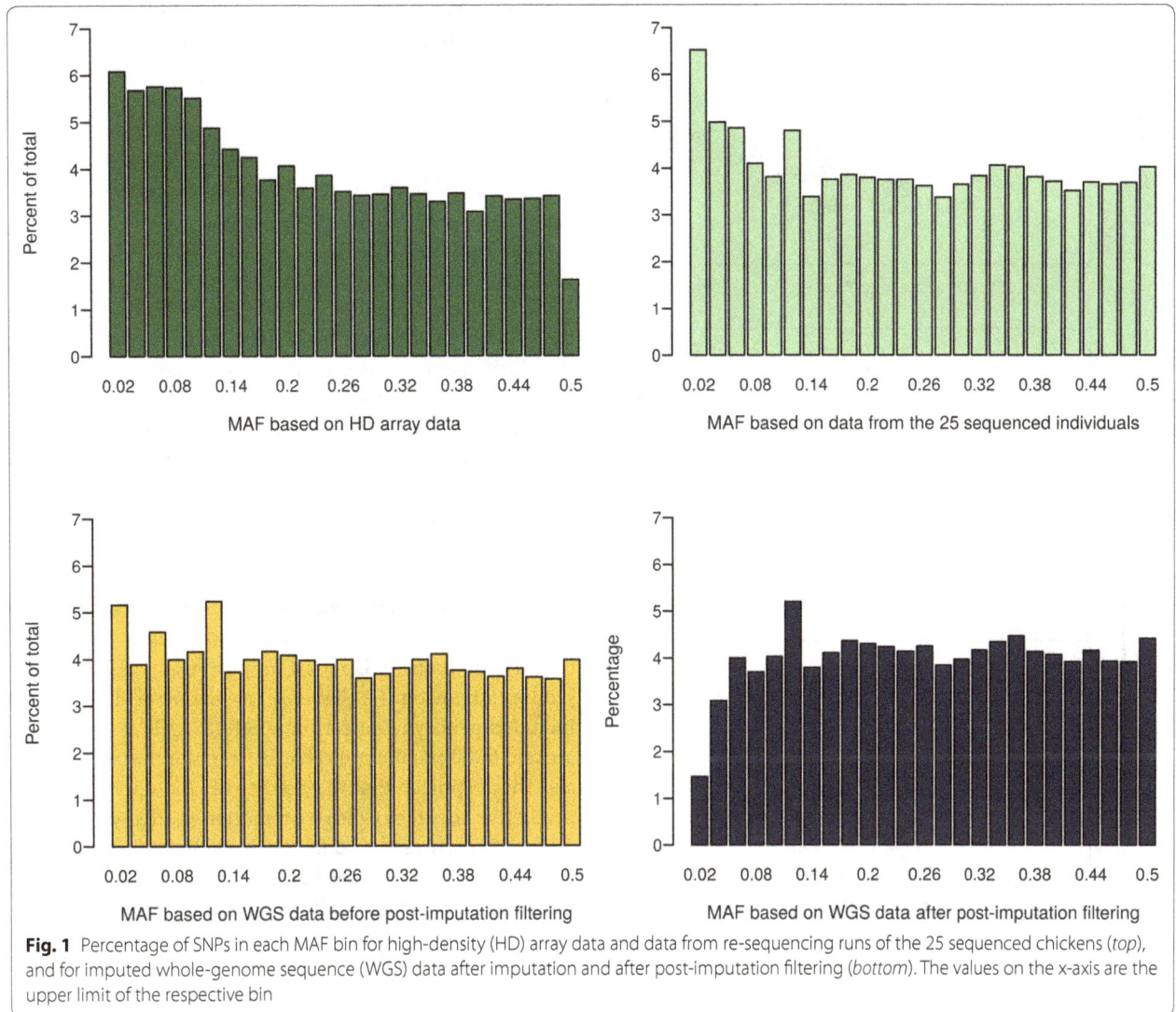

Fig. 1 Percentage of SNPs in each MAF bin for high-density (HD) array data and data from re-sequencing runs of the 25 sequenced chickens (*top*), and for imputed whole-genome sequence (WGS) data after imputation and after post-imputation filtering (*bottom*). The values on the x-axis are the upper limit of the respective bin

U-shaped distribution of MAF in the imputed WGS data. This means that some of the rare SNPs in the re-sequenced individuals were either not present in all the other individuals of the population or got lost during the imputation process, partly because of the poor imputation accuracy for SNPs with a low MAF [35, 36]. Starting from more than 9 million SNPs after imputation (monomorphic SNPs excluded), 200,679 SNPs were filtered out due to a low MAF, and 85% of these filtered SNPs had low imputation accuracy (Rsq of minimac3 <0.8) as well, which means that SNPs with a low MAF are even less represented in the SNP set. Furthermore, 1.3 million SNPs among the imputed SNP set, which passed the MAF criteria, were filtered out due to low imputation accuracy only; these were evenly distributed over all MAF bins. In total, more than 50% of SNPs were filtered out due to low imputation accuracy in the leftmost

three MAF bins ($0 < MAF \leq 0.06$). The fact that we found high rates of low Rsq values within the set of SNPs with a low MAF could be due to low LD between these SNPs and adjacent SNPs, which can result in lower imputation accuracy [for imputation accuracies in different MAF bins (see Additional file 2: Figure S1)] [37–41]. Filtering out a large number of SNPs with a low MAF—in many cases, because imputation accuracy is too low—could weaken the advantage of imputed WGS data, which contain a large number of rare SNPs [6], although GP with all imputed SNPs without quality-based filtering did not improve the prediction ability in our case (results not shown). In addition, LD pruning was not performed in our study, because in a preliminary study we found that predictive ability based on the pruned dataset was the same as that based on data without pruning (results not shown).

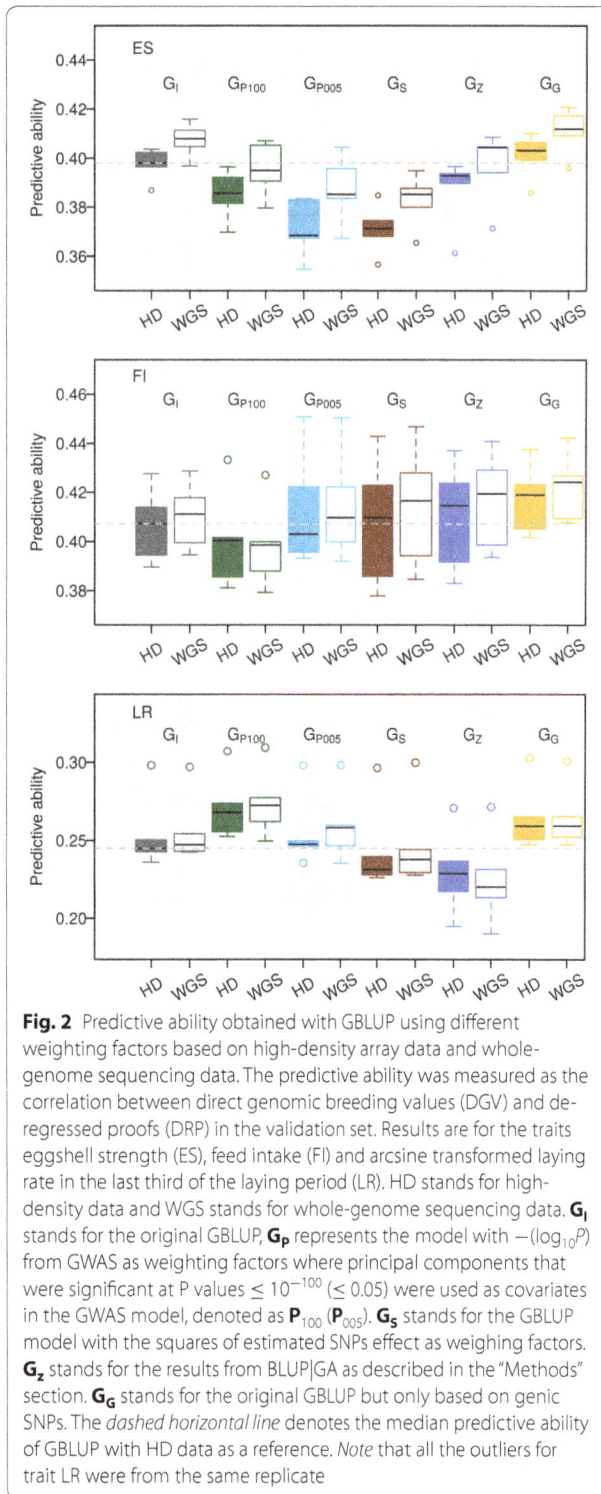

Fig. 2 Predictive ability obtained with GBLUP using different weighting factors based on high-density array data and whole-genome sequencing data. The predictive ability was measured as the correlation between direct genomic breeding values (DGV) and de-regressed proofs (DRP) in the validation set. Results are for the traits eggshell strength (ES), feed intake (FI) and arcsine transformed laying rate in the last third of the laying period (LR). HD stands for high-density data and WGS stands for whole-genome sequencing data. $\mathbf{G_I}$ stands for the original GBLUP, $\mathbf{G_P}$ represents the model with $-(\log_{10}P)$ from GWAS as weighting factors where principal components that were significant at P values $\leq 10^{-100}$ (≤ 0.05) were used as covariates in the GWAS model, denoted as $\mathbf{P_{100}}$ ($\mathbf{P_{005}}$). $\mathbf{G_S}$ stands for the GBLUP model with the squares of estimated SNPs effect as weighing factors. $\mathbf{G_z}$ stands for the results from BLUP|GA as described in the "Methods" section. $\mathbf{G_G}$ stands for the original GBLUP but only based on genic SNPs. The *dashed horizontal line* denotes the median predictive ability of GBLUP with HD data as a reference. *Note* that all the outliers for trait LR were from the same replicate

Comparison between HD array data and WGS data using different weighting factors

Predictive abilities obtained with GBLUP using different weighting factors based on HD array data and WGS

data are in Fig. 2 for the traits ES, FI, and LR, respectively. Predictive ability was defined as the correlation between DGV and DRP of individuals in the validation set. Generally speaking, predictive ability could not be clearly increased when using WGS data compared to HD array data regardless of the different weighting factors studied. Using genic SNPs from WGS data had a positive effect on prediction ability in our study design.

Averaging over the three traits analyzed here, the predictive ability $\pm$ standard deviation for the original GBLUP was 0.353 ± 0.074 based on HD array data and 0.358 ± 0.076 based on WGS data. When $-(\log_{10}P)$ (with P values from GWAS with different covariates in the model) were used as weighting factors, predictive abilities for G_{P100} (G_{005}) were 0.352 ± 0.062 (0.347 ± 0.072) based on HD array data and 0.356 ± 0.062 (0.354 ± 0.073) based on WGS data. Unlike the SNP effects that were estimated from RRBLUP, in which effects are assessed simultaneously, SNP effects were estimated independently in GWAS. Thus, effects of a group of SNPs which represent the same QTL could not be fitted simultaneously, and thus the overall weighting of a region might depend on the marker density. De los Campos et al. [15] studied a public human type-2 diabetes case–control dataset that included genotype data from a 500 K SNP platform and around 6000 phenotype records from unrelated individuals. They reported that the predictive reliability (square of predictive accuracy) with a prediction model weighted by $-(\log_{10}P)$ increased by a factor of 110% compared to that with the original GBLUP. Similarly, Su et al. [28] reported that predictive ability using $-(\log_{10}P)$ as weighting factors was higher than that obtained with the original GBLUP, based on more than 5000 Nordic Holstein bulls that were genotyped with the Illumina Bovine SNP50 BeadChip. However, the improvement in predictive ability by using $-(\log_{10}P)$ as weighting factors in GP was not observed in our dataset.

Furthermore, using the squares of SNP effects as weighting factors in GBLUP ($\mathbf{G_S}$) resulted in slightly lower predictive abilities compared to the original GBLUP, in both analyses based on HD array data and on WGS data, respectively, as shown in Fig. 2. For $\mathbf{G_S}$, averaging over the three traits, predictive ability was 0.341 ± 0.076 based on HD data and 0.348 ± 0.078 based on WGS array data, compared to 0.353 ± 0.074 (for HD array data) and 0.358 ± 0.076 (for WGS data) with the original GBLUP. These results are in agreement with Su et al. [28], who reported that GBLUP with the squares of SNP effects as weighting factors did not improve predictive ability compared to the original GBLUP or to the model with $-(\log_{10}P)$ as weighting factors. The lack of improvement in predictive ability when using the squares of SNP effects as weighting factors might be due to two

Whole-genome sequence-based genomic prediction in laying chickens with different genomic relationship...

79

reasons. One reason is the occurrence of sequencing or imputation errors, i.e. in our study, the most probable genotypes imputed from Minimac3 were used as WGS data rather than genotype probabilities, which does not account for the uncertainty of imputation. The second reason is that the noise and uncertainty of estimated SNP effects could also bias predictive ability [28]. In our study, DGV of the training population were assigned to millions of SNPs [Figs. 3, 4; (Additional file 3: Figure S2)], so that the effect of each SNP was very small. However, the prediction error of a SNP effect might be even larger than the SNP effect itself. In addition, the size of the training set was relatively small, which could further enhance the uncertainty of SNP effects. Thus, the combination of both mentioned reasons could lead to lower predictive ability, since the DGV of individual i is the summation of estimated SNP effects times its genotypes (i.e. $DGV_i = \sum_{k=1}^{m} X_{ik}\beta_k$).

With BLUP|GA, predictive ability was 0.342 (± 0.085) based on HD array data and 0.346 (± 0.091) based on WGS data averaged over the three traits analyzed (Fig. 2). Generally speaking, BLUP|GA did not improve predictive ability with WGS or HD data, compared to the original GBLUP. Zhang et al. [17] reported that BLUP|GA outperformed the original GBLUP for production traits (i.e. fat percentage and milk yield) in a German Holstein cattle population, while its performance was similar to that of GBLUP for SCS. A well-known candidate gene *DGAT1* has a strong influence on fat percentage [42, 43], while for SCS no major genes are known. This suggests that BLUP|GA is especially useful when QTL regions that heavily influence the trait are present in the genome. The genetic architecture of ES, FI, and LR seems to be more similar to that of SCS than of fat percentage which might explain why no strong candidate genes have been identified to date and also that no strong SNP effects have been detected in the GWAS runs performed in this study (see Additional file 4: Figure S3). The SNP effects estimated from RRBLUP based on HD array (WGS) data are in Fig. 3 (Fig. 4) and further illustrate that ES, FI, and LR are controlled by numerous SNPs with very small effects.

When focusing on the training stage of BLUP|GA, the burden of calculation to identify the optimal combination for parameters top% and ω with a grid strategy was huge. Prediction abilities of BLUP|GA in the training stage are in Fig. 5 for each parameter combination exemplarily for the first fold of the first replicate. The combination of large ω and small top% tended to give lower predictive ability. As top% increased and ω decreased, predictive ability tended to increase. In most cases, the optimal option for ω based on HD data and WGS data was 0.1 in our study, which is the minimal ω we analyzed. The optimal option for top% was 10%, which is the maximal top% we analyzed, and is different from the findings of Zhang et al. [17]. These authors tended to select a smaller top% while there was no obvious pattern in the selection of ω. Those 10% SNPs explained approximately 23% of the total variance of SNP effects for ES. Optimal combinations in each fivefold cross-validation of each replicate for each trait are in Additional file 5: Table S2 and Additional file 6: Table S3. It should be noted that, as described in Zhang et al. [17], accuracy of GP based on the optimal parameters obtained in the training stage by cross-validation may not lead to the highest accuracy in the application stage.

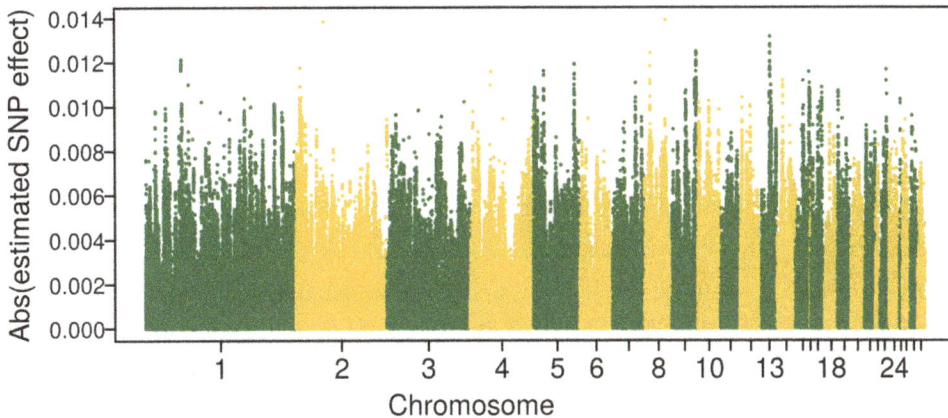

Fig. 3 Manhattan plot of absolute estimated SNP effects for trait eggshell strength based on high-density (HD) array data. SNP effects were obtained from RRBLUP in the training set of the first replicate

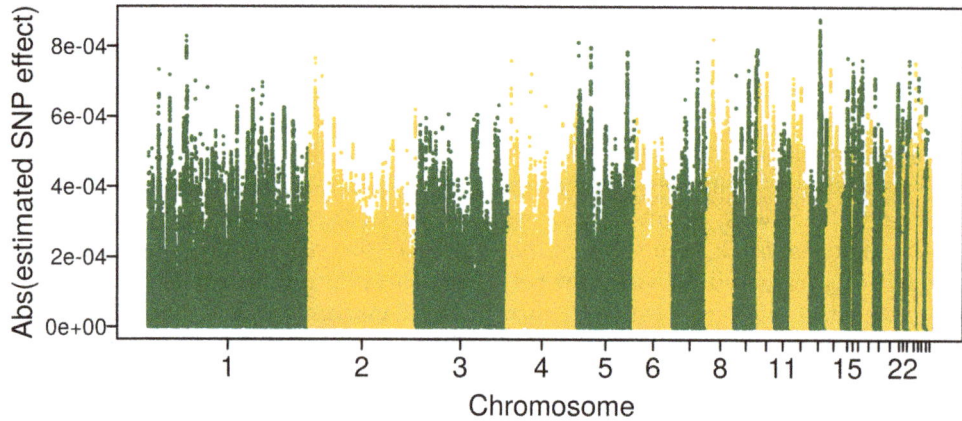

Fig. 4 Manhattan plot of absolute estimated SNP effects for trait eggshell strength based on whole-genome sequence (WGS) data. SNP effects were obtained from RRBLUP in the training set of the first replicate

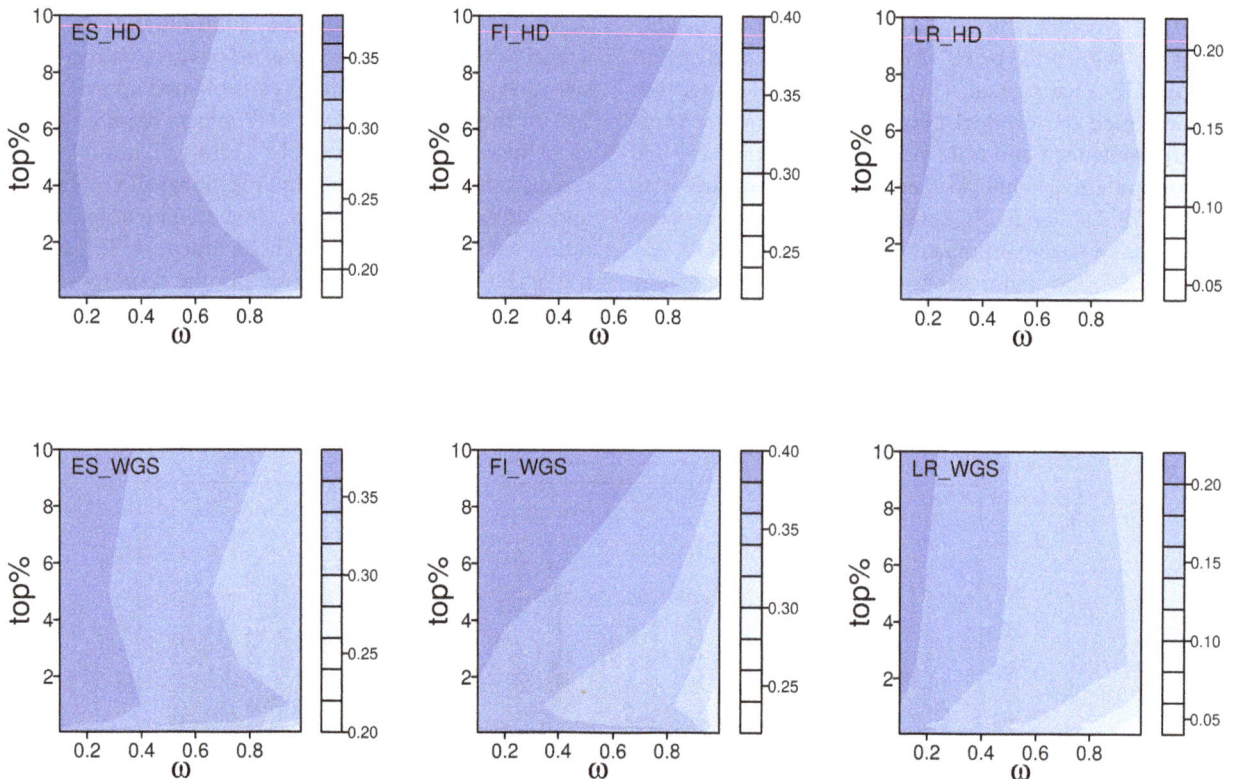

Fig. 5 Predictive ability of the best linear unbiased prediction given the genetic architecture (BLUP|GA) in the training stage to select the optimal parameter combination for the application stage. Predictive ability in this figure is the mean correlation between direct genomic breeding values (DGV) and de-regressed proofs (DRP). The first row is for high-density (HD) array data, while the second row is for whole-genome sequence (WGS) data. The x-axis stands for the overall weighting factor; y-axis stands for the percentage of SNPs selected based on the SNP effects (top%); different colors stand for different levels of predictive ability

Averaging over the three traits analyzed here, predictive ability ± standard deviation was 0.366 ± 0.075 based on the WGS_genic data and 0.361 ± 0.072 based on HD_genic data, compared to 0.353 (HD array data) and 0.358 (WGS data), which means that GP with WGS_genic resulted in the highest predictive ability in our study. Similarly, Do et al. [44] reported that predictive ability increased only when SNPs in genes were considered for residual feed intake based on 1272 Duroc pigs, which were genotyped with the 60 K SNP chip, although the increase was not significantly different from that obtained with 1000 randomly SNPs. In chicken, Morota et al. [12] studied predictive ability with 1351 commercial broiler chickens genotyped with the Affymetrix 600 K chip, and found that prediction based on SNPs in or around genes did not result in a higher accuracy using kernel-based Bayesian ridge regression. In our dataset, predictive ability with HD_genic data was slightly higher than that with all HD data. Furthermore, the benefit was observed when using WGS_genic, which could be due to the fact that using only genic SNPs reduces the noise in WGS data and might increase the chance to identify the potential causal mutations. Koufariotis et al. [45] found that significant SNPs in the GWAS were enriched in coding regions based on 17,425 Holstein or Jersey bulls and cows, which were genotyped with the 777 K Illumina Bovine HD array. The enrichment of significant SNPs could further imply that using genic SNPs can help us to achieve higher predictive ability.

The bias of DGV was assessed as the slope coefficient of the linear regressions of DRP on DGV within the validation sets of random fivefold cross-validation. The averaged regression coefficient ranged from 0.520 (G_{P005} of HD dataset) to 0.871 (G_I of WGS dataset) for the trait ES (see Additional file 7: Figure S4). No major differences were observed between using HD and WGS datasets within different methods. Generally, regression coefficients were all smaller than 1, which means that the variance of the breeding values tends to be overestimated. However, the regression coefficients were closer to 1 when the identity matrix was used in the prediction model (i.e. G_I, G_G). The overestimation could be due to the fact that those analyses were based on cross-validation where the relationship between training and validation populations might cause a bias. Another possible reason for the overestimation could be that, in this chicken population, individuals were under strong within-line selection. The same tendency was observed for traits FI and LR (results not shown).

Comparison within a full-sib family

To get an insight into the ranking of 12 full-sibs within a family according to DRP and DGV, DGV that were predicted in the validation sets with different **G** matrices in the first of the five replicates of the cross-validation runs are in Figs. 6 (HD data) and 7 (WGS data) for ES, and Additional file 8: Figure S5 and Additional file 9: Figure S6 for traits FI and LR, respectively. The higher the rank correlation is, the higher is the possibility to select the same candidates. Based on HD array data, DGV from different weighting models had a relatively high rank correlation with those from G_I (from 0.88 to 0.97 for ES). This suggested that the same candidate tended to be selected in different models. Likewise, the rank correlations based on WGS data were relatively high as well, with minimal values of 0.91 between G_G and G_{P005}. In addition, the Spearman's rank correlation between G_I based on HD array data and that based on WGS data was 0.98. Spearman's rank correlation between G_G with WGS_genic data and G_I with WGS data was 0.99, which indicated that there was hardly any difference in selecting candidates based on HD array data, or WGS data, or WGS_genic data with GBLUP. Generally, the same set of candidates tended to be selected regardless of the dataset (HD array data or WGS data) and weighting factors (identity weights, squares of SNPs effect, or P values from GWAS) used in the model. When comparing the DGV from different models with DRP, the Spearman's rank correlations were modest (from 0.38 to 0.54 with HD data and from 0.31 to 0.50 with WGS data) and within the expected range considering the overall predictive ability obtained in the cross-validation study (see Fig. 2). Although DGV from different models were highly correlated, Spearman's rank correlation of the respective DGV to DRP clearly varied. This fact, however, should not be overvalued regarding the small sample size that was used here (n = 12) and the fact that the DGV of the full-sib family were estimated from different CV folds. Thus, a forward prediction was performed with 146 individuals from the last two generations as validation set. In this case the same tendency was observed, namely that DGV from different models were highly correlated within a large half-sib family. However, in this forward prediction scenario, the predictive ability with genic SNPs was slightly lower than that with all SNPs (results not shown).

Perspectives and implications

Using WGS data in GP was expected to lead to higher predictive ability, since WGS data should include most of the causal mutations that influence the trait and

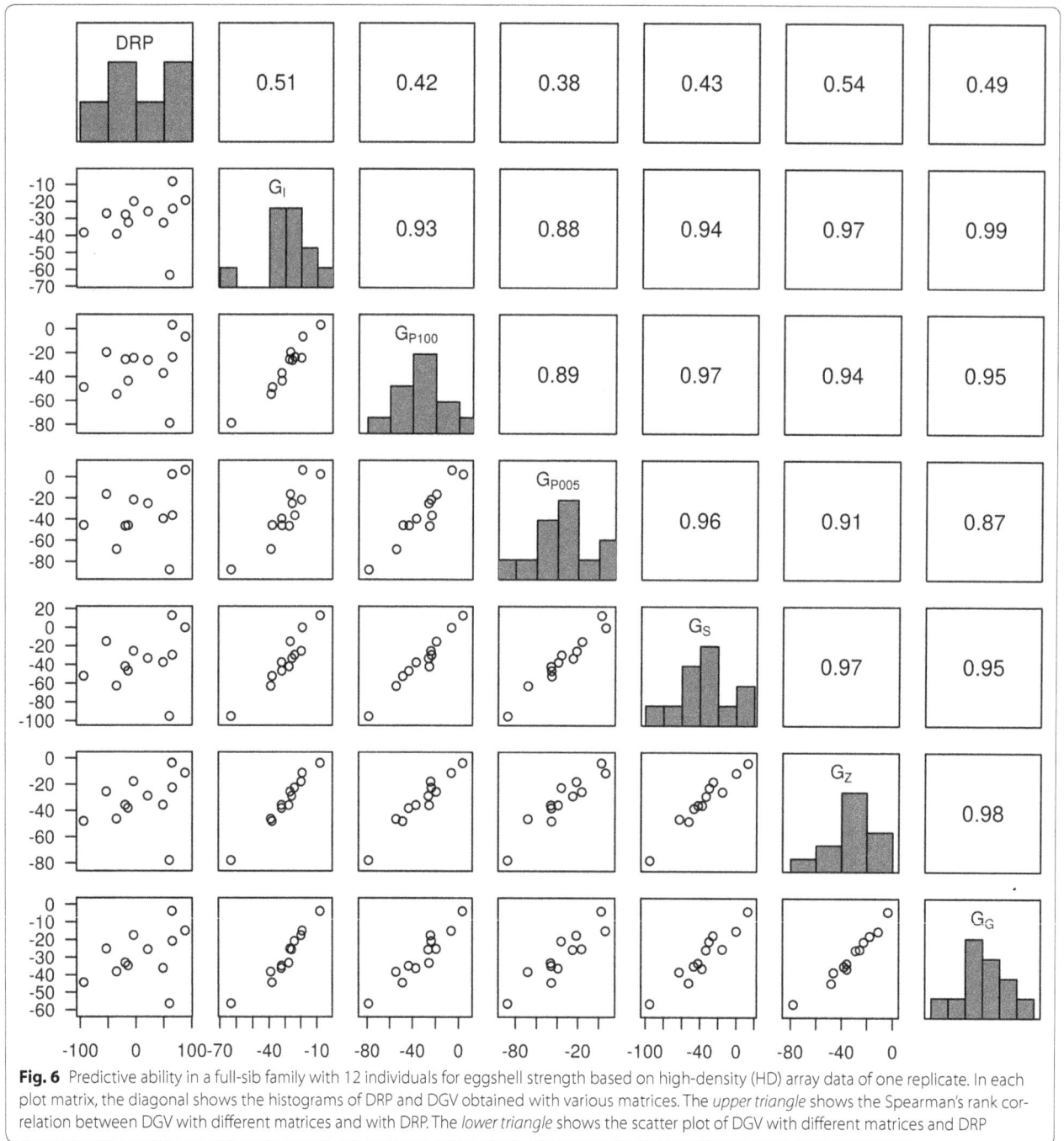

Fig. 6 Predictive ability in a full-sib family with 12 individuals for eggshell strength based on high-density (HD) array data of one replicate. In each plot matrix, the diagonal shows the histograms of DRP and DGV obtained with various matrices. The *upper triangle* shows the Spearman's rank correlation between DGV with different matrices and with DRP. The *lower triangle* shows the scatter plot of DGV with different matrices and DRP

prediction is much less limited by LD between SNPs and causal mutations. Contrary to this expectation, little gain was found in our study. One possible reason could be that QTL effects were not estimated properly, due to the relatively small dataset (892 chickens) with imputed WGS data [18]. Imputation has been widely used in many livestock [38, 46–48], however, the magnitude of the potential imputation errors remains difficult to detect. In fact, Van

Binsbergen et al. [10] reported from a study based on data of more than 5000 Holstein–Friesian bulls that predictive ability was lower with imputed HD array data than with the actual genotyped HD array data, which confirms our assumption that imputation could lead to lower predictive ability. In addition, discrete genotype data were used as imputed WGS data in this study, instead of genotype probabilities which can account for the uncertainty of

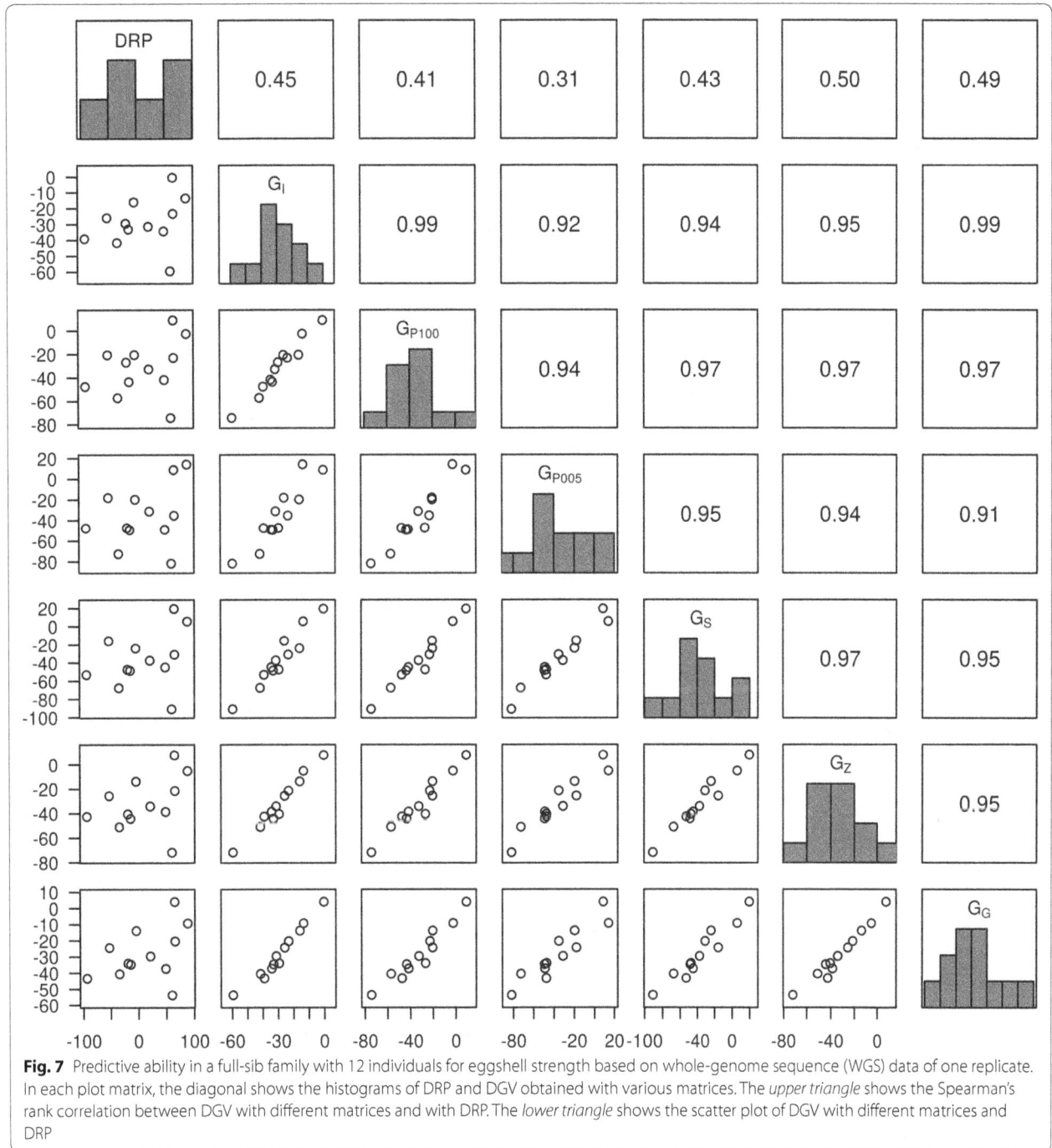

Fig. 7 Predictive ability in a full-sib family with 12 individuals for eggshell strength based on whole-genome sequence (WGS) data of one replicate. In each plot matrix, the diagonal shows the histograms of DRP and DGV obtained with various matrices. The *upper triangle* shows the Spearman's rank correlation between DGV with different matrices and with DRP. The *lower triangle* shows the scatter plot of DGV with different matrices and DRP

imputation and may be more informative [49]. At present, sequencing all individuals in a population is not realistic. In practice, there is a trade-off between predictive ability and cost efficiency. When focusing on the post-imputation filtering criteria, the threshold for imputation accuracy was 0.8 in our study to guarantee the high quality of the imputed WGS data. Numerous rare SNPs, however,

were filtered out due to the low imputation accuracy as shown in Fig. 1 and Additional file 2: Figure S1. This could increase the risk of excluding rare causal mutations. However, Ober et al. [8] did not observe an increase in predictive ability for starvation resistance when rare SNPs were included in the GBLUP based on ~2.5 million SNPs that had been identified from 192 *D. melanogaster*. Further investigation needs to be done in chicken, especially when more founder sequences become available.

Another reason why we did not observe any increase in predictive ability when using WGS data could be that we did not apply variable selection. The density of WGS data was around 15 times higher than that of HD array data, which increased LD between SNPs. Thus, QTL effects were assigned to more SNPs in WGS data than in HD array data, which could be overcome by variable selection. Su et al. [28] reported that reliability of GP increased by more than 5% when grouping 30 adjacent SNPs. In each group, a common weight was assigned which reflected the mean over the SNP effects in the same group. In addition, Brøndum et al. [11] reported that the reliability of GP can be improved by adding several significant QTL into the regular bovine 54 K array data. In our study, 20 top SNPs were selected according to their estimated effects from RRBLUP or $-(\log_{10}P)$ of GWAS and used as fixed effects in GBLUP, but it did not improve predictive ability (results not shown). GP with genic SNPs from WGS (the WGS_genic data) provided the highest predictive ability compared to that obtained when all SNPs from WGS data were used. This implies that selecting the proper variables could help us to reduce noise and increase predictive ability. Using a variable selection model may also help. Based on a simulated WGS data, Wimmer et al. [50] reported that feature selection methods (e.g. the least absolute shrinkage and selection operator) have limitations when the ratio between sample size and number of SNPs is unfavorable, which was the case in our study. A similar conclusion was drawn by Heidaritabar et al. [14] who found that there was no advantage in genomic prediction with a BayesC model compared to GBLUP based on imputed WGS data of 1244 white layer chickens. In addition, Ober et al. [8] observed no differences in predictive ability with BayesC and GBLUP based on a dataset with ~2.5 million SNPs that were identified for a *D. melanogaster* population. In addition, with the increasing knowledge about gene networks, pathways and suitable prediction models, blending biological knowledge based on gene annotations and complex interactions may provide insights to guide GP [51].

Our fourth possible explanation for the small improvement in predictive ability with WGS data refers to the population structure. Commercial chickens have been subject to intensive within-line selection, which has a strong effect on the population structure. MacLeod et al. [52] studied the accuracy of GP based on WGS data for two simulated populations with a different demographic history. They found that in a highly selected population with a small effective population size there was almost no gain in prediction accuracy when using WGS data compared to HD data, which is in agreement with our findings. The way the data were split for the cross-validation strategy might enhance this effect compared to a forward prediction scenario.

The use of incomplete WGS information could also weaken its predictive ability. First, in most studies, sex chromosomes were disregarded in the GP scheme, considering that the transmission of sex chromosomes and that of autosomes differ and that the density of SNPs and LD structure on the sex chromosomes is lower than on autosomes in commercial SNP chips. However, recent studies have discovered an increasing number of genes on the sex chromosomes that affect economic traits. For example, Su et al. [53] found that including the sex chromosomes in the GP scheme could increase the predictive ability averaged over 15 traits that were included in the Nordic Total Merit index (e.g. milk yield and fat yield). Second, WGS data, technically include all DNA variants [e.g. copy number variations (CNV) and InDels], but the studies on GP in livestock have so far mostly focused on SNPs. However, according to previous studies [54, 55], CNV and other types of structural variations play an important role in gene expression and phenotypic variation. Third, although the chicken karyotype consists of 39 chromosomes, data from re-sequencing represent only 30 chromosomes and two linkage groups since the reference genome was not available for some of the microchromosomes which are also assumed to be gene-rich [56, 57]. Beyond that, chromosome 16, which hosts the chicken major histocompatibility complex, is included in the reference sequence but has a low marker density [58] and the quality of the reference sequence is expected to be inferior due to the high genetic variability. Furthermore, non-nuclear DNA present in the mitochondria is not accounted for. In general, further work is necessary to assess the importance of the entire DNA variation on the predictive ability of GP in chicken.

Conclusions

In this study, we compared the ability of genomic prediction using both high-density array data and imputed whole-genome sequencing data. More comparisons were performed based on GBLUP with different genomic relationship matrices to account for the genetic architecture of the three traits analyzed: eggshell strength, feed intake, and laying rate. Our results show that little or no benefit was gained when using all imputed WGS data compared to HD array data with different weighting approaches in the GBLUP model. However, our results suggest that using genic SNPs for genomic prediction has the potential to improve the predictive ability both with HD and WGS data. Overall, the same candidates tend to be selected from a full-sib family of interest regardless of the genotype data and weighting factors used.

Additional files

Additional file 1: Table S1. Number of individuals in each generation.

Additional file 2: Figure S1. Imputation accuracy (Rsq of Minimac3) in each minor allele frequency (MAF) interval.

Additional file 3: Figure S2. Manhattan plot of absolute estimated SNP effects for traits FI and LR based on high-density (HD) array data and whole-genome sequence (WGS) data, respectively.

Additional file 4: Figure S3. Manhattan plots of $-(\log_{10}P)$ for the three traits based on high density array data (panels 1–3) and the whole-genome sequence (WGS) data (panels 4–6). Significance among principal components (PC) was tested in advance with a Tracy Widom test and PC with P values less than 0.05 were used as fixed covariates in single-SNP GWAS runs.

Additional file 5: Table S2. The optimal parameter in the training stage of BLUP|GA based on HD array data for each fold of fivefold cross-validation in each replicate.

Additional file 6: Table S3. The optimal parameter in the training stage of BLUP|GA based on WGS data for each fold of fivefold cross-validation in each replicate.

Additional file 7: Figure S4. Regression coefficient of DGV on genomic prediction using different weighting factors based on high-density array data and whole-genome sequencing data.

Additional file 8: Figure S5. Predictive ability in a full-sib family with 12 individuals for feed intake based on high-density (HD) array data (top) and whole-genome sequence (WGS) data (bottom) of one replicate. In each plot matrix, the diagonal shows the histograms of DRP and DGV obtained with various **G** matrices. The upper triangle shows the Spearman's rank correlation between DGV with different **G** matrices and DRP. The lower triangle shows the scatter plot of DGV with different **G** matrices and DRP.

Additional file 9: Figure S6. Predictive ability in a full-sib family with 12 individuals for laying rate based on high-density (HD) array data (top) and whole-genome sequence (WGS) data (bottom) of one replicate. In each plot matrix, the diagonal shows the histograms of DRP and DGV obtained with various **G** matrices. The upper triangle shows the Spearman's rank correlation between DGV with different **G** matrices and DRP. The lower triangle shows the scatter plot of DGV with different **G** matrices and DRP.

Authors' contributions

GN, HS and ME participated in the design of this study, GN carried out the data analysis, and GN, HS and ME drafted and edited the manuscript. DC contributed pedigree information and materials. AF and ME participated in the data analysis. All authors read and approved the final manuscript.

Author details
[1] Animal Breeding and Genetics Group, Georg-August-Universität, Göttingen, Germany. [2] Lohmann Tierzucht GmbH, Cuxhaven, Germany. [3] Institute for Animal Breeding, Bavarian State Research Centre for Agriculture, Grub, Germany.

Acknowledgements
We thank Annett and Steffen Weigend (Friedrich-Loeffler-Institute Mariensee) for the management and DNA extraction of the samples. We also thank Tim M. Strom (Helmholtz Zentrum München) for sequencing and Hubert Pausch (Technische Universität München) for genotyping. We are indebted to those researchers who programmed the software used in this study. This research was funded by the German Federal Ministry of Education and Research (BMBF) within the AgroClustEr "Synbreed—Synergistic plant and animal breeding" (Grant ID 0315528C). GN personally thanks the China scholarship council for financial support.

Competing interests
The authors declare that they have no competing interests.

References
1. Meuwissen TH, Hayes BJ, Goddard ME. Prediction of total genetic value using genome-wide dense marker maps. Genetics. 2001;157:1819–29.
2. Hayes BJ, Bowman PJ, Chamberlain AJ, Goddard ME. Invited review: genomic selection in dairy cattle: progress and challenges. J Dairy Sci. 2009;92:433–43.
3. VanRaden PM, Van Tassell CP, Wiggans GR, Sonstegard TS, Schnabel RD, Taylor JF, et al. Invited review: reliability of genomic predictions for North American Holstein bulls. J Dairy Sci. 2009;92:16–24.
4. Daetwyler HD, Hickey J, Henshall JM, Dominik S, Gredler B, van der Werf JHJ, et al. Accuracy of estimated genomic breeding values for wool and meat traits in a multi-breed sheep population. Anim Prod Sci. 2010;50:1004–10.
5. Daetwyler HD, Calus MPL, Pong-Wong R, de los Campos G, Hickey JM. Genomic prediction in animals and plants: simulation of data, validation, reporting, and benchmarking. Genetics. 2013;193:347–65.
6. Georges M. Towards sequence-based genomic selection of cattle. Nat Genet. 2014;46:807–9.
7. Pérez-Enciso M, Rincón JC, Legarra A. Sequence- vs. chip-assisted genomic selection: accurate biological information is advised. Genet Sel Evol. 2015;47:43.
8. Ober U, Ayroles JF, Stone EA, Richards S, Zhu D, Gibbs R, et al. Using whole-genome sequence data to predict quantitative trait phenotypes in *Drosophila melanogaster*. PLoS Genet. 2012;8:e1002685.
9. Hayes BJ, MacLeod IM, Daetwyler HD, Bowman PJ, Chamberlian AJ, Vander Jagt CJ, et al. Genomic prediction from whole genome sequence in livestock: the 1000 bull genomes project. In: Proceedings of the 10th world congress on genetics applied to livestock production. Vancouver; 2014. 17–22 Aug 2014. https://asas.org/docs/default-source/wcgalp-proceedings-oral/183_paper_10441_manuscript_1644_0.pdf?sfvrsn=2.
10. van Binsbergen R, Calus MPL, Bink MCAM, van Eeuwijk FA, Schrooten C, Veerkamp RF. Genomic prediction using imputed whole-genome sequence data in Holstein Friesian cattle. Genet Sel Evol. 2015;47:71.
11. Brøndum RF, Su G, Janss L, Sahana G, Guldbrandtsen B, Boichard D, et al. Quantitative trait loci markers derived from whole genome sequence data increases the reliability of genomic prediction. J Dairy Sci. 2015;98:4107–16.
12. Morota G, Abdollahi-Arpanahi R, Kranis A, Gianola D. Genome-enabled prediction of quantitative traits in chickens using genomic annotation. BMC Genomics. 2014;15:109.
13. Abdollahi-Arpanahi R, Morota G, Valente BD, Kranis A, Rosa GJM, Gianola D. Assessment of bagging GBLUP for whole-genome prediction of broiler chicken traits. J Anim Breed Genet. 2015;132:218–28.
14. Heidaritabar M, Calus MPL, Megens HJ, Vereijken A, Groenen MAM, Bastiaansen JWM. Accuracy of genomic prediction using imputed whole-genome sequence data in white layers. J Anim Breed Genet. 2016;133:167–79.
15. de Los Campos G, Vazquez AI, Fernando R, Klimentidis YC, Sorensen D. Prediction of complex human traits using the genomic best linear unbiased predictor. PLoS Genet. 2013;9:e1003608.
16. Zhou L, Lund MS, Wang Y, Su G. Genomic predictions across Nordic Holstein and Nordic Red using the genomic best linear unbiased prediction model with different genomic relationship matrices. J Anim Breed Genet. 2014;131:249–57.
17. Zhang Z, Erbe M, He J, Ober U, Gao N, Zhang H, et al. Accuracy of whole-genome prediction using a genetic architecture-enhanced variance-covariance matrix. G3 (Bethesda). 2015;5:615–27.
18. Druet T, Macleod IM, Hayes BJ. Toward genomic prediction from whole-genome sequence data: impact of sequencing design on genotype imputation and accuracy of predictions. Heredity (Edinb). 2014;112:39–47.
19. Goddard ME, Hayes BJ. Genomic selection based on dense genotypes infererred from sparse genotypes. Proc Assoc Advmt Anim Breed Genet. 2009;18:26–9.

20. Li H, Durbin R. Fast and accurate short read alignment with Burrows–Wheeler transform. Bioinformatics. 2009;25:1754–60.

21. McKenna A, Hanna M, Banks E, Sivachenko A, Cibulskis K, Kernytsky A, et al. The genome analysis toolkit: a MapReduce framework for analyzing next-generation DNA sequencing data. Genome Res. 2010;20:1297–303.

22. Howie BN, Donnelly P, Marchini J. A flexible and accurate genotype imputation method for the next generation of genome-wide association studies. PLoS Genet. 2009;5:e1000529.

23. Browning SR, Browning BL. Rapid and accurate haplotype phasing and missing-data inference for whole-genome association studies by use of localized haplotype clustering. Am J Hum Genet. 2007;81:1084–97.

24. Ni G, Strom TM, Pausch H, Reimer C, Preisinger R, Simianer H, et al. Comparison among three variant callers and assessment of the accuracy of imputation from SNP array data to whole-genome sequence level in chicken. BMC Genomics. 2015;16:824.

25. Wang K, Li M, Hakonarson H. ANNOVAR: functional annotation of genetic variants from high-throughput sequencing data. Nucleic Acids Res. 2010;38:e164.

26. Curwen V, Eyras E, Andrews TD, Clarke L, Mongin E, Searle SMJ, et al. The Ensembl automatic gene annotation system. Genome Res. 2004;14:942–50.

27. Garrick DJ, Taylor JF, Fernando RL. Deregressing estimated breeding values and weighting information for genomic regression analyses. Genet Sel Evol. 2009;41:55.

28. Su G, Christensen OF, Janss L, Lund MS. Comparison of genomic predictions using genomic relationship matrices built with different weighting factors to account for locus-specific variances. J Dairy Sci. 2014;97:6547–59.

29. VanRaden PM. Efficient methods to compute genomic predictions. J Dairy Sci. 2008;91:4414–23.

30. Price AL, Patterson NJ, Plenge RM, Weinblatt ME, Shadick NA, Reich D. Principal components analysis corrects for stratification in genome-wide association studies. Nat Genet. 2006;38:904–9.

31. Gilmour AR, Gogel BJ, Cullis BR, Thompson R. ASReml User Guide 3.0. Hemel Hempstead: VSN International Ltd; 2009.

32. Eynard SE, Windig JJ, Leroy G, van Binsbergen R, Calus MP. The effect of rare alleles on estimated genomic relationships from whole genome sequence data. BMC Genet. 2015;16:24.

33. Fujimoto A, Nakagawa H, Hosono N, Nakano K, Abe T, Boroevich KA, et al. Whole-genome sequencing and comprehensive variant analysis of a Japanese individual using massively parallel sequencing. Nat Genet. 2010;42:931–6.

34. Muir WM, Wong GK, Zhang Y, Wang J, Groenen MAM, Crooijmans RPMA, et al. Genome-wide assessment of worldwide chicken SNP genetic diversity indicates significant absence of rare alleles in commercial breeds. Proc Natl Acad Sci USA. 2008;105:17312–7.

35. Calus MPL, Bouwman AC, Hickey JM, Veerkamp RF, Mulder HA. Evaluation of measures of correctness of genotype imputation in the context of genomic prediction: a review of livestock applications. Animal. 2014;8:1743–53.

36. Hickey JM, Crossa J, Babu R, de los Campos G. Factors affecting the accuracy of genotype imputation in populations from several maize breeding programs. Crop Sci. 2012;52:654.

37. Daetwyler HD, Capitan A, Pausch H, Stothard P, van Binsbergen R, Brøndum RF, et al. Whole-genome sequencing of 234 bulls facilitates mapping of monogenic and complex traits in cattle. Nat Genet. 2014;46:858–65.

38. Ma P, Brøndum RF, Zhang Q, Lund MS, Su G. Comparison of different methods for imputing genome-wide marker genotypes in Swedish and Finnish Red Cattle. J Dairy Sci. 2013;96:4666–77.

39. Deelen P, Menelaou A, van Leeuwen EM, Kanterakis A, van Dijk F, Medina-Gomez C, et al. Improved imputation quality of low-frequency and rare variants in European samples using the "Genome of The Netherlands". Eur J Hum Genet. 2014;22:1321–6.

40. Liu Q, Cirulli ET, Han Y, Yao S, Liu S, Zhu Q. Systematic assessment of imputation performance using the 1000 Genomes reference panels. Brief Bioinform. 2014;16:549–62.

41. Zheng HF, Rong JJ, Liu M, Han F, Zhang XW, Richards JB, et al. Performance of genotype imputation for low frequency and rare variants from the 1000 genomes. PLoS One. 2015;10:e0116487.

42. Grisart B, Coppieters W, Farnir F, Karim L, Ford C, Berzi P, et al. Positional candidate cloning of a QTL in dairy cattle: identification of a missense mutation in the bovine DGAT1 gene with major effect on milk yield and composition. Genome Res. 2002;12:222–31.

43. Thaller G, Kühn C, Winter A, Ewald G, Bellmann O, Wegner J, et al. DGAT1, a new positional and functional candidate gene for intramuscular fat deposition in cattle. Anim Genet. 2003;34:354–7.

44. Do DN, Janss LLG, Jensen J, Kadarmideen HN. SNP annotation-based whole genomic prediction and selection: an application to feed efficiency and its component traits in pigs. J Anim Sci. 2015;93:2056–63.

45. Koufariotis L, Chen YPP, Bolormaa S, Hayes BJ. Regulatory and coding genome regions are enriched for trait associated variants in dairy and beef cattle. BMC Genomics. 2014;15:436.

46. Chen L, Li C, Sargolzaei M, Schenkel F. Impact of genotype imputation on the performance of GBLUP and Bayesian methods for genomic prediction. PLoS One. 2014;9:e101544.

47. Segelke D, Chen J, Liu Z, Reinhardt F, Thaller G, Reents R. Reliability of genomic prediction for German Holsteins using imputed genotypes from low-density chips. J Dairy Sci. 2012;95:5403–11.

48. Mulder HA, Calus MPL, Druet T, Schrooten C. Imputation of genotypes with low-density chips and its effect on reliability of direct genomic values in Dutch Holstein cattle. J Dairy Sci. 2012;95:876–89.

49. Kutalik Z, Johnson T, Bochud M, Mooser V, Vollenweider P, Waeber G, et al. Methods for testing association between uncertain genotypes and quantitative traits. Biostatistics. 2011;12:1–17.

50. Wimmer V, Lehermeier C, Albrecht T, Auinger HJ, Wang Y, Schön CC. Genome-wide prediction of traits with different genetic architecture through efficient variable selection. Genetics. 2013;195:573–87.

51. Snelling WM, Cushman RA, Keele JW, Maltecca C, Thomas MG, Fortes MRS, et al. Networks and pathways to guide genomic selection. J Anim Sci. 2013;91:537–52.

52. MacLeod IM, Hayes BJ, Goddard ME. The effects of demography and long-term selection on the accuracy of genomic prediction with sequence data. Genetics. 2014;198:1671–84.

53. Su G, Guldbrandtsen B, Aamand GP, Strandén I, Lund MS. Genomic relationships based on X chromosome markers and accuracy of genomic predictions with and without X chromosome markers. Genet Sel Evol. 2014;46:47.

54. Redon R, Ishikawa S, Fitch KR, Feuk L, Perry GH, Andrews TD, et al. Global variation in copy number in the human genome. Nature. 2006;444:444–54.

55. McCarroll SA, Hadnott TN, Perry GH, Sabeti PC, Zody MC, Barrett JC, et al. Common deletion polymorphisms in the human genome. Nat Genet. 2006;38:86–92.

56. McQueen HA, Siriaco G, Bird AP. Chicken microchromosomes are hyperacetylated, early replicating, and gene rich. Genome Res. 1998;8:621–30.

57. International Chicken Genome Sequencing Consortium. Sequence and comparative analysis of the chicken genome provide unique perspectives on vertebrate evolution. Nature. 2004;432:695–716.

58. Kranis A, Gheyas AA, Boschiero C, Turner F, Yu L, Smith S, et al. Development of a high density 600 K SNP genotyping array for chicken. BMC Genomics. 2013;14:59.

7

Rapid evolution of genetic and phenotypic divergence in Atlantic salmon following the colonisation of two new branches of a watercourse

Arne Johan Jensen[1*] ⓘ, Lars Petter Hansen[2], Bjørn Ove Johnsen[1] and Sten Karlsson[1]

Abstract

Background: Selection acts strongly on individuals that colonise a habitat and have phenotypic traits that deviate from the local optima. Our objective was to investigate the evolutionary rates in Atlantic salmon (*Salmo salar*) in a river system (the Vefsna watershed in Norway), fewer than 15 generations after colonisation of two new branches of the watercourse for spawning, which were made available by construction of fish ladders in 1889.

Methods: Differences in age and size were analysed using scale samples collected by anglers. Age and size of recaptures from a tagging experiment were compared between the three branches. Furthermore, genetic analyses of scale samples collected in the three river branches during two periods were performed to evaluate whether observed differences evolved by genetic divergence over this short period, or were the result of phenotypic plasticity.

Results: We demonstrate that evolution can be rapid when fish populations are subjected to strong selection, in spite of sympatry with their ancestral group, no physical barriers to hybridisation, and natal homing as the only reproductive isolating barrier. After fewer than 15 generations, there was evidence of genetic isolation between the two branches based on genetic variation at 96 single nucleotide polymorphism loci, and significant differences in several life history traits, including size and age at maturity. Selection against large size at maturity appears to have occurred, since large individuals were reluctant to ascend the branch with less abundant water. The estimated evolutionary rate of change in life history traits is within the upper 3 to 7% reported in other fish studies on microevolutionary rates.

Conclusions: These findings suggest that with sufficient genetic diversity, Atlantic salmon can rapidly colonise and evolve to new accessible habitats. This has profound implications for conservation and restoration of populations and habitats in order to meet evolutionary challenges, including alterations in water regime, whether altered by climate change or anthropogenic factors.

Background

Selection acts strongly on individuals that colonise a habitat and have phenotypic traits that are far from the local optima. Different environments may directly induce differences in the behaviour, morphology, and physiology of an individual [1, 2]. In most cases of natural colonisation, information on the source population and time since colonisation is unavailable. However, exotic species may be useful for observing evolutionary processes in real time [3, 4]. Rates of change have been measured for many such populations [5]; however, these metrics have been recorded often without examining whether this change is based entirely upon phenotypic plasticity, or if it involves genetic divergence [6, 7].

The introduction of salmonid fishes to new locations has provided the opportunity to study the rates and patterns of evolution [6, 8, 9]. Anadromous salmonids are known for their long feeding migrations at sea, where they

*Correspondence: arne.jensen@nina.no
[1] Norwegian Institute for Nature Research (NINA), P.O. Box 5685, 7485 Sluppen, Trondheim, Norway
Full list of author information is available at the end of the article

gain most of their body mass, followed by returning, or homing, to their natal river to spawn. This is a general feature of Pacific salmon [10, 11] as well as Atlantic salmon [12, 13]. Homing is important for the evolution of local adaptive genetic characteristics in populations [14, 15]. Some straying may occur with the ability to colonise new habitats, and may—depending on the strength of selection—constrain local adaptation [16, 17]. Usually, individuals from donor populations have been stocked in new habitats isolated from their ancestors, and their rates have been estimated decades later, including chinook salmon (*Oncorhynchus tshawytscha*) in New Zealand [18–20], European grayling (*Thymallus thymallus*) in Norway [21–24], and sockeye salmon (*Oncorhynchus nerka*) in Lake Washington [8, 25]. Quantitative estimates are needed to determine how rapidly populations can evolve to meet new evolutionary challenges as a function of key population parameters [26]. Recent research has indicated that evolution can occur over relatively short (10 to 20 generations) periods of time [27]. For example, Hendry et al. [8] studied two adjacent sockeye salmon (*O. nerka*) populations in Lake Washington of common ancestry that colonised divergent reproductive environments (a river and a lake beach), and found evidence for the evolution of reproductive isolation after fewer than 13 generations.

In the current study, evolutionary rates for Atlantic salmon (*Salmo salar*) were investigated in a river system (the Vefsna watershed in Norway), fewer than 15 generations from the colonisation of two new branches of the watercourse for spawning, which were made available by construction of fish ladders in 1889. All Atlantic salmon in this river system occur in sympatry, with natal homing being the only barrier to genetic exchange between spawning areas. However, because the environmental conditions in these spawning areas differ considerably in terms of annual water flow and nutrients, we hypothesised that natural selection has been sufficient for the evolution of local genetic adaptation. Hence, life history, marine growth, size and age at maturity, and genetic divergence between Atlantic salmon from the two upper branches were compared to similar data from individuals captured in the main river further downstream (i.e., descendants of the ancestor population). Differences in age and size were analysed using scale samples collected by anglers between 1969 and 1981. Moreover, the age and size of recaptures in the three branches from a tagging experiment performed in 1979 were compared. Furthermore, genetic analyses of scale samples collected in the three river areas during two periods (1972 and 1979) were performed to evaluate whether observed differences evolved as the consequence of genetic divergence during this short period, or were entirely due to phenotypic plasticity.

Methods

Study site

The Vefsna watershed is located in Norway in the northern reach of the Boreal Uplands [28]. The river mouth is located in the innermost part of the Vefsnfjorden fjord at 65°51′N, 13°11′E. The catchment area is 4231 km^2 in size, and the mean annual water discharge at the outlet to the sea is 181 m^3 s^{-1}. Two main branches (the Austervefsna and Svenningdalselva rivers) meet at Trofors, 42 km from the sea. Downstream of this confluence, the name of the river changes to Vefsna. Starting at the Swedish border, the Austervefsna River flows mainly westwards to Trofors, where it merges with Svenningdalselva from the south, from where the name changes to Vefsna, and flows northwards to the sea at the city of Mosjøen (Fig. 1). Svenningdalselva has less water than Austervefsna, with mean annual discharges of 35 and 98 m^3 s^{-1}, respectively. The watercourse is rather steep, with several waterfalls, and an average gradient of 2.6 m km^{-1} [28]. The western part of the catchment (the Svenningdalen valley) consists of strongly transformed bedrocks from the Cambro-Silurian period, while the eastern catchment (Austervefsna) has little transformed bedrock from the same period, and a wide limestone belt. This influences the water quality, with higher values of total hardness, calcium, alkalinity, pH, and conductivity reported [28]. Hence, Austervefsna is slightly more productive than Svenningdalselva.

The predominant fish species at the study site are Atlantic salmon and brown trout (*Salmo trutta*). The Vefsna watershed used to be one of the largest inhabited by Atlantic salmon in Norway. In the late 1970s, however, the non-native parasite *Gyrodactylus salaris* was introduced to the river, and caused a severe decline in the Atlantic salmon population [29, 30]. Although the river was treated with rotenone in 2011 and 2012 to exterminate the parasite, and a recovery program was initiated to rebuild the Atlantic salmon population, this study is based on data collected before the invasion of the parasite, i.e., mainly during the 1970s.

Atlantic salmon reproduce in fresh water, where their offspring usually stay for 1 to 5 years before they migrate, at a length of 10 to 20 cm, to sea to feed and gain most of their weight. After 1 to 4 years at sea, they mature and return to their natal river with high precision, and even to the section of the river where they were born, to spawn [11, 31, 32]. The size of an individual at spawning, and hence its reproductive success, is dependent on the duration of its stay at sea (sea age). After one winter at sea, Atlantic salmon are typically 1 to 3 kg (50 to 65 cm), compared to 7 to 20 kg (>90 cm) after 3 years at sea [33].

Originally in the Vefsna watercourse, Atlantic salmon could only ascend to the Laksforsen waterfall, about 29 km from the sea. In 1889, fish ladders were established

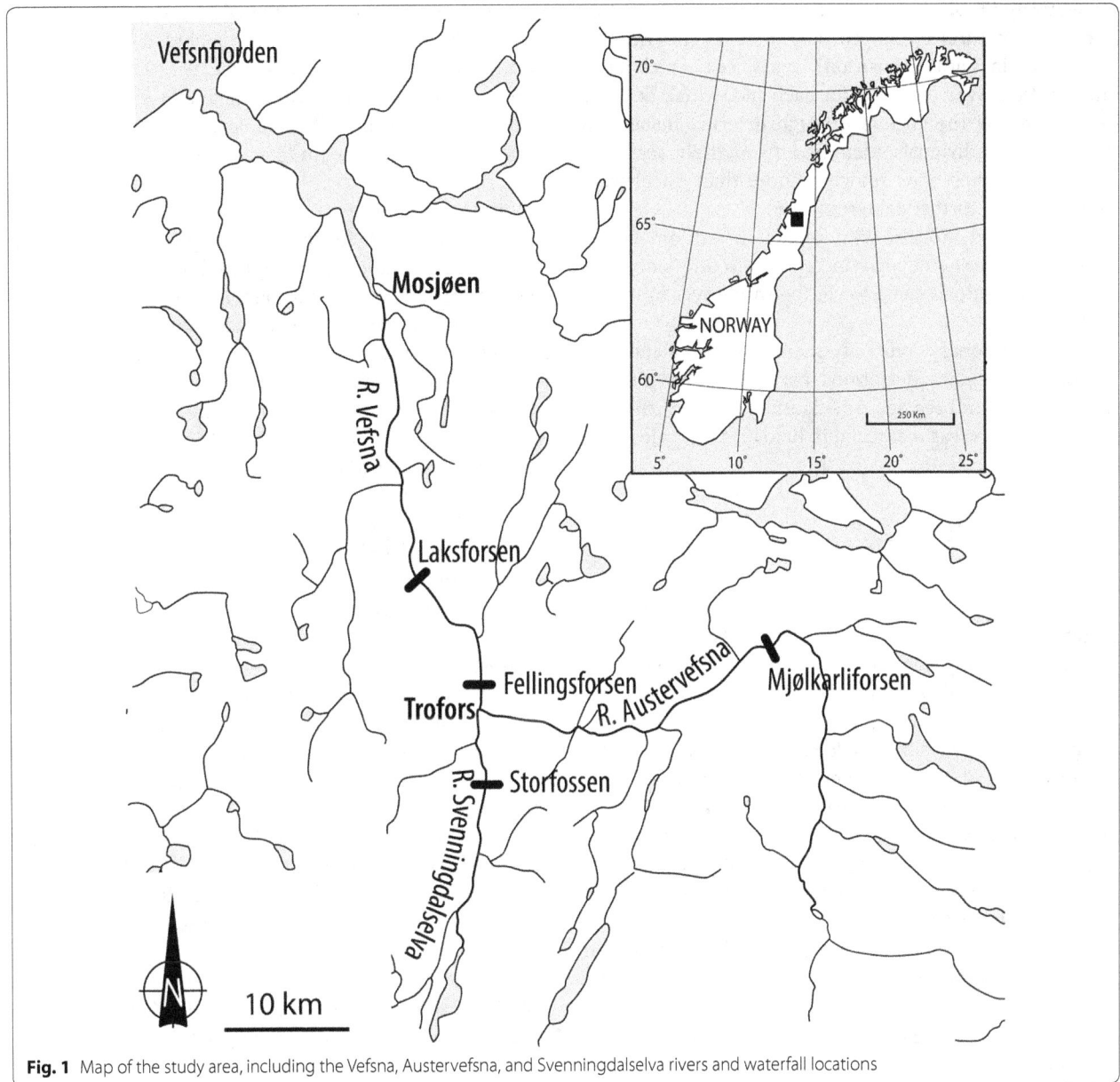

Fig. 1 Map of the study area, including the Vefsna, Austervefsna, and Svenningdalselva rivers and waterfall locations

in the 16-m high Laksforsen and in another waterfall (Fellingforsen, 5-m high) further upstream in Vefsna. This provided access to the upper parts of Vefsna, and a further 24 and 4 km upstream, the Austervefsna (to the Mjølkarliforsen waterfall) and Svenningdalselva (to the Storfossen waterfall) rivers, respectively [34]. In 1903, another 23 km of the Svenningdalselva River was opened for salmon, when a new fish ladder was constructed in the 7-m high Storfossen waterfall [34]. In 1922, fish ladders were built in four different waterfalls further upstream in the Austervefsna River, including Mjølkarliforsen, allowing salmonids to migrate a further 22 km upstream. In addition, particularly in the 1950s, old fish ladders were

improved and new ladders were constructed in Austervefsna [34]. Hence, today, 126 km of the watercourse up to 330 m above sea level is accessible to this species. All ladders are of the same type (pool and weir ladders), and all new stretches have been colonised naturally, i.e., without stocking of fish.

Age and growth
Age and growth were analysed from scale samples of adult Atlantic salmon collected by anglers during the period between 1969 and 1981. In total, 3089 samples were available, of which 2060, 595, and 434 were collected in Vefsna, Austervefsna, and Svenningdalselva,

respectively (Table 1; Additional file 1: Table S1). Information on natural tip length (i.e., total length with the caudal fin in natural position), mass, sex, catch date, and location were recorded for each individual. Sex was determined by the anglers through external inspection of the fish. Individuals identified from their scales as multiple spawners (i.e., spawned more than once) were excluded from further analyses (0.7%).

Linear mixed models, with year as a random factor, were used to test similarity in fish length and smolt age between individuals captured in the three river branches [35].

The evolutionary rate of change (i.e., phenotypic change in standard deviations per generation [36]) in length, mass, and sea age of Atlantic salmon in the two upper branches was measured in haldanes (h), calculated as:

$$h = \frac{(x_2/S_p) - (x_1/S_p)}{g},$$

where, x_1 is the mean natural log measurements from either the Svenningdalselva or Austervefsna, and x_2 is the corresponding measurement from the Vefsna River; g is the number of generations since the upper branches of the watercourse were opened to anadromous fish (i.e., since 1889), and S_p is the pooled standard deviation (SD) from the natural log measurements for both populations [6]. Data from Vefsna prior to establishing the first fish ladders in 1889 were not available; therefore, corresponding data obtained from samples collected in the part of Vefsna where anadromous fish were naturally distributed (i.e., downstream from the Laksforsen waterfall) between 1969 and 1981 were used as a proxy. This assumption may lead to some uncertainties regarding evolutionarily

Table 1 Number of scale samples from Atlantic salmon collected in the three river stretches, sorted by collection year

Year	Vefsna	Austervefsna	Svenningdalselva	Total
1969	5	0	40	45
1971	4	4	69	77
1972	148	52	50	250
1973	240	51	55	346
1974	52	187	126	365
1975	104	29	19	152
1977	305	0	0	305
1978	308	109	15	432
1979	375	63	46	484
1980	283	59	9	351
1981	236	41	5	282
Total	2060	595	434	3089

rates per se [6]. Some individuals from the two upper branches were likely to have been included in the catches in the lower part of the watercourse; hence, underestimation of mean length, mass, and sea-age of Atlantic salmon in the lower part of the Vefsna, and, by extension, evolutionary rates of change (in h), is probable.

Tagging experiment
In the period from 25 June to 20 September 1979, 1130 adult Atlantic salmon were captured in the fish ladder in the Laksforsen waterfall during their migration upstream, individually tagged with Lea tags, and released again. During tagging, the water temperature varied between 7 and 16 °C. Most individuals were tagged in July at temperatures between 11 and 16 °C. In a note included in each Lea tag, the angler was asked to report the length and mass of the fish, the day and exact locality of recapture, and a scale sample was requested (see Additional file 2: Table S2). Individuals recovered in Austervefsna and Svenningdalselva were assumed to have homed, whereas those recovered in Vefsna might also belong to the upper sections.

Genetic analyses
Overall, 285 individuals, 95 from each of Austervefsna, Svenningdalselva, and Vefsna, were assayed for genetic variation at 96 single nucleotide polymorphisms (SNPs). Half of the samples were collected in 1972 and the remainder were collected in 1979. Eighteen individuals had a genotyping success lower than 80% and were excluded from further analyses. The remaining 267 successfully genotyped individuals were included in analyses of temporal and spatial genetic variation (see Additional file 3: Table S3).

Total genomic DNA was extracted from dried scale samples using the DNeasy kit from Qiagen (Hombrechtikon, Switzerland). Ninety-six SNPs previously described by Bourret et al. [37] were genotyped with an EP1™ 96.96 Dynamic array IFCs (Fluidigm, San Francisco, CA, USA). Fifteen of the 96 SNPs were located within the mitochondrial genome [38].

Observed and expected heterozygosity levels at the 81 nuclear SNPs within individuals from the Austervefsna, Svenningdalselva, and Vefsna rivers were estimated using GENALEX 6.0 [39], and deviation from Hardy–Weinberg equilibrium was tested in Genepop v. 4.1.4 [40]. The 15 SNPs in the mitochondrial DNA (mtDNA) were compiled into haplotypes, and standard genetic indices, including number of haplotypes, haplotype diversity, and nucleotide diversity, were estimated in Arlequin version 3.5 [41].

Estimates of pairwise F_{ST} and tests for differences in allele frequencies between samples from Austervefsna,

Svenningdalselva, and Vefsna sampled in 1972 and 1979 were performed in Genepop v. 4.1.4 [40]. Genetic differences between sampling localities and sampling years were visualised in a principal coordinate analysis (PCoA) plot based on pairwise F_{ST} as implemented in GENALEX 6.0 [39]. Estimates of pairwise F_{ST} and tests for significance variance in the mitochondrial SNPs were performed in Arlequin [41]. To investigate possible migrants, individual genetic assignment to the three branches of the watershed, so called self-assignment, was conducted using the direct assignment approach and the Bayesian method [42], as implemented in GeneClass2 [43]. Relatedness coefficients between individuals were estimated using the Coancestry program [44], and the level of relatedness between individuals within Austervefsna was compared against that within Svenningdalselva using the same program.

Effective population size (N_e) was estimated using the linkage disequilibrium (LD) method [45, 46] implemented in LDNE [47] and the temporal method with Fs estimator implemented in the TempoFS software [48], applying sample plan 2.

Results

Age and growth

As for most Atlantic salmon populations containing multi-sea-winter (MSW) fish, sea age at maturity in the Vefsna watercourse differed between sexes, with a majority of males returning after one winter at sea (1SW), and females after two or three winters at sea (2SW and 3SW, respectively). Sea age at maturity differed; however, among the three river stretches, with the youngest individuals (both males and females) found in Svenningdalselva and the oldest in Vefsna (Fig. 2). The mean sea age ($\pm$95% confidence interval, CI) of males captured in Svenningdalselva was 1.10 ± 0.04 years, while the corresponding means in Austervefsna and Vefsna were 1.24 ± 0.06 and 2.10 ± 0.10 years, respectively. Females were usually older than males (one-way ANOVA; Svenningdalselva: $F_{1,429} = 59.7$, P < 0.001; Austervefsna: $F_{1,602} = 217.4$, P < 0.001; Vefsna: $F_{1,1991} = 806.9$, P < 0.001), with mean sea ages of 1.47 ± 0.10, 2.01 ± 0.08, and 2.55 ± 0.04 years in Svenningdalselva, Austervefsna, and Vefsna, respectively. The sex ratio differed significantly between Vefsna and the two other river stretches ($X^2 = 1563$, P < 0.001), since males predominated in Svenningdalselva and Austervefsna (62 and 57%, respectively), while females comprised the majority of salmon caught in Vefsna (53%).

The length of Atlantic salmon differed significantly between the three river branches (one-way ANOVA, $F_{2,3468} = 173.4$, P < 0.001), with the smallest individuals captured in Svenningdalselva (mean length 613.7 ± 10.7 mm, $n = 433$), intermediate-sized individuals were caught in

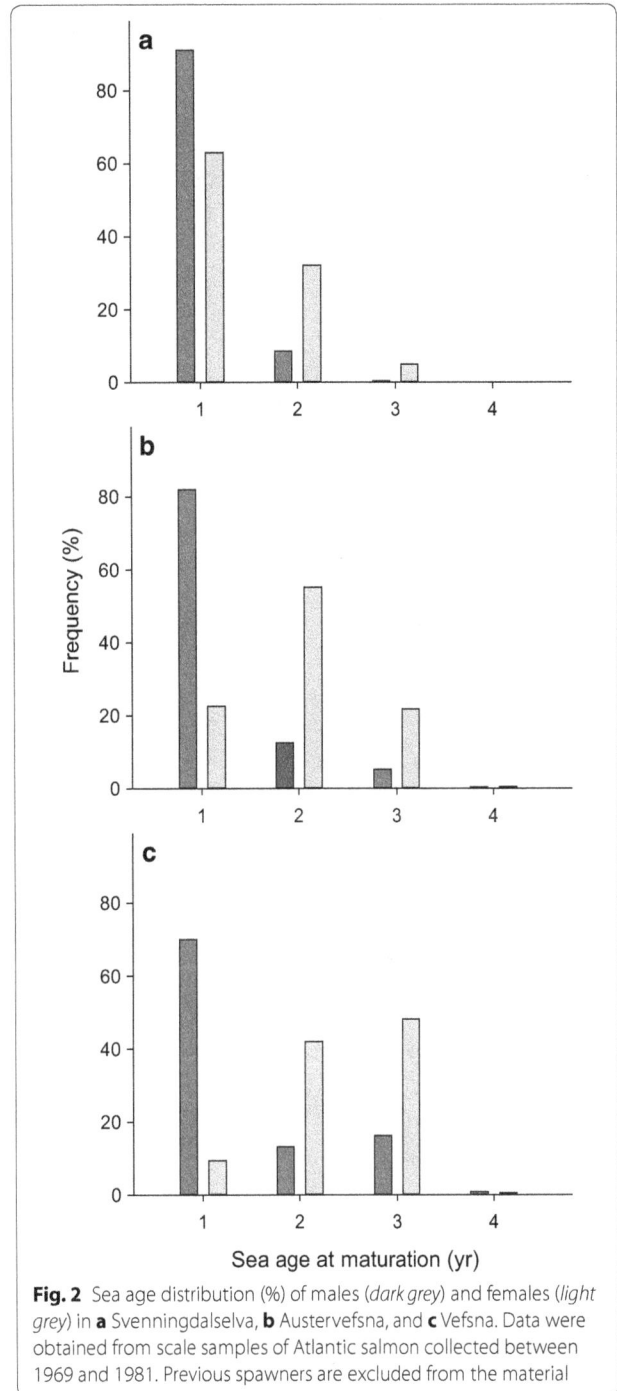

Fig. 2 Sea age distribution (%) of males (*dark grey*) and females (*light grey*) in **a** Svenningdalselva, **b** Austervefsna, and **c** Vefsna. Data were obtained from scale samples of Atlantic salmon collected between 1969 and 1981. Previous spawners are excluded from the material

Austervefsna (714.3 ± 12.6 mm, $n = 595$), and the largest individuals were caught in Vefsna (765.2 ± 7.7 mm, $n = 2046$), partly because of differences in sea age when they returned to the river. When grouping by sea age, the mean lengths of both 1SW and 2SW fish remained the smallest in Svenningdalselva, both among males and females (Table 2). Controlling for sex, sea age, and river branch, and with year as a random factor in a linear mixed model, individuals captured in Svenningdalselva were significantly smaller than in both Austervefsna (−25.5 ± 8 mm, t = 6.18, df = 157, P < 0.001) and Vefsna (−20.6 ± 8.8 mm, t = 4.61, df = 1413, P < 0.001).

The mean (±SD) age of adult Atlantic salmon when they returned to the watercourse as mature individuals, was 5.68 ± 1.04 years; they stayed 3.91 ± 0.65 years in fresh water (smolt age) until they migrated to sea, and remained 1.78 ± 0.83 years at sea (sea age at maturity) before they returned to the river to spawn. The mean length of adult fish after a stay at sea of 1, 2, and 3 years was 568 ± 46, 820 ± 67, and 947 ± 61 mm, respectively.

The predominating smolt age was 4 years (2- to 6-year range) in all three branches of the watercourse. The mean smolt age (±95% CI) was greater in Svenningdalselva (3.97 ± 0.06 years, n = 420) than in Austervefsna (3.88 ± 0.05 years, n = 580) and Vefsna (3.90 ± 0.03 years, n = 1838). Controlling for sex and river branch, and with year as a random factor in a linear mixed model, individuals captured in Svenningdalselva were significantly older when they migrated to sea as smolts than in both Austervefsna (0.10 ± 0.09 year, t = −2.13, df = 886, P < 0.05)

and Vefsna (0.20 ± 0.09 year, t = −4.39, df = 1659, P < 0.001).

Estimated evolutionary rates of change in the length, mass, and sea age of Atlantic salmon between 1889 and the 1970s, were higher in the Svenningdalselva population than in the Austervefsna population, and higher for females than for males, with values varying between −0.054 and −0.075 h for Austervefsna and −0.086 to −0.187 h for Svenningdalselva (Table 3).

Tagging experiment

Among the 315 recaptures reported in the same year that they were tagged, 82 were reported in Austervefsna, 44 in Svenningdalselva, and 117 in Vefsna, upstream of the tagging location (Fig. 3). The length distribution of individuals recaptured in Svenningdalselva differed from that of individuals captured from the two other river stretches (Fig. 4). In Svenningdalselva, only individuals less than 650-mm long were caught, while in the two other river branches, a significant portion of the catch was larger (between 800 and 1100 mm in length). Such patchy length distributions reflect different sea age groups, where individuals smaller than 700 mm from this watercourse stayed for only one winter at sea, while larger individuals were a mixture of 2SW, 3SW, and 4SW fish (Fig. 4). Thus, recaptures smaller than 700 mm were assumed to be 1SW fish, and these were significantly smaller in Svenningdalselva compared within Austervefsna [mean lengths 541.8 ± 10.5 (±95% CI) mm and 572.3 ± 9.8 mm in Svenningdalselva and Austervefsna, respectively (one-way ANOVA: $F_{1,95}$ = 18.09, P < 0.001)]. The mean length of 1SW fish recaptured in Vefsna upstream of the tagging location was 567.9 ± 7.3 mm, i.e., intermediate between that of the two other river stretches (significantly larger than that in Svenningdalselva [one-way ANOVA: $F_{1,147}$ = 15.43, P < 0.001], but not significantly different from that in Austervefsna [one-way ANOVA: $F_{1,156}$ = 0.49, P < 0.05]). No MSW fish from the tagging experiment was captured in Svenningdalselva, while 30 individuals (36%) captured in Austervefsna and 59 individuals (36%) captured in Vefsna were MSW fish.

Table 2 Mean length (mm ± 95% CI) of adult Atlantic salmon captured in three river stretches of the Vefsna watercourse between 1969 and 1981, maturing after one, two, and three winters at sea, respectively, and sorted between males and females

River	Vefsna	Austervefsna	Svenningdalselva
Males			
1SW	589.8 ± 9.0 (118)	592.1 ± 5.1 (261)	573.3 ± 5.9 (205)
2SW	848.3 ± 11.2 (97)	849.1 ± 29.5 (41)	777.4 ± 44.4 (23)
3SW	1003.5 ± 11.9 (138)	951.7 ± 41.0 (17)	NA
Females			
1SW	575.0 ± 25.3 (17)	575.0 ± 13.3 (52)	560.5 ± 9.1 (85)
2SW	809.3 ± 5.1 (345)	827.0 ± 11.7 (135)	794.2 ± 19.4 (52)
3SW	928.2 ± 4.4 (476)	952.7 ± 12.2 (54)	945.0 ± 34.3 (8)

Individuals identified as previous spawners were excluded. Sample sizes are in parentheses

1SW one sea winter, *2SW* two sea winters, *3SW* three sea winters, *NA* not available

Table 3 Evolutionary rate of change in haldanes (*h*) in length, mass, and sea age at maturity in male and female Atlantic salmon estimated from scale samples collected between 1969 and 1981

River branch	Length (*h*)		Mass (*h*)		Sea age (*h*)	
	Males	Females	Males	Females	Males	Females
Austervefsna	−0.060	−0.054	−0.075	−0.071	−0.075	−0.074
Svenningdalselva	−0.086	−0.161	−0.106	−0.187	−0.094	−0.162

Data obtained from scale samples collected in the lower part of Vefsna (where anadromous fish were naturally distributed, i.e., downstream of the Laksforsen waterfall) were used as a proxy for corresponding data prior to the first fish ladder being constructed in 1889

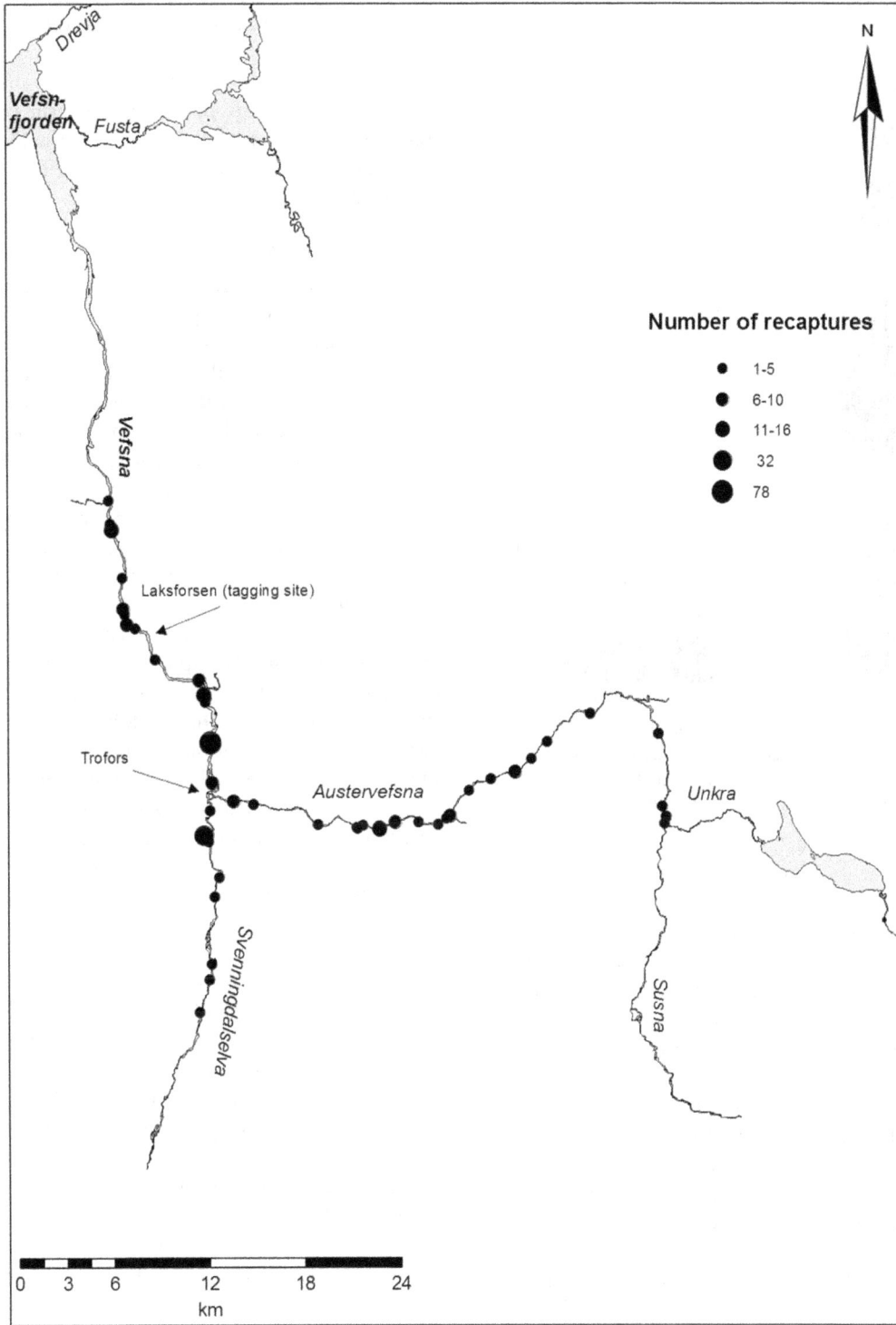

Fig. 3 Geographic distribution of adult Atlantic salmon recaptures within the Vefsna watercourse tagged in the Laksforsen waterfall during 1979

Genetic analyses

No significant deviations from Hardy–Weinberg equilibrium were detected within individuals from the Austervefsna, Svenningdalselva, and Vefsna rivers, and the average observed and expected heterozygosities in the three localities were 0.332 to 0.352 and

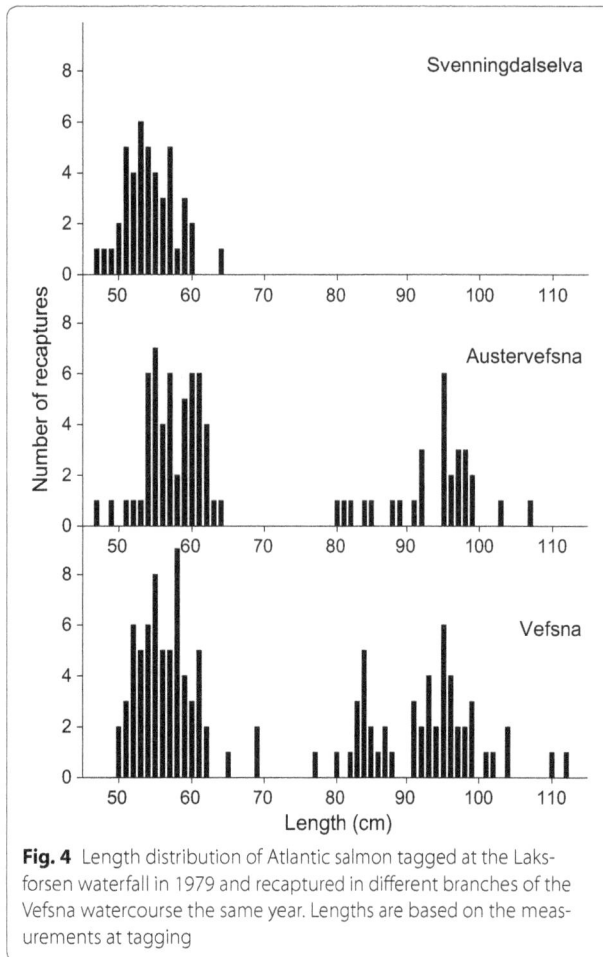

Fig. 4 Length distribution of Atlantic salmon tagged at the Laksforsen waterfall in 1979 and recaptured in different branches of the Vefsna watercourse the same year. Lengths are based on the measurements at tagging

0.332 to 0.339, respectively, with the lowest level in Svenningdalselva. Svenningdalselva also had fewer mitochondrial haplotypes (4), and the lowest levels of nucleotide diversity ($\pi = 0.102$) and haplotype diversity (hd = 0.446) compared to Vefsna (six haplotypes, $\pi = 0.144$, and hd = 0.167).

There were no significant genetic differences between sampling years within localities ($P > 0.305$). Based on 81 nuclear SNPs, a temporally stable genetic structure was observed (Fig. 5); the subpopulation from Svenningdalselva was significantly different (P < 0.001) from those from Austervefsna and Vefsna, with estimated F_{ST} values of 0.012 and 0.011, respectively. The subpopulation from Austervefsna was not significantly different from that from Vefsna ($F_{ST} = -0.001$, $P = 0.802$). Genetic variation at 15 mitochondrial SNPs revealed that fish collected at Svenningdalselva were significantly genetically distinct from those in the Vefsna ($F_{ST} = 0.038$, $P < 0.001$), but not from those in the Austervefsna ($F_{ST} = -0.008$, $P = 0.766$). In terms of the genetic assignment of individuals from the Austervefsna and Svenningdalselva rivers,

8.5% of the individuals from Austervefsna were assigned to Svenningdalselva and not to Austervefsna, while 12.7% of the individuals from Svenningdalselva were assigned to Austervefsna and not Svenningdalselva (Fig. 6). This suggests an exchange of individuals, and possible gene flow between Austervefsna and Svenningdalselva, with a higher rate of gene flow observed from Austervefsna to Svenningdalselva than from Svenningdalselva to Austervefsna.

According to the Wang estimator, average relatedness between individuals from Svenningdalselva was 0.017, and was significantly higher at the 2.5 percentile than that between individuals from Austervefsna (-0.010). This finding, combined with the lower genetic variation and significant genetic differentiation of fish in the Austervefsna and Vefsna branches, indicates the presence of a lower effective population size (N_e) in Svenningdalselva compared to Austervefsna and Vefsna.

Estimates of N_e using the temporal method were 71 in Svenningdalselva, with a 95% confidence interval (CI) from 31 to ∞, while no reasonable estimates were obtained for Austervefsna and Vefsna (negative values). A negative value indicates that most of the observed shift in allele frequency is explained by the sampling variance and not by the genetic drift, and that large sample sizes are needed [47, 49]. In addition, temporal samples were only one generation apart, which also lowers the precision of this method. The LD method for estimating current N_e gave reasonable estimates (with 95% CI) for Austervefsna 1972, 171 (94–656); Svenningdalselva 1979, 140 (79–472); Vefsna 1972, 249 (123–3210); and Vefsna 1979, 1931 (221–∞).

Discussion

In accordance with previous studies [5, 8, 50, 51], the findings reported here demonstrate that evolution can occur rapidly when fish populations encounter strong selection. In addition, this can occur even in sympatry with an ancestral population, isolated by natal homing only, and with no physical barriers to hybridisation. In the Vefsna watercourse, strong selection for smaller and younger individuals at maturity may be due to the lower water discharge in the two upper branches compared to the main river further downstream, and hence reluctance of the largest individuals to ascend these branches, particularly the branch that has less water (Svenningdalselva).

An alternative, albeit less likely, hypothesis to explain the smaller size of Atlantic salmon ascending Svenningdalselva compared with that from the other river stretches could be difficulties for MSW fish in crossing the fish ladder in Storfossen in Svenningdalselva. However, with a fall of 7 m, Storfossen is not considered more

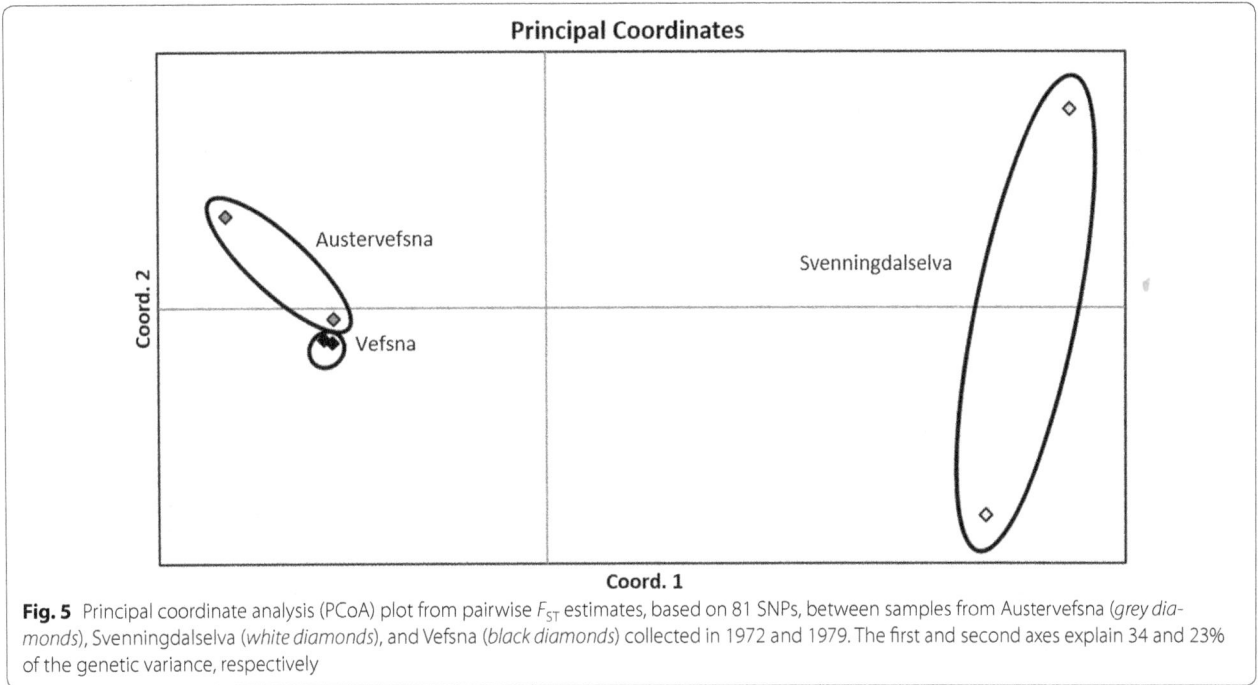

Fig. 5 Principal coordinate analysis (PCoA) plot from pairwise F_{ST} estimates, based on 81 SNPs, between samples from Austervefsna (*grey diamonds*), Svenningdalselva (*white diamonds*), and Vefsna (*black diamonds*) collected in 1972 and 1979. The first and second axes explain 34 and 23% of the genetic variance, respectively

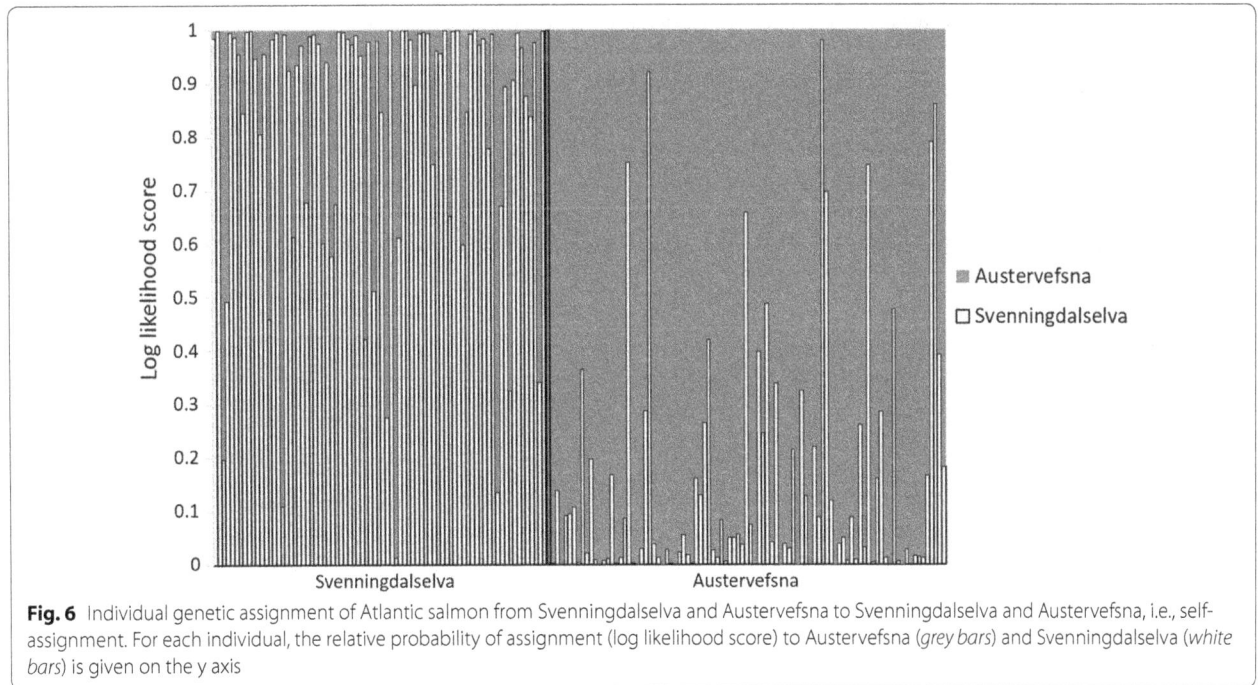

Fig. 6 Individual genetic assignment of Atlantic salmon from Svenningdalselva and Austervefsna to Svenningdalselva and Austervefsna, i.e., self-assignment. For each individual, the relative probability of assignment (log likelihood score) to Austervefsna (*grey bars*) and Svenningdalselva (*white bars*) is given on the y axis

difficult to cross than the other fish ladders in the watercourse. Furthermore, this fall is located 4 km upstream from the confluent with Austervefsna, and the lower 4 km stretch of Svenningdalselva, which was already open for Atlantic salmon in 1889, is as suited to produce young fish as the remainder of the watercourse.

Atlantic salmon captured in the two upper branches, which were both made available from 1889 following fish ladder construction in Vefsna, were a genetically distinct lineage with unique population demographics, including smolt age, and sea age and size at maturity. In both river branches, sea age and size at maturity were lower than those in the main river further downstream. Individuals captured in the Svenningdalselva section genetically diverged from those captured in the other two river branches; in addition, individuals of the same sea age were smaller. Both upper branches had been available for Atlantic salmon for a maximum of 83 years before the first genetic sample was collected (in 1972), which corresponds to fewer than 15 generations.

In this study, large individuals appeared reluctant to ascend Svenningdalselva, and some also hesitated to ascend Austervefsna due to the lower level of water compared to the downstream of their confluence. In Atlantic salmon, there is a large increase in mass due to each additional summer at sea, compared to the variance in mass among fish of the same sea age. Hence, this selection pressure also affects duration at sea and age at maturity. In general, small rivers contain smaller and younger Atlantic salmon than larger rivers [52]. In a study of Atlantic salmon in 18 Norwegian rivers, mean body length and mean sea age at maturity increased with increasing annual discharge of the spawning river until 40 $m^3 s^{-1}$, while at higher discharges, no such correlation was found [53]. Jonsson et al. [53] argued that selection in such rivers may act against large salmon because of low water flow, which increases the risk of successful ascent and breeding by large salmon, counteracting size advantages such as fecundity and competitive ability. This study supports the hypothesis of Jonsson et al. [53], who proposed that body size and sea age at maturity increase with increasing annual discharge. From both Austervefsna and Svenningdalselva, lower values were found for these life history traits compared with those from Vefsna, suggesting that Atlantic salmon with larger body sizes may also hesitate to ascend rivers at discharges higher than 40 $m^3 s^{-1}$.

Mean smolt age was greater in Svenningdalselva than in Austervefsna, demonstrating high precision homing to each river branch. Growth differences of juvenile parr—and hence, differences in smolt age—are expected to be caused by the slightly higher productivity in Austervefsna than in Svenningdalselva, because of differences in geology and hence water quality [28]. Water temperature may also have affected growth [54], although data on the temperature regimes of these branches are not available.

Atlantic salmon from Svenningdalselva have diverged from their source population in the lower part of Vefsna, presumably because of interplay between adaptation and ecologically-mediated reproductive isolation [55]. Because of the precise homing behaviour of salmonids to the river section where they are born, colonising habitats upstream of new fish ladders might be a slow process, depending mainly on strayers from downstream stretches or other rivers [17, 56]. After their first sea sojourn, offspring of these strayers are expected to ascend to the location of the river where they were born to spawn. To the upper stretches of the Vefsna watershed, most strayers are expected to belong to downstream sections of the Vefsna watershed, rather than to other rivers. This is because Vefsna housed the largest Atlantic salmon population within 200 km before the fish ladders were constructed, with reported annual catches in the 1880s of 1.5 to 4 tons [57].

The Atlantic salmon population in the Vefsna watercourse is a multi-sea-winter (MSW) population, with a high proportion of large individuals, such that approximately 35% of individuals captured in Vefsna (i.e., downstream Trofors) were longer than 900 mm, and had stayed three or more winters at sea, with a predominance of females. Results from both the tagging experiment and catches in different parts of the watercourse demonstrate that most of the large individuals (3SW or older) were captured in Vefsna. Furthermore, males comprised more of the population in Svenningdalselva and Austervefsna, whereas females predominated in Vefsna. These results suggest that large individuals (predominately females) born in the two upper sections hesitated to ascend these sections, and hence stayed in Vefsna for longer than their smaller conspecifics. In this way, the probability of being caught by anglers in Vefsna was higher than that for their smaller conspecifics, and may have resulted in a predominance of females in catches from Vefsna, and a predominance of males further upstream. This suggests that there is strong selection for smaller body size of fish in the two upper branches, especially in the section with the lowest water flow (Svenningdalselva). This hypothesis is supported by the differences observed between Svenningdalselva and Austervefsna in nuclear and mitochondrial SNPs. Significant differences are found between Svenningdalselva and Austervefsna in nuclear SNPs, but no such differences are observed in maternally inherited mtDNA, despite an expected higher evolutionary rate (lower effective population size) in mtDNA. This indicates a higher rate of gene flow in females than in males. The age and size at which an individual reproduces is an important life history trait. In Atlantic salmon, body size (and hence the number of years spent at sea before maturity) is important for reproductive success, and is more important for females than for males [58]. There is evidence for sex-specific selection patterns on sea age at maturity, as life-history strategies differ considerably between females and males [58]. A recent discovery that a single locus controls up to 39% of the variation in age at maturity in Atlantic salmon in a sex-specific manner [59] substantiates such rapid selection for adaptation in size-related traits under a changing environment.

The estimated evolutionary rates for life history traits in this study were in the upper range of values (for

Svenningdalselva the upper 3 to 7%) found in other fish studies reporting microevolutionary rates [5]. Furthermore, they were similar to the highest rates observed in fish in response to stream impoundment [60], but lower than those for exploited fish stocks [61, 62]. Genetic differences in this study appeared within fewer than 15 generations, which is similar to the 10 to 20 generations reported as the shortest time observed for other vertebrates [27].

Conclusions

The results of this study suggest that with sufficient genetic diversity, Atlantic salmon can rapidly colonise and evolve to new habitats if made accessible. This has profound implications for conservation and restoration of populations and habitats to meet evolutionary challenges, including alterations in water regime, whether in response to climate change or anthropogenic factors.

Additional files

Additional file 1: Table S1. Information on scale samples for length, mass, sex, smolt age, and sea age of adult Atlantic salmon captured in different sections of the Vefsna watershed during the period 1969 to 1981.

Additional file 2: Table S2. Recaptures of adult Atlantic salmon from a tagging experiment performed in Vefsna in 1979. Only recaptures upstream of the tagging site during the year of tagging are included.

Additional file 3: Table S3. Genotype data for 96 SNPs in Atlantic salmon from the Vefsna watershed.

Authors' contributions

AJJ and BOJ collected scale samples, LPH conducted the tagging experiment, and SK conducted the genetic analyses. AJJ and SK prepared the data and wrote the paper. All authors read and approved the final manuscript.

Author details

[1] Norwegian Institute for Nature Research (NINA), P.O. Box 5685, 7485 Sluppen, Trondheim, Norway. [2] Norwegian Institute for Nature Research (NINA), Gaustadalléen 21, 0349 Oslo, Norway.

Acknowledgements

The authors are grateful to J.G. Jensås and the late P.I. Møkkelgjerd for analysing the Atlantic salmon scales and to O. Diserud for providing advice on statistical analyses.

Competing interests

The authors declare that they have no competing interests.

References

1. Price TD, Qvarnström A, Irwin DE. The role of phenotypic plasticity in driving genetic evolution. Proc R Soc Lond B Biol Sci. 2003;270:1433–40.

2. Colautti RI, Lau JA. Contemporary evolution during invasion: evidence for differentiation, natural selection, and local adaptation. Mol Ecol. 2015;24:1999–2017.

3. Mathys BA, Lockwood JL. Contemporary morphological diversification of passerine birds introduced to the Hawaiian archipelago. Proc R Soc Lond B Biol Sci. 2011;278:2393–400.

4. Sax DF, Stachowicz JJ, Brown JH, Bruno JF, Dawson MN, Gaines SD, et al. Ecological and evolutionary insights from species invasions. Trends Ecol Evol. 2007;22:465–71.

5. Hendry AP, Farrugia TJ, Kinnison MT. Human influences on rates of phenotypic change in wild animal populations. Mol Ecol. 2008;17:20–9.

6. Hendry AP, Kinnison MT. Pespective: the pace of modern life: measuring rates of contemporary microevolution. Evolution. 1999;53:1637–53.

7. Lucek K, Sivasundar A, Seehausen O. Disentangling the role of phenotypic plasticity and genetic divergence in contemporary ecotype formation during a biological invasion. Evolution. 2014;68:2619–32.

8. Hendry AP, Wenburg JK, Berntzen P, Volk EC, Quinn TP. Rapid evolution of reproductive isolation in the wild: evidence from introduced salmon. Science. 2000;290:516–8.

9. Hutchings JA. Unintentional selection, unanticipated insights: introductions, stocking and the evolutionary ecology of fishes. J Fish Biol. 2014;85:1907–26.

10. Quinn TP, Fresh K. Homing and straying in Chinook salmon (Oncorhynchus tshawytscha) from Cowlitz River Hatchery, Washington. Can J Fish Aquat Sci. 1984;41:1078–82.

11. Quinn TP. A review of homing and straying of wild and hatchery-produced salmon. Fish Res. 1993;18:29–44.

12. Hansen LP, Jonsson N, Jonsson B. Oceanic migration in homing Atlantic salmon. Anim Behav. 1993;45:927–41.

13. Hansen LP, Quinn TP. The marine phase of the Atlantic salmon (Salmo salar) life cycle, with comparisons to Pacific salmon. Can J Fish Aquat Sci. 1998;55:104–18.

14. Taylor EB. A review of local adaptation in Salmonidae, with particular reference to Pacific and Atlantic salmon. Aquaculture. 1991;98:185–207.

15. Garcia de Leaniz C, Fleming IA, Einum S, Verspoor E, Jordan WC, Consuegra S, et al. A critical review of adaptive genetic variation in Atlantic salmon: implications for conservation. Biol Rev Camb Philos Soc. 2007;82:173–211.

16. Milner AM, Bailey RG. Salmonid colonization of new streams in Glacier Bay National Park, Alaska. Aquac Res. 1989;20:179–92.

17. Hendry AP, Castric V, Kinnison MT, Quinn TP. The evolution of philopatry and dispersal: homing versus straying in salmonids. In: Hendry AP, Stearns SC, editors. Evolution illuminated: Salmon and their relatives. Oxford: Oxford University Press; 2004. p. 52–91.

18. Kinnison MT, Unwin MJ, Hendry AP, Quinn TP. Migratory costs and the evolution of egg size and number in introduced and indegenous salmon populations. Evolution. 2001;55:1656–67.

19. Unwin MJ, Kinnison MT, Boustead NC, Quinn TP. Genetic control over survival in Pacific salmon (Oncorhynchus spp.): experimental evidence between and within populations of New Zealand chinook salmon (O. tshawytscha). Can J Fish Aquat Sci. 2003;60:1–11.

20. Quinn TP, Unwin MJ, Kinnison MT. Contemporary divergence in migratory timing of naturalized populations of chinook salmon, Oncorhynchus tshawytscha, in New Zealand. Evol Ecol Res. 2011;13:45–54.

21. Koskinen MT, Haugen TO, Primmer CR. Contemporary fisherian life-history evolution in small salmonid populations. Nature. 2002;419:826–30.

22. Haugen TO, Vøllestad LA. Population differences in early life-history traits in grayling. J Evol Biol. 2000;13:897–905.

23. Gregersen F, Haugen TO, Vøllestad LA. Contemporary egg size divergence among sympatric grayling demes with common ancestors. Ecol Freshw Fish. 2008;17:110–8.

24. Junge C, Vøllestad LA, Barson NJ, Haugen TO, Otero J, Sætre G-P, et al. Strong gene flow and lack of stable population structure in the face of rapid adaptation to local temperature in a spring-spawning salmonid, the European grayling (Thymallus thymallus). Heredity (Edinb). 2011;106:460–71.

25. Hendry AP. Adaptive divergence and the evolution of reproductive isolation in the wild: an empirical demonstration using introduced sockeye salmon. Genetica. 2001;112:515–34.

26. Stearns SC, Hendry AP. Introduction. The salmonid contribution to key issues in evolution. In: Stearns SC, Hendry AP, editors. Evolution

illuminated: Salmon and their relatives. Oxford: Oxford University Press; 2004. p. 3–19.

27. Hendry AP, Nosil P, Rieseberg LH. The speed of ecological speciation. Funct Ecol. 2007;21:455–64.

28. L'Abée-Lund JH, Haugland S, Melvold K, Saltveit SJ, Eie JA, Hvidsten NA, et al. Rivers of Boreal Uplands. In: Tockner K, Robinson CT, Uehlinger U, editors. Rivers of Europe. Amsterdam: Elsevier Ltd.; 2009. p. 577–606.

29. Johnsen BO, Jensen AJ. Introduction and establishment of *Gyrodactylus salaris* Malmberg, 1957, in Atlantic salmon, *Salmo salar* L., fry and parr in the River Vefsna, northern Norway. J Fish Dis. 1988;11:35–45.

30. Johnsen BO, Jensen AJ. The *Gyrodactylus* story in Norway. Aquaculture. 1991;98:289–302.

31. Heggberget TG, Lund RA, Ryman N, Ståhl G. Growth and genetic variation of Atlantic salmon (*Salmo salar*) from different sections of the River Alta, North Norway. Can J Fish Aquat Sci. 1986;43:1828–35.

32. Thorpe JE. Salmon migration. Sci Prog. 1988;72:345–70.

33. Hutchings JA, Jones MEB. Life history variation and growth rate thresholds for maturity in Atlantic salmon, *Salmo salar*. Can J Fish Aquat Sci. 1998;55(Suppl. 1):22–47.

34. Berg M. North Norwegian salmon rivers. Oslo: Johan Grundt Tanum Forlag; 1964. p. 299 **(in Norwegian, with English summary)**.

35. McCulloch CE, Searle SR. Generalized, linear, and mixed models. New York: Wiley; 2000.

36. Gingerich PD. Quantification and comparison of evolutionary rates. Am J Sci. 1993;293A:453–78.

37. Bourret V, Kent MP, Primmer CR, Vasemägi A, Karlsson S, Hindar K, et al. SNP-array reveals genome-wide patterns of geographical and potential adaptive divergence across the natural range of Atlantic salmon (*Salmo salar*). Mol Ecol. 2013;22:532–51.

38. Karlsson S, Moen T, Hindar K. Contrasting patterns of gene diversity between microsatellites and mitochondrial SNPs in farm and wild Atlantic salmon. Conserv Genet. 2010;11:571–82.

39. Peakall R, Smouse PE. GENALEX 6: genetic analysis in Excel. Population genetic software for teaching and research. Mol Ecol Notes. 2006;6:288–95.

40. Raymond M, Rousset F. Genepop (version 2.1): population genetics software for exact tests and ecumenicism. J Hered. 1995;86:248–9.

41. Schneider S, Roessli D, Exoffier L. ARLEQUIN, version 2.000: a software for population genetic data analysis. Geneva: Genetics and Biometry Laboratory, University of Geneva; 2000.

42. Rannala B, Mountain JL. Detecting immigration by using multilocus genotypes. Proc Natl Acad Sci USA. 1997;94:9197–201.

43. Piry S, Alapetite A, Cornuet JM, Paetkau D, Baudouin L, Estoup A. GeneClass2: a software for genetic assignment and first-generation migrant detection. J Hered. 2004;95:536–9.

44. Wang J. COANCESTRY: a program for simulating, estimating and analysing relatedness and inbreeding coefficients. Mol Ecol Res. 2011;11:141–5.

45. Wapes RS. A bias correction for estimates of effective population size based on linkage disequilibrium at unlinked gene loci. Conserv Genet. 2006;7:167–84.

46. Hill WG. Estimation of effective population size from data on linkage disequilibrium. Genet Res. 1981;38:209–16.

47. Waples RS, Do C. LDNE: a program for estimating effective population size from data on linkage disequilibrium. Mol Ecol Resour. 2008;8:753–6.

48. Jorde PE, Ryman N. Unbiased estimator for genetic drift and effective population size. Genetics. 2007;177:927–35.

49. Waples RS. A generalized approach for estimating effective population size from temporal changes in allele frequency. Genetics. 1989;121:379–91.

50. Endler JA. Natural selection on color patterns in *Poecilia reticulata*. Evolution. 1980;34:76–91.

51. Haugen TO. Growth and survival effects on maturation pattern in populations of grayling with recent common ancestors. Oikos. 2000;90:107–18.

52. Power G. Stock characteristics and catches of Atlantic salmon (*Salmo salar*) in Quebec, and Newfoundland and Labrador in relation to environmental variables. Can J Fish Aquat Sci. 1981;38:1601–11.

53. Jonsson N, Hansen LP, Jonsson B. Variation in age, size and repeat spawning of adult Atlantic salmon in relation to river discharge. J Anim Ecol. 1991;60:937–47.

54. Jonsson B, Forseth T, Jensen AJ, Næsje TF. Thermal performance of juvenile Atlantic salmon, *Salmo salar* L. Funct Ecol. 2001;15:701–11.

55. Kinnison MT, Hendry AP. From macro- to micro-evolution: tempo and mode in salmonid evolution. In: Stearns SC, Hendry AP, editors. Evolution illuminated Salmon and their relatives. Oxford: Oxford University Press; 2004. p. 208–31.

56. Pess GR, McHenry ML, Beechie TJ, Davies J. Biological impacts of the Elwha River Dams and potential salmonid responses to dam removal. Northwest Sci. 2008;82(sp1):72–90.

57. Dahl K, Dahl E. Norges lakseelver. Deres utbytte i tabeller og grafer (Norwegian salmon rivers. Their catches in tables and grafs). Oslo: Landbruksdepartementet, Fiskerikontoret; 1942.

58. Fleming IA, Einum S. Reproductive ecology: a tale of two sexes. In: Aas Ø, Einum S, Klemetsen A, Skurdal J, editors. Atlantic salmon ecology. Oxford: Blackwell; 2011. p. 33–65.

59. Barson NJ, Aykanat T, Hindar K, Baranski M, Bolstad GH, Fiske P, et al. Sex-dependent dominance at a single locus maintains variation in age at maturity in salmon. Nature. 2015;528:405–8.

60. Cureton JC II, Broughton RE. Rapid morphological divergence of a stream fish in response to changes in water flow. Biol Lett. 2014;10:20140352.

61. Devine JA, Wright PJ, Pardoe HE, Heino M. Comparing rates of contemporary evolution in life-history traits for exploited fish stocks. Can J Fish Aquat Sci. 2012;69:1105–20.

62. van Wijk SJ, Taylor MI, Creer S, Dreyer C, Rodrigues FM, Ramnarine IW, et al. Experimental harvesting of fish populations drives genetically based shifts in body size and maturation. Front Ecol Environ. 2013;11:181–7.

8

Multiple-trait structured antedependence model to study the relationship between litter size and birth weight in pigs and rabbits

Ingrid David[1]* iD, Hervé Garreau[1], Elodie Balmisse[2], Yvon Billon[3] and Laurianne Canario[1]

Abstract

Background: Some genetic studies need to take into account correlations between traits that are repeatedly measured over time. Multiple-trait random regression models are commonly used to analyze repeated traits but suffer from several major drawbacks. In the present study, we developed a multiple-trait extension of the structured antedependence model (SAD) to overcome this issue and validated its usefulness by modeling the association between litter size (LS) and average birth weight (ABW) over parities in pigs and rabbits.

Methods: The single-trait SAD model assumes that a random effect at time t_j can be explained by the previous values of the random effect (i.e. at previous times). The proposed multiple-trait extension of the SAD model consists in adding a cross-antedependence parameter to the single-trait SAD model. This model can be easily fitted using ASReml and the OWN Fortran program that we have developed. In comparison with the random regression model, we used our multiple-trait SAD model to analyze the LS and ABW of 4345 litters from 1817 Large White sows and 8706 litters from 2286 L-1777 does over a maximum of five successive parities.

Results: For both species, the multiple-trait SAD fitted the data better than the random regression model. The difference between AIC of the two models (AIC_random regression-AIC_SAD) were equal to 7 and 227 for pigs and rabbits, respectively. A similar pattern of heritability and correlation estimates was obtained for both species. Heritabilities were lower for LS (ranging from 0.09 to 0.29) than for ABW (ranging from 0.23 to 0.39). The general trend was a decrease of the genetic correlation for a given trait between more distant parities. Estimates of genetic correlations between LS and ABW were negative and ranged from −0.03 to −0.52 across parities. No correlation was observed between the permanent environmental effects, except between the permanent environmental effects of LS and ABW of the same parity, for which the estimate of the correlation was strongly negative (ranging from −0.57 to −0.67).

Conclusions: We demonstrated that application of our multiple-trait SAD model is feasible for studying several traits with repeated measurements and showed that it provided a better fit to the data than the random regression model.

Background

In genetic studies, many traits of interest are repeatedly measured over time, which gives rise to longitudinal data. The main issue when modeling such data is to account for the covariance structure of the repeated records with a limited number of parameters. A good approach to reduce the number of parameters that need to be estimated (compared with a multiple-trait model) is to use a repeatability model, which is often used because of its simplicity. However, this model assumes, too narrowly, that repeated records are expressions of the same genetic trait and that the phenotypic correlation between repeated measures is uniform. More flexible approaches have been proposed, such as random regression (RR) [1–3], character process (CP), and structured antedependence (SAD) models [4,

*Correspondence: ingrid.david@inra.fr
[1] GenPhySE, INRA, INPT, ENVT, Université de Toulouse, 31326 Castanet-Tolosan, France
Full list of author information is available at the end of the article

5]. The most commonly used approach is the RR model [6], although this approach suffers from various drawbacks, the main one being the well-known "border effect" problem [7]. CP and SAD models have been shown to fit the covariance structures better than RR models [4, 5, 8]. However, they are less often used in genetic analyses due to the lack of user-friendly software for SAD models and to the difficulty in accounting for nonstationary longitudinal data without inflation of the number of parameters or without appropriate software for CP models.

When the correlation between several longitudinal traits needs to be taken into account, modeling the covariance structure of repeated records becomes even more complicated. Once again, RR models are more often used than CP or SAD models because their extension to multivariate situations is straightforward. However, such extensions can require a large number of parameters and the drawbacks described for univariate RR models remain [9]. Multiple-trait extension of the CP model is more complex and was discussed in Jaffrézic et al. [10]. The same authors also proposed an extension of the SAD model to accommodate multiple-trait situations [9]. Nonetheless, the use of these multiple-trait CP and SAD models is still challenging because of the lack of user-friendly and readily available software. The application of the multiple-trait RR, CP and SAD approaches is relevant for a wide range of traits that are repeatedly measured over time, such as traits related to milk production (milk yield, milk composition, and somatic cell counts [11]), animal growth (body weight and feed intake) or reproduction traits (litter size and birth weight [12]).

Breeding programs in polytocous species aim at increasing the number of young weaned per female. To reach this goal, they have primarily focused on litter size [13]. However, in some selection programs, response to selection using a simple repeatability model has been low for two reasons: litter sizes at different parities were not considered as different traits [14, 15], and larger litter sizes were correlated with decreased survival of the young [16], probably because of a smaller weight and a reduced level of maturity at birth [17].

The objective of our study was to propose a new multiple-trait SAD model (freely available software) that can take correlations among several longitudinal traits into account. To illustrate the functionality of our model, we used it to analyze litter size and average birth weight in two species, pigs and rabbits.

Methods
Multiple-trait SAD models
Let $y_i(t_j)$ be the observation of animal i at time t_j. All linear mixed models used to study repeated measures of y_i over time can be decomposed as follows:

$$y_i(t_j) = \mu_i(t_j) + u_i(t_j) + p_i(t_j),$$

where $\mu_i(t_j)$ represents the fixed effects at time t_j, and $u_i(t_j)$ and $p_i(t_j)$ the genetic and pseudo-permanent animal effect random functions, with covariance functions $\mathbf{U}(t_j, t_{j'})$ and $\mathbf{P}(t_j, t_{j'})$, respectively. Note that this model does not include a residual term in order to help convergence and avoid identifiability problems between structured permanent and classical residual covariance matrices [18]. Thus, the residual variance was, by definition, included in the (co)variance matrix of the pseudo-permanent effects included in the model. In the single-trait situation, possible non-null covariance between random effects at different times [i.e. $\mathbf{U}(t_j, t_{j'})$ and $\mathbf{P}(t_j, t_{j'})$] are taken into account in the SAD approach by modeling the form of the random-effects functions. Specifically, it assumes that a random effect at time t_j can be explained by the previous random effects (i.e. at time $t_k, k < j$). For instance, for a given random effect $\mathbf{p}(t)$, the general form of the SAD model of order α is:

$$\mathbf{p}(t_j) = \sum_{s=1}^{\alpha} \theta_{sj} \mathbf{p}(t_{j-s}) + \mathbf{e}(t_j),$$

where θ_{sj} is the sth antedependence parameter for time t_j, and $\mathbf{e}(t_j)$ is a random normally distributed effect (error term) with mean 0 and innovation variance $\sigma_p^2(t_j)$. To reduce the number of parameters in the SAD model, θ_{sj} and $\sigma_p^2(t_j)$ are assumed to be continuous functions of time: $\theta_{sj} = \sum_{q=0}^{\beta_s} a_{sq} t_j^q$ for a function of degree β_s and $\sigma_p^2(t_j) = \exp\left(\sum_{q=0}^{\gamma} b_q t_j^q\right)$ for a function of degree γ. A single-trait SAD model is then defined by the order of the antedependence (α), the degree of the polynomial for each antedependence parameter (β_1 to β_α), and the degree of the polynomial for the innovation variance (γ) for each random effect. We will refer to single-trait SAD models as SAD $\alpha\beta_1 \ldots \beta_\alpha\gamma$ [8]. For instance SAD 111 stands for a SAD model with:

$$\mathbf{p}(t_j) = \theta_{1j} \mathbf{p}(t_{j-1}) + \mathbf{e}(t_j), \quad \theta_{1j} = a_{1,0} + a_{1,1} t_j$$
$$\text{and } \sigma_p^2(t_j) = \exp(b_0 + b_1 t_j).$$

We propose an extension of the single-trait SAD model to the multiple-trait situation by assuming that, in addition to the within-trait antedependence relationship, a random effect of one trait can be a function of the same random effect of the other traits considered. For two traits $\mathbf{y}_1$ and $\mathbf{y}_2$, the general form of the multiple-trait SAD model of order $\alpha, \alpha', \eta, \eta'$, for a given random effect $\mathbf{p}$, can be written as (for $j > \max(\alpha, \alpha', \eta, \eta')$):

$$\mathbf{p}_1(t_j) = \sum_{s=1}^{\alpha} \theta_{sj}\mathbf{p}_1(t_{j-s}) + \sum_{s=c}^{\eta} \delta_{sj}\mathbf{p}_2(t_{j-s}) + \mathbf{e}_1(t_j),$$

$$\mathbf{p}_2(t_j) = \sum_{s=1}^{\alpha'} \theta'_{sj}\mathbf{p}_2(t_{j-s}) + \sum_{s=c'}^{\eta'} \delta'_{sj}\mathbf{p}_1(t_{j-s}) + \mathbf{e}_2(t_j), \tag{1}$$

where θ_{sj} and θ'_{sj} are the sth antedependence parameters at time j for traits 1 and 2, respectively, and δ_{sj} and δ'_{sj} are the $(s-c+1)$th or $(s-c'+1)$th cross-antedependence parameters at time j for traits 1 and 2, respectively. Note that, in contrast to the antedependence relationship that starts at time t_{j-1}, the cross antedependence relationships show greater flexibility and start at time t_{j-c} ($t_{j-c'}$), with $c(c')$ greater or equal to 0. Here, $\mathbf{e}_1(t_j)$ and $\mathbf{e}_2(t_j)$ are normally distributed random effects with mean 0 and innovation variance $\sigma_{p1}^2(t_j)$ and $\sigma_{p2}^2(t_j)$, respectively. Error terms $\mathbf{e}_1$ and $\mathbf{e}_2$ are assumed to be independent, except if $c > 0$ and $c' > 0$ when a correlation between the two can be considered at time t_1. This constraint on the correlation between error terms ensures the identifiability of the parameters in the multiple-trait SAD model, as has been demonstrated for structural equation models (SEM). Indeed, the multiple-trait SAD model can be considered to be a specific kind of SEM [19] in which the structural parameters are functions of time. Consider, for the sake of simplicity, the simple case of no repetition per subject. A SEM with a recursive relationship between two traits $\mathbf{y}_1$ and $\mathbf{y}_2$ for animal i is:

SAD model. Following recommendations for identifiability in the SEM, identifiability in the SAD model is achieved by assuming independency between the error terms in Eq. (1) for both genetic and pseudo-permanent environmental effects. In the more general case of repeated measurements over time for both traits, if $c > \eta$ and $c' = 1$ (recursive cross-antedependence) for random effects, the SAD model at time t_1 is equivalent to a classical multiple-trait model (where parameters are identifiable even if random effects are correlated between traits). Correlation between the error terms of random effects at time t_1 is thus permitted in the SAD model without adversely affecting parameter identifiability. Although the previous considered a simple situation with two traits, the same SAD models for each random effect, and a recursive relationship, it is straightforward to extend the reasoning to more complicated SAD models.

As for the single-trait model, (cross-)antedependence parameters and innovation variances were assumed to be continuous functions of time. The multiple-trait SAD model is then defined for two traits by the order of the antedependence for each trait (α, α'), the starting points (c, c'), the order of the cross-antedependence $(\eta - c + 1, \eta' - c' + 1)$, the degree of the polynomial for each (cross-)antedependence parameter, and the degree of the polynomial for the innovation variance of each trait (γ, γ'), as well as an indicator of the presence of an initial correlation between $\mathbf{e}_1(t_1)$, $\mathbf{e}_2(t_1)$. An interesting computational property of this multiple-trait SAD model is that, as is the case for the single-trait model,

$$\begin{cases} y_{1i} = \mathbf{x}'_{1i}\boldsymbol{\beta}_1 + u_{1i} + \varepsilon_{1i} \\ y_{2i} = \lambda y_{1i} + \mathbf{x}'_{2i}\boldsymbol{\beta}_2 + u_{2i} + \varepsilon_{2i} \end{cases} \Leftrightarrow \begin{cases} y_{1i} = \mathbf{x}'_{1i}\boldsymbol{\beta}_1 + u_{1i} + \varepsilon_{1i} \\ y_{2i} = \lambda\mathbf{x}'_{1i}\boldsymbol{\beta}_1 + \mathbf{x}'_{2i}\boldsymbol{\beta}_2 + \lambda u_{1i} + u_{2i} + \lambda\varepsilon_{1i} + \varepsilon_{2i} \end{cases} \tag{2}$$

Rosa et al. [20] showed that parameter identifiability in Eq. (2) is possible by assuming independency between residuals $\boldsymbol{\varepsilon}_1$ and $\boldsymbol{\varepsilon}_2$. Varona et al. [12] proposed an extension of the SEM that allows for unequal recursive relationships between random terms:

$$\begin{cases} y_{1i} = \mathbf{x}'_{1i}\boldsymbol{\beta}_1 + u_{1i} + \varepsilon_{1i} \\ y_{2i} = \lambda\mathbf{x}'_{1i}\boldsymbol{\beta}_1 + \mathbf{x}'_{2i}\boldsymbol{\beta}_2 + \lambda_u u_{1i} + u_{2i} + \lambda_p\varepsilon_{1i} + \varepsilon_{2i} \end{cases} \tag{3}$$

Identifiability of the parameters in Eq. (3) is achieved by assuming independency between $\mathbf{u}_1$ and $\mathbf{u}_2$ and between $\boldsymbol{\varepsilon}_1$ and $\boldsymbol{\varepsilon}_2$ [12]. If we discard antedependency because no repetition occurs and focus on the cross-antedependency, model (3) [and (2)] is equivalent to a multiple-trait SAD model with $c > \eta$ (recursive but not simultaneous relationship), $c' = 0$ and $\eta' = 0$ and a polynomial cross-antedependence function of degree 0 for both the genetic and pseudo-permanent environmental effects. Terms $\mathbf{u}_1$, $\mathbf{u}_2$ and $\boldsymbol{\varepsilon}_1$, $\boldsymbol{\varepsilon}_2$ are equivalent to the error terms for the genetic and pseudo-permanent environmental effects of a multiple-trait

the inverse of the covariance matrix $\mathbf{P}$ can be easily calculated by the following Cholesky decomposition [21]: $\mathbf{P}^{-1} = \mathbf{L}'\mathbf{D}^{-1}\mathbf{L}$, where $\mathbf{D}$ is a diagonal matrix with innovation variances as components, $\mathbf{L}$ is a lower triangular matrix with 1s on the diagonal and the negatives of the (cross-)antedependence parameters and the initial correlation (ρ) between $\mathbf{e}_1$, $\mathbf{e}_2$ as diagonal entries. For instance, for two traits, three time points, all (cross-)antedependence of order 1, $c = c' = 1$ and an initial correlation ρ between $\mathbf{e}_1(t_1)$ and $\mathbf{e}_2(t_1)$, the $\mathbf{L}$ matrix for the vector $[\mathbf{p}_{y1}(1) \ \mathbf{p}_{y2}(1) \ \mathbf{p}_{y1}(2) \ \mathbf{p}_{y2}(2) \ \mathbf{p}_{y1}(3) \ \mathbf{p}_{y2}(3)]$ is:

$$\mathbf{L} = \begin{bmatrix} 1 & & & & \\ -\rho & 1 & & 0 & \\ -\theta_{12} & -\delta_{12} & 1 & & \\ -\delta'_{12} & -\theta'_{12} & 0 & 1 & \\ 0 & 0 & -\theta_{13} & -\delta_{13} & 1 \\ 0 & 0 & -\delta'_{13} & -\theta'_{13} & 0 & 1 \end{bmatrix}.$$

This multiple-trait SAD model can be easily fitted to data using ASReml [22], as well as the OWN Fortran program that we have developed (available online at https://zenodo.org/record/192036#.WEAYLdLhBaQ).

The orders of (cross-)antedependence and the degrees of the polynomial functions of time for (cross-)antedependence parameters and innovation variances can be selected by comparing nested models using the likelihood ratio test and by comparing non-nested models using the Bayesian or Akaike information criterion (AIC) [23, 24]. To reduce the number of models that need to be compared, we suggest selecting the specification (order and degree) of the antedependence and innovation variances by first using a single-trait SAD model and then the cross-antedependence specification by using a multiple-trait SAD model. For the single-trait SAD model step of the selection process, we suggest to start with a SAD 100, then test an increase of the degree of the function of time (γ) for the innovation variance (i.e. SAD 101), and finally test an increase of the degree of the function of time for the antedependence parameter (β_1) (i.e. SAD 111). This procedure is repeated until there is no additional significant improvement of the model. Then, an increase of the order of the dependence (α) can be tested, starting with a constant second order antedependence parameter. For instance, if the last model of order 1 selected is SAD 121, the next model tested is SAD 2201. The increase of the polynome of time for the second order antedependence parameter can then be tested (i.e. SAD 2211) and so on. Using this step-by-step selection procedure, all models are nested and can be compared using the likelihood ratio test.

Data application

The multiple-trait SAD model was applied to data from two polytocous species (pigs and rabbits) to study the relationship between the number of young born alive per litter (litter size, LS) and the average birth weight (ABW) per litter, calculated as the sum of the weight at birth of all the individuals of a litter divided by LS. The pig dataset included 4345 litters from 1817 Large White sows over a maximum of five successive farrowings. The mean LS was 12.1 ± 3.6. Piglets were weighed at birth in all litters, the mean ABW was 1514 ± 31 g. The rabbit dataset included 8706 litters from 2286 L-1777 does [25] over a maximum of five successive kindlings. Kittens were weighed at birth for 3490 litters. The mean LS and ABW were 9.5 ± 3.1 and 83 ± 14 g, respectively. The descriptive statistics of LS and ABW per parity are in Table 1.

Fixed effects included in the model were initially selected separately for each trait using a step-by-step descending procedure. To do this, simple models that did not take relationships between animals into account were applied to the data, nested models were then compared using the likelihood ratio test. The same within-species

Table 1 Litter size and average birth weight by parity in pigs and rabbits

Parity	Number of litters	LS (s.e.)	Number of litters	Average BW (s.e.)
Pigs				
1	1508	11.7 (3.4)	1508	1419 (29)
2	1134	12.0 (3.6)	1134	1565 (31)
3	861	12.6 (3.6)	861	1564 (30)
4	515	12.5 (3.5)	515	1566 (31)
5	327	12.4 (3.3)	327	1566 (31)
Rabbits				
1	2013	8.1 (2.7)	115	84 (17)
2	1868	9.5 (3.0)	955	85 (15)
3	1812	10.0 (3.2)	1174	83 (14)
4	1638	10.3 (3.2)	907	82 (13)
5	1375	10.0 (3.0)	339	82 (11)

fixed effects were included in the genetic multiple-trait SAD model for both traits: parity (five classes), proportion of females in the litter (covariate), and the combination of year and month of delivery (121 levels) for rabbits; and parity (five classes), season of farrowing (four classes), sire breed (seven classes) and sow weight when entering the farrowing unit (covariates) for pigs. It should be noted that the contemporary group effect (constant over parities) was included in the model as a random effect for pig data, in addition to the genetic and pseudo-permanent environmental effects.

ABW and LS were defined as traits of the sow/doe and analyzed using the previously described multiple-trait SAD model with successive time points at each farrowing/kindling. Data were also analyzed using a multiple-trait RR model in order to provide a comparison of the SAD multiple-trait model with the most currently used method. Selection of the degree of the Legendre polynomials for the permanent and genetic effects was performed for each trait by comparing nested single-trait RR models using the log likelihood ratio test. Then, multiple-trait RR models were applied to the data using the selected degree of polynomials for each trait. Goodness-of-fit of the SAD and RR models to the data were compared using the AIC.

Results

In rabbits, model selection for the antedependence relationship using single-trait analysis showed that the most appropriate SAD models were SAD 111 for the animal genetic effect on both traits (LS and ABW), and SAD 100 and SAD 111 for the pseudo-permanent environmental effect on LS and ABW, respectively. These antedependence characteristics were retained for the multiple-trait model. Using the multiple-trait SAD model, the same cross-antedependence characteristics were selected for the genetic and permanent environmental effects, that is a recursive relationship (i.e. $c > \eta$, in other words, LS random effects are not functions of ABW random effects), $c' = \eta' = 0$ (i.e. ABW random effects at parity j are functions of LS random effects at parity j) and a degree 2 for the cross-antedependence function of time, which resulted in 20 parameters to model $\mathbf{U}$ and $\mathbf{P}$. Let $LS_i(j)$ and $ABW_i(j)$ be the LS and ABW of dam i at parity j, the selected multiple-trait SAD model then is:

$$LS_i(j) = \mathbf{x}'_{ij}\boldsymbol{\beta}_{LS} + u_i(j) + p_{LSi}(j),$$

$$ABW_i(j) = \mathbf{x}'_{ij}\boldsymbol{\beta}_{ABW} + v_i(j) + p_{ABWi}(j), \text{ with}$$

$$u_i(j) = \theta_{u,1j}u_i(j-1) + e_{u,i}(j),$$

$$v_i(j) = \theta'_{v,1j}v_i(j-1) + \delta'_{v,0j}u_i(j) + e_{v,i}(j), \text{ and}$$

$$p_{LS,i}(j) = \theta_{p,1j}p_{LS,i}(j-1) + e_{pLS,i}(j),$$

$$p_{ABW,i}(j) = \theta'_{p,1j}p_{ABW,i}(j-1) + \delta'_{p,0j}p_{LS,i}(j) + e_{pABW,i}(j).$$

In pigs, the SAD 101 model was retained for the within-trait antedependency for the genetic and pseudo-permanent environmental effects for LS, and SAD 111 and SAD 101 for the genetic and pseudo-permanent environmental effects for ABW, respectively. Recursive cross-antedependence functions of order 1 with $c' = 0$, degree 0 and 1 were selected for the genetic and pseudo-permanent environmental effects, respectively. The random contemporary group effects of the two traits were independent.

After model selection, the single trait RR models selected in pigs included a Legendre polynomial of degree 1 for the genetic effects and a constant permanent effect over time for both traits. In rabbits, the best single trait RR model was of degree 1 for both the genetic and permanent environmental effects for both traits. Unfortunately, the multiple-trait extension of these models did not converge in rabbits. Thus, the multiple-trait RR model considered for rabbits consisted, as for pigs, of a polynomial function of degree 1 for genetic effects only.

In pigs, the AIC for the multiple-trait SAD and RR models were 66,319 and 66,326, respectively. For rabbits, the AIC were 70,684 and 70,911 for the multiple-trait SAD and RR models, respectively.

Heritability estimates with the SAD model for the two traits are in Table 2. Heritability estimates were moderate for all traits at all parities, with the exception of LS

Table 2 Heritability estimates obtained with the SAD model in pigs and rabbits

Parity	Pigs		Rabbits	
	Litter size	Average BW	Litter size	Average BW
1	0.21	0.29	0.09	0.23
2	0.25	0.33	0.19	0.26
3	0.24	0.35	0.27	0.32
4	0.22	0.33	0.29	0.37
5	0.19	0.30	0.24	0.39

at parity 1 in rabbits, which had a lower heritability estimate, i.e. 0.09. Heritability estimates for ABW were generally higher than for LS. In pigs, heritability estimates were quite stable across parities, ranging from 0.19 to 0.25 for LS and from 0.29 to 0.35 for ABW. In rabbits, heritability estimates tended to increase with parity for both traits, ranging from 0.09 to 0.29 for LS and from 0.23 to 0.39 for ABW. In pigs, heritability estimates obtained with the RR model tended to be slightly lower than those obtained with the SAD model, ranging from 0.18 to 0.28 for LS and from 0.22 to 0.31 for ABW. In rabbits, compared to the SAD model, heritability estimates obtained with the RR model were slightly lower for LS (ranging from 0.08 to 0.25) and higher for ABW (ranging from 0.24 to 0.42).

Genetic correlation matrices estimated with the multiple-trait SAD models are shown in Figs. 1 and 2 for pigs and rabbits, respectively. Estimates of the genetic correlation between LS tended to decrease as the distance between parities increased, the decrease being slightly more pronounced for pigs than for rabbits (0.84 versus 0.68 between parities 1 and 5). The same general pattern, i.e. a decrease of the estimated genetic correlation between more distant parities, was observed for ABW in rabbits. In pigs, estimates of the genetic correlations between ABW at different parities were high, the minimum being 0.96 between parities 1 and 5. Estimates of the genetic correlation between LS and ABW were negative in both species. In general, we observed an increase of the genetic antagonism between the two traits with the parity level for ABW regardless of the parity level for LS.

The genetic correlation matrix estimated with the RR model for pigs is in Fig. 3. The general pattern of these correlation estimates was similar to the one obtained with the SAD model for both species but with a more pronounced genetic antagonism between the two traits in pigs and less pronounced in rabbits (result not shown, the lowest correlation between traits in rabbits was equal to −0.34).

Correlation matrices for the pseudo-permanent environmental effects obtained with the multiple-trait SAD

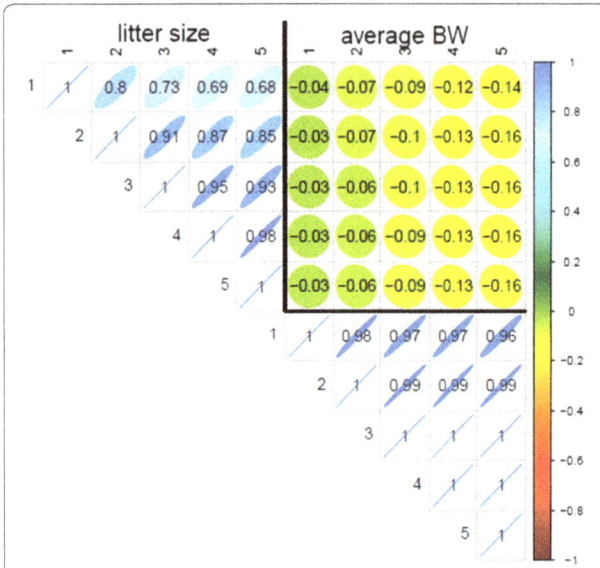

Fig. 1 Genetic correlations estimated with the SAD model in pigs

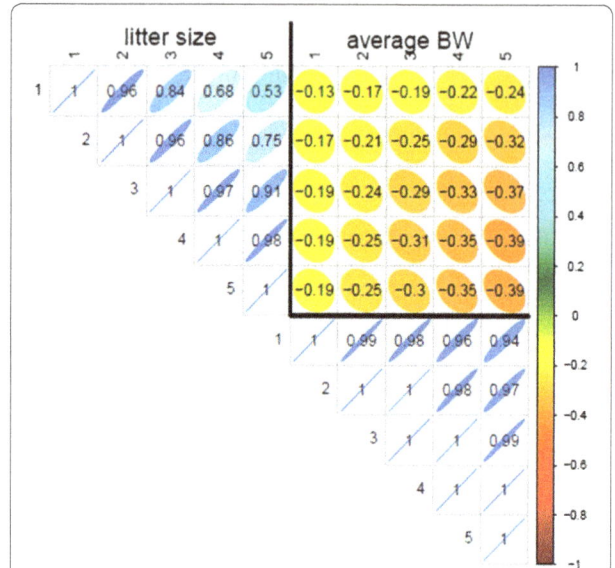

Fig. 3 Genetic correlations estimated with the RR model in pigs

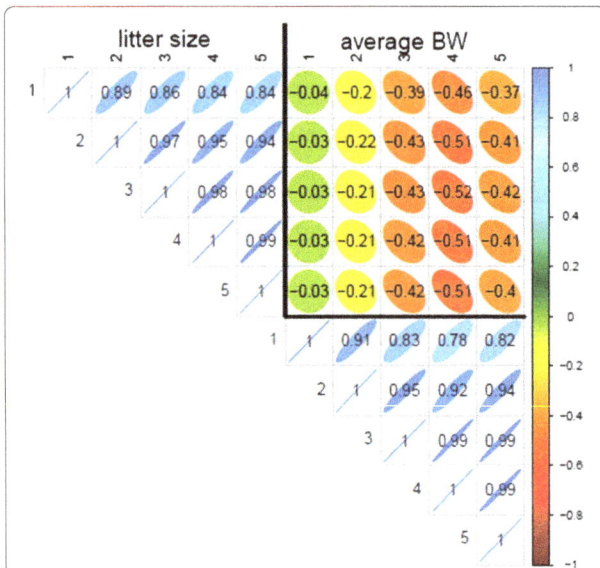

Fig. 2 Genetic correlations estimated with the SAD model in rabbits

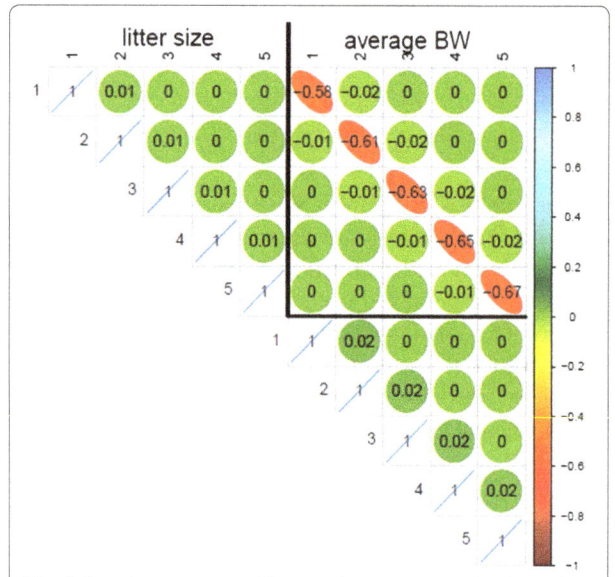

Fig. 4 Pseudo-permanent effect correlations estimated with the SAD model in pigs

model are in Figs. 4 and 5 for pigs and rabbits, respectively. Estimates of the correlations between pseudo-permanent environmental effects at different parities within-trait were close to 0. The same across-trait pattern of estimates of correlations between pseudo-permanent environmental effects was observed for both species: a high negative correlation between the pseudo-permanent environmental effects of the two traits for the same parity level (ranging from −0.57 to −0.67) and correlations close to 0 otherwise. The same general pattern of correlations was obtained with the RR model for pigs (Fig. 6) and rabbits (result not shown).

Discussion

We propose in this article a multiple-trait SAD model that takes the within-trait and cross-trait correlations over time into account. A first multiple-trait extension of the SAD model was proposed a few years back by Jaffrézic et al. [9], however our multiple-trait SAD model

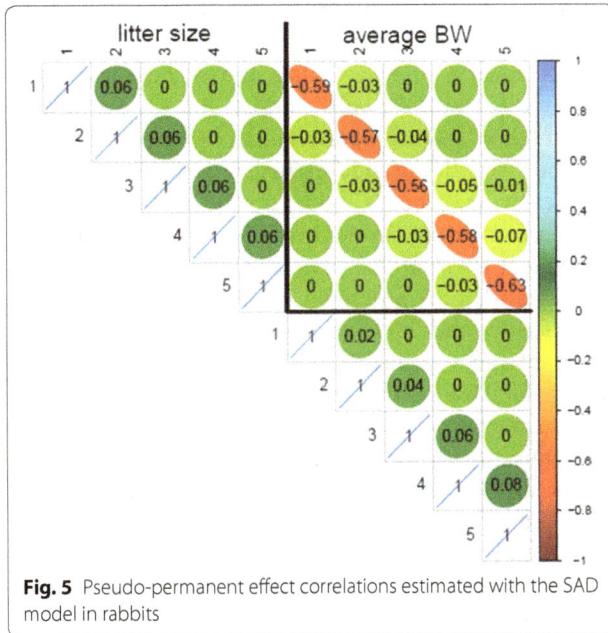

Fig. 5 Pseudo-permanent effect correlations estimated with the SAD model in rabbits

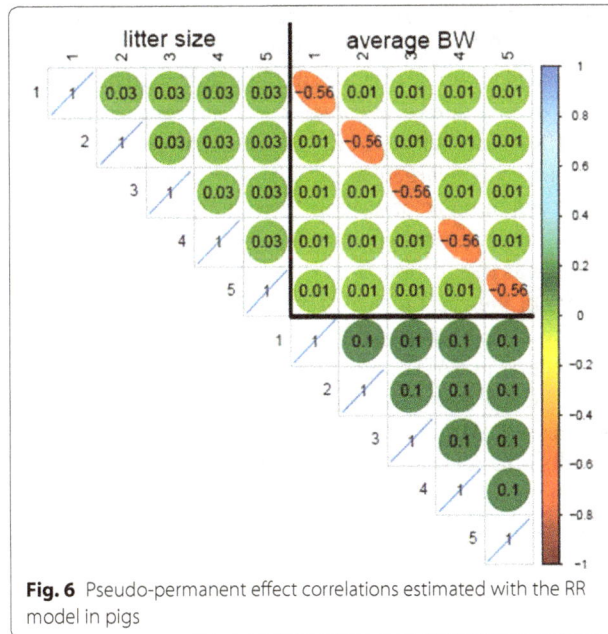

Fig. 6 Pseudo-permanent effect correlations estimated with the RR model in pigs

functions: within-trait selection of the antedependence parameters using a single-trait SAD model, followed by selection of the cross-antedependence parameters using the multiple-trait SAD model. This same two-step approach was performed by Jaffrézic et al. [9]. In the multiple-trait SAD model, selection of the order of cross-antedependence may differ according to the goal of the analysis. If the goal is to estimate the correlation between traits, our experience showed that a recursive cross-antedependence with zero as starting point and order 1 (i.e. $c' = \eta' = 0$) is generally sufficient to model all forms of correlation across traits. In such cases, only the degree of the cross-antedependence function needs to be selected by comparing nested models using the likelihood ratio test. Conversely, if the goal of the analysis is to study causal relationships between traits, it seems necessary to define, a priori, the general pathway between traits, as in SEM (resursiveness, simultaneity), to fit the order of the cross-antedependence in the multiple-trait SAD model to match the pathway. Then, the degree of the polynomial function for each cross-antedependence parameter is selected by comparing nested models using the likelihood ratio test. In such situations, antedependence parameter values are of interest in addition to the **U** and **P** covariance matrices.

Similar to a RR model, the number of breeding values predicted per animal with a SAD model equals the number of time points. Eigen decomposition (eigenvalues, eigenvectors) of the genetic covariance matrix obtained with the SAD model can help to determine which linear combination of breeding values can be used for genetic selection purposes. Nonetheless, given the way the SAD covariance matrix is structured, a decomposition in eigenfunctions (continuous function of time) similar to that proposed for RR models [26] is certainly not feasible.

We applied the multiple-trait SAD model to study the correlation between LS and ABW over parities in two species. For this purpose, we used a recursive cross-antedependence relationship. To study the causal relationship between these two traits with a SEM, Varona et al. [12] proposed a one-way causal path that established an effect of LS on ABW within parity as the most likely general pathway between LS and ABW. If our goal was to study the causal relationship between traits, similar to Varona et al. [12]. we would have needed to add cross-antedependence parameters to the SAD model that links the random parameters of ABW at time t to those of LS at time $t + 1$. As in previous studies, we considered LS and ABW as traits of the dam [12, 13].

Heritability estimates for LS were higher than the values of about 0.10 that were reported in previous studies [27–31] but were consistent with the total heritability for LS reported for the same pig breed by Kaufmann et al.

differs from theirs regarding the assumption made about the innovation covariance matrix. For their extension, Jaffrézic et al. [9] assumed that the error terms $\mathbf{e}_1(t)$ and $\mathbf{e}_2(t)$ are correlated as a function of time. To insure identifiability of the parameters, our assumption is that these error terms are independent, except at time t_1, when they can be correlated if $c > 0$ and $c' > 0$ in Eq. (1).

We suggest a two-step procedure for selecting the order of (cross-)antedependence and degree of the polynomial

(0.24) [32] and for rabbits by Nagy et al. (0.11–0.31) [33]. The higher heritabilities obtained in the current study can be explained by the fact that heritabilities estimated with a model that assumes different traits over time are generally higher than those obtained by using a simple repeatability model [34]. An increase of the heritability of LS with parity similar to that observed here was previously reported for both species [28–30], however it was not observed in pigs by Lukovic et al. [35] using records over 10 parities. Our heritability estimates for ABW were consistent with those reported by Hermesch in pigs (0.31) [13] but higher than those reported by Varona et al. in pigs (0.23) [12] and Bolet et al. [36] (0.04) or Garreau et al. [37] (0.06) in rabbits.

On the one hand, the decrease in the within-trait genetic correlation with distance between measurements (parities) is a result frequently reported in the literature [38, 39] and the genetic correlation matrix reported by Hanenberg et al. [28] for LS in Dutch Landrace pigs is close to the estimates obtained in our study. On the other hand, estimates of the genetic correlation for ABW reported for Australian pigs tended to be lower than obtained in the current study, but their standard errors were high [40]. It is generally recommended to consider LS and ABW at first parity as different traits from the performances at later parities [28, 40]. The same conclusion can be drawn from our results, except for ABW in pigs, which can be considered as a repeatable trait, the lowest genetic correlation value being 0.95 between parity 1 and 5.

Regarding cross-correlations, our results showed that the general trend was an unfavorable genetic correlation between LS and ABW (slightly negative in pigs and more strongly negative in rabbits). Several studies have reported the same negative genetic correlation between these traits in pigs [31, 40, 41]. Even if the trend was less clear in pigs, we observed the same cross-correlation tendency over parities: ABW tended to be more negatively correlated to LS in late parities than in first parity. In other words, if one considered that LS occurs before ABW and studies the relationship between successive traits (i.e. ABW at time $t - 1$ has an "effect" on LS at time t which has an "effect" on ABW at time t and so on), our results show that the antagonism between successive traits increases with parity. This could be considered as an increasing adaptation of traits to their environmental conditions with time, LS at time t being part of the environment for ABW at time t, which in turn is the environment to which the animal has to adapt its LS at time $t + 1$ and so on. Hermesch et al. [40] also studied cross-correlations over parities in pigs and did not find any trend, but their estimates had large standard errors.

The close to 0 correlations that we found between pseudo-permanent effects of the same trait were previously reported between first and second parities in rabbits [30] and pigs [40]. Consistent with our results, Hermesch et al. [40] reported a strong negative correlation between the pseudo-permanent effects of the two traits at the same parity and no correlation between traits at different parities. In our model, the so-called pseudo-permanent environmental effect combines the dam characteristics that are not under genetic control and persist over parities and the environmental factors that are not taken into account in the model because not measured/observable. The correlation between pseudo-permanent environmental effects at different parities was extremely low. Thus, the pseudo-permanent effects probably mainly reflect specific animal parity-related factors that are not taken into account in the model or environmental factors that affect traits differently during the reproductive period. Given the high negative correlation between the pseudo-permanent environmental effects between traits, these unobservable factors have opposite effects on LS and ABW.

We compared results obtained with the multiple-trait SAD and RR models. The latter is the approach most often used for analyzing longitudinal data in genetic studies. Probably due to the small number of repetitions, the best RR model selected in pigs and the multiple-trait RR model that converged in rabbits only considered a Legendre polynomial of degree 1 for genetic effects. Thus, with the same number of parameters in pigs (16), the RR model was less flexible than the multiple-trait SAD model to model the variance covariance matrix of environmental effects. Comparison of AIC values showed that the multiple-trait SAD model provided a better fit to the data than the RR model in both species. The same conclusion has been drawn in previous studies on other traits [8, 9]. Results obtained with RR models were essentially consistent with those of the multiple-trait SAD model, which enables us to be confident in the estimations obtained with the SAD approach.

Conclusions

In this paper, we have outlined a multiple-trait SAD model to simultaneously analyze repeated measurements of several traits. This flexible approach was developed to provide an advantageous approach to model the covariance matrix of random terms with few parameters. When used to study relationship between LS and ABW over five parities in two species, the multiple-trait SAD model showed that a total of 16 or 20 parameters was sufficient to model the random term covariance matrices. This is much less than the number of parameters required to obtain estimates when unstructured covariance matrices are assumed for **U** and **P** (up to 110 parameters). Furthermore, we showed that multiple-trait SAD models provide a better fit to the data than multiple-trait RR models. We offer a freely-available online Fortran program that can be used to implement this SAD model in ASReml.

Authors' contributions
YB and EB provided pig and rabbit data, respectively. HG and LC prepared the pig and rabbit data, respectively. They all took part in the interpretation of the results. ID developed the multiple-trait SAD model and wrote the Fortran program, performed data analysis and wrote the paper. All authors read and approved the final manuscript.

Author details
[1] GenPhySE, INRA, INPT, ENVT, Université de Toulouse, 31326 Castanet-Tolosan, France. [2] Pectoul, INRA, 31326 Castanet-Tolosan, France. [3] GenESI, INRA, 17700 Surgères, France.

Competing interests
The authors declare that they have no competing interests.

References
1. Schnyder U, Hofer A, Labroue F, Kunzi N. Genetic parameters of a random regression model for daily feed intake of performance tested French Landrace and Large White growing pigs. Genet Sel Evol. 2001;33:635–58.
2. Lorenzo Bermejo J, Roehe R, Schulze V, Rave G, Looft H, Kalm E. Random regression to model genetically the longitudinal data of daily feed intake in growing pigs. Livest Prod Sci. 2003;82:89–99.
3. Manzanilla Pech CIV, Veerkamp RF, Calus MPL, Zom R, van Knegsel A, Pryce JE, et al. Genetic parameters across lactation for feed intake, fat- and protein-corrected milk, and liveweight in first-parity Holstein cattle. J Dairy Sci. 2014;97:5851–62.
4. Jaffrézic F, Venot E, Laloe D, Vinet A, Renand G. Use of structured antedependence models for the genetic analysis of growth curves. J Anim Sci. 2004;82:3465–73.
5. Jaffrézic F, Pletcher SD. Statistical models for estimating the genetic basis of repeated measures and other function-valued traits. Genetics. 2000;156:913–22.
6. Schaeffer L. Application of random regression models in animal breeding. Livest Prod Sci. 2004;86:35–45.
7. Druet T, Jaffrézic F, Boichard D, Ducrocq V. Modeling lactation curves and estimation of genetic parameters for first lactation test-day records of french Holstein cows. J Dairy Sci. 2003;86:2480–90.
8. David I, Ruesche J, Drouilhet L, Garreau H, Gilbert H. Genetic modeling of feed intake. J Anim Sci. 2015;93:965–77.
9. Jaffrézic F, Thompson R, Hill WG. Structured antedependence models for genetic analysis of repeated measures on multiple quantitative traits. Genet Res. 2003;82:55–65.
10. Jaffrézic F, Thompson R, Pletcher SD. Multivariate character process models for the analysis of two or more correlated function-valued traits. Genetics. 2004;168:477–87.
11. Jamrozik J, Bohmanova J, Schaeffer L. Relationships between milk yield and somatic cell score in Canadian Holsteins from simultaneous and recursive random regression models. J Dairy Sci. 2010;93:1216–33.
12. Varona L, Sorensen D, Thompson R. Analysis of litter size and average litter weight in pigs using a recursive model. Genetics. 2007;177:1791–9.
13. Hermesch S. Avenues for genetic improvement of litter size and litter mortality. In: Proceedings of the AGBU Pig Genetics Workshop: 7–8 November 2001; Armidale; 2001. p. 36–43.
14. Irgang R, Fávero JA, Kennedy BW. Genetic parameters for litter size of different parities in Duroc, Landrace, and large white sows. J Anim Sci. 1994;72:2237–46.
15. Baselga M, Gomez E, Cifre P, Camacho J. Genetic diversity of litter size traits between parities in rabbits. J Appl Rabbit Res. 1992;15:198–205.
16. Johnson RK, Nielsen MK, Casey DS. Responses in ovulation rate, embryonal survival, and litter traits in swine to 14 generations of selection to increase litter size. J Anim Sci. 1999;77:541–57.
17. Canario L, Père M, Tribout T, Thomas F, David C, Gogué J, et al. Estimation of genetic trends from 1977 to 1998 of body composition and physiological state of Large White pigs at birth. Animal. 2007;1:1409–13.
18. Wang W. Identifiability of linear mixed effects models. Electron J Stat. 2013;7:244–63.
19. Gianola D, Sorensen D. Quantitative genetic models for describing simultaneous and recursive relationships between phenotypes. Genetics. 2004;167:1407–24.
20. Rosa GJM, Valente BD, de los Campos G, Wu X-L, Gianola D, Silva MA. Inferring causal phenotype networks using structural equation models. Genet Sel Evol. 2011;43:6.
21. Pourahmadi M. Joint mean-covariance models with applications to longitudinal data: unconstrained parameterisation. Biometrika. 1999;86:677–90.
22. Gilmour AR, Gogel BJ, Cullis BR, Thompson R, ASReml User guide Release 3.01, Hemel Hempstead: VSN International Ltd; 2009.
23. Schwarz GE. Estimating the dimension of a model. Ann Stat. 1978;6:461–4.
24. Akaike H. A new look at the statistical model identification. IEEE Trans Autom Control. 1974;19:716–23.
25. Garreau H, Duzert R, Tudela F, Baillot C, Ruesche J, Grauby G, et al. Gestion et sélection de la souche INRA 1777: résultats de trois générations de sélection. In: Proceedings of Les 11èmes Journées de la Recherche Cunicole: 29–30 November 2005; Paris; 2005. p. 19–22.
26. Van Der Werf JHJ, Goddard ME, Meyer K. The use of covariance functions and random regressions for genetic evaluation of milk production based on test day records. J Dairy Sci. 1998;81:3300–8.
27. Putz A, Tiezzi F, Maltecca C, Gray K, Knauer M. Variance component estimates for alternative litter size traits in swine. J Anim Sci. 2015;93:5153–63.
28. Hanenberg EHAT, Knol EF, Merks JWM. Estimates of genetic parameters for reproduction traits at different parities in Dutch Landrace pigs. Livest Prod Sci. 2001;69:179–86.
29. Roehe R, Kennedy BW. Estimation of genetic parameters for litter size in Canadian Yorkshire and Landrace swine with each parity of farrowing treated as a different trait. J Anim Sci. 1995;73:2959–70.
30. Piles M, García M, Rafel O, Ramon J, Baselga M. Genetics of litter size in three maternal lines of rabbits: repeatability versus multiple-trait models. J Anim Sci. 2006;84:2309–15.
31. Mérour I, Bernard E, Bidanel JP, Canario L. Genetic parameters for litter traits including farrowing duration and piglet survival up to weaning in the French Large White and Landrace sows. In: Proceedings of the 16th Annual Meeting of the European Association of Animal Production: 23–27 August 2010; Heraklios. 2010.
32. Kaufmann D, Hofer A, Bidanel JP, Künzi N. Genetic parameters for individual birth and weaning weight and for litter size of Large White pigs. J Anim Breed Genet. 2000;117:121–8.
33. Nagy I, Radnai I, Nagyné-Kiszlinger H, Farkas J, Szendrő Z. Genetic parameters and genetic trends of reproduction traits in synthetic Pannon rabbits using repeatability and multi-trait animal models. Arch Tierz. 2011;54:297–307.
34. Mekkawy W, Roehe R, Lewis R, Davies M, Bünger L, Simm G, et al. Comparison of repeatability and multiple trait threshold models for litter size in sheep using observed and simulated data in Bayesian analyses. J Anim Breed Genet. 2010;127:261–71.
35. Luković Z, Uremović M, Konjačić M, Uremović Z, Vincek D. Genetic parameters for litter size in pigs using a random regression model. Asian Aust J Anim Sci. 2007;20:160–5.
36. Bolet G, Zerrouki N, Gacem M, Brun JM, Lebas F. Genetic parameters and trends for litter and growth traits in a synthetic line of rabbits created in

Algeria. In: Proceedings of the 10th World Rabbit Congress: 3 6 August 2012; Sharm El-Sheikh. 2012.

37. Garreau H, Bolet G, Larzul C, Robert Granié C, Saleil G, San Cristobal M, et al. Results of four generations of a canalising selection for rabbit birth weight. Livest Sci. 2008;119:55–62.

38. Sawalha RM, Keown JF, Kachman SD, Van Vleck LD. Evaluation of autoregressive covariance structures for test-day records of Holstein cows: estimates of parameters. J Dairy Sci. 2005;88:2632–42.

39. Nunez-Anton V, Zimmerman DL. Modeling non-stationary longitudinal data. Biometrics. 2000;56:699–705.

40. Hermesch S, Luxford BG, Graser HU. Genetic parameters for lean meat yield, meat quality, reproduction and feed efficiency traits for Australian pigs-3. Genetic parameters for reproduction traits and genetic correlations with production, carcase and meat quality traits. Livest Prod Sci. 2000;65:261–70.

41. Huby M, Canario L, Tribout T, Caritez J, Billon Y, Gogué J, et al. Genetic correlations between litter size and weights, piglet weight variability and piglet survival from birth to weaning in Large White pigs. In: Proceedings of the 54th Annual Meeting of the European Association for Animal Production: 31 August–3 September 2003; Roma. 2003.

9

A hybrid method for the imputation of genomic data in livestock populations

Roberto Antolín[1], Carl Nettelblad[2], Gregor Gorjanc[1], Daniel Money[1] and John M. Hickey[1*]

Abstract

Background: This paper describes a combined heuristic and hidden Markov model (HMM) method to accurately impute missing genotypes in livestock datasets. Genomic selection in breeding programs requires high-density geno-typing of many individuals, making algorithms that economically generate this information crucial. There are two common classes of imputation methods, heuristic methods and probabilistic methods, the latter being largely based on hidden Markov models. Heuristic methods are robust, but fail to impute markers in regions where the thresholds of heuristic rules are not met, or the pedigree is inconsistent. Hidden Markov models are probabilistic methods which typically do not require specific family structures or pedigree information, making them very flexible, but they are computationally expensive and, in some cases, less accurate.

Results: We implemented a new hybrid imputation method that combined heuristic and HMM methods, AlphaImpute and MaCH, and compared the computation time and imputation accuracy of the three methods. AlphaImpute was the fastest, followed by the hybrid method and then the HMM. The computation time of the hybrid method and the HMM increased linearly with the number of iterations used in the hidden Markov model, however, the computation time of the hybrid method increased almost linearly and that of the HMM quadratically with the number of template haplotypes. The hybrid method was the most accurate imputation method for low-density panels when pedigree information was missing, especially if minor allele frequency was also low. The accuracy of the hybrid method and the HMM increased with the number of template haplotypes. The imputation accuracy of all three methods increased with the marker density of the low-density panels. Excluding the pedigree information reduced imputation accuracy for the hybrid method and AlphaImpute. Finally, the imputation accuracy of the three methods decreased with decreasing minor allele frequency.

Conclusions: The hybrid heuristic and probabilistic imputation method is able to impute all markers for all individuals in a population, as the HMM. The hybrid method is usually more accurate and never significantly less accurate than a purely heuristic method or a purely probabilistic method and is faster than a standard probabilistic method.

Background

This paper describes a combined heuristic and hidden Markov model (HMM) method to accurately impute missing genotypes in livestock datasets. Methods for imputing genotypes are essential for modern livestock breeding because they help to facilitate genomic selection, which has become the dominant method for genetic evaluation of livestock. Imputation can cost-effectively generate the high-density genotypes of many individuals required for genomic selection [1, 2]. Typically, the genotyping strategies used in livestock breeding involve genotyping a small number of individuals with expensive high-density marker panels and large numbers with cheaper low-density panels, then using imputation to infer the untyped high-density markers in the individuals genotyped at low-density. Imputation methods work by identifying haplotypes shared between individuals. The methods used generally fall into two broad categories: (1) heuristic methods that are designed to identify and propagate linkage information about long haplotypes

*Correspondence: john.hickey@roslin.ed.ac.uk
[1] The Roslin Institute and Royal (Dick) School of Veterinary Studies, The University of Edinburgh, Easter Bush Research Centre, Midlothian EH25 9RG, Scotland, UK
Full list of author information is available at the end of the article

(e.g., >10 cM), which is typically shared between closely related individuals; and (2) probabilistic methods that are designed to identify and propagate linkage disequilibrium information about short haplotypes (e.g., <1 cM), which is typically shared between distantly related individuals.

Heuristic methods use the basic principles of inheritance and are fast and accurate in many of the circumstances that are common to livestock applications [3]. Heuristic methods make explicit use of pedigree information and make inferences from information on closely related individuals from large families and the large portions of the genome shared between pairs of related individuals. However, heuristic methods do not impute alleles if such data is lacking or unreliable. The AlphaImpute program [3] is an example that combines several heuristic methods, such as basic rules of Mendelian inheritance, long-range phasing, and haplotype library imputation algorithm [4]. Other examples that are based on heuristic methods include Findhap [5] and FImpute [6].

Probabilistic methods mainly use HMM approaches to model genotype variation along chromosomes and the sharing of genomic segments between nominally unrelated individuals. HMM-based imputation methods are computationally more demanding, slower, and inherently less accurate than heuristic methods when they do not take pedigree information into account. HMM methods commonly used in livestock applications were primarily developed for application in human populations where pedigree information is typically lacking and pedigree structures, such as small family sizes, are not well-suited for exploiting heuristic algorithms.

HMM methods are used to describe the variation of an observable variable in a sequence, as a function of an underlying sequence of hidden variables that each have a set of K distinct states [7]. When HMM methods are applied to genotype imputation, the observable variable is a marker genotype, the sequence is a set of M markers along the chromosome, and the hidden variable represents the possible haplotypes that underlie the genotype. Given the number of markers, M, and the number of hidden states, K, the computational time of hidden Markov models scale as $O(M \times K^2)$, which limits the effectiveness of genomic applications with large numbers of markers and many possible haplotypes.

Distinct HMM algorithms with different representation of hidden states and computational considerations have been developed to alleviate the computational burden when analysing dense genomic data, such as, PHASE [8]; fastPHASE [9]; Beagle [10]; SHAPE-IT [11]; Impute2 [12]; MaCH [13]; MERLIN [14]; cnF2freq [15]. PHASE uses a Markov chain Monte Carlo algorithm to estimate the actual pair of hidden gametes of each individual as a mosaic of haplotypes given the observed genotypes

and the underlying recombination rates [8]. PHASE is very accurate but computationally intractable for large datasets. fastPHASE, Beagle, SHAPE-IT, Impute2, and MaCH are computationally tractable HMM methods for phasing and imputation. fastPHASE uses an expectation–maximisation approach and infers the most likely hidden states by clustering similar haplotypes [9]. Albeit faster, fastPHASE is still computationally expensive and its expectation–maximisation algorithm can get stuck in a local maximum. Beagle relies on a similar concept as fastPHASE, but clusters haplotypes locally [10]. SHAPE-IT follows the HMM of PHASE but collapses all the haplotypes into a graph structure and uses this structure to divide the haplotypes into disjoint segments of J distinct haplotypes used as hidden states [11]. SHAPE-IT samples pairs of haplotypes with a Markov chain Monte Carlo algorithm that is linear in the number of the distinct haplotypes, J, to reduce computational intensity even further. SHAPE-IT is also able to integrate pedigree information for disjoint sets of duos and trios directly in the model, as well as adding a proof-reading step based on the separate local duo-HMM model [16]. Impute2 approximates PHASE but, instead of conditioning on all haplotypes of all individuals, it restricts the number of haplotypes to the effective population size to decrease computational intensity [12]. Impute2 selects haplotypes that are similar to the haplotypes of the individual being imputed as the hidden states, which can lead to local minima [17].

MaCH is close to PHASE in the use of a mosaic of haplotypes to explain the observed genotypes. However, MaCH uses a model parameterised by recombination and mutation rates to iteratively improve the phasing of each individual in a Markov chain Monte Carlo framework [13]. In this regard, MaCH is similar to Impute2, but the method selects a user specified number of template haplotypes at random instead of selecting those that are expected to be similar to haplotypes carried by the individual being imputed as in the case of Impute2. To reduce computational intensity, MaCH limits the template haplotypes, to a number specified by the user. MERLIN models the state space as the combination of haplotypes of all individuals that are included in the same pedigree [14]. This is successful for small nuclear families, but it is not feasible for livestock applications where the number of hidden states increases exponentially with the number of individuals. However, in the cases it can handle, the optimum solution achieved is the globally preferable solution based on the modelling assumptions, while most other HMM approaches only provide approximations. cnF2freq models the state of multiple individuals at once, but maintains several separate local pedigrees of a single individual and its immediate ancestors, making the problem tractable, but again introducing the risk of getting stuck in a local optimum [15].

Heuristic and HMM imputation methods have different strengths and weaknesses. Combining the two approaches in a single algorithm may improve performance. Errors and missing genotypes generated by heuristics can be resolved by exploiting the probabilistic nature of Markov models. For example, if one parent of an individual is known, genotyped, and can be fully phased using heuristics, while the other parent is unknown, not genotyped, or cannot be phased using heuristics, a heuristic method can be used to impute the alleles on the gamete inherited from the first parent, while a HMM can be used to impute the alleles on the gamete inherited from the other parent.

Heuristic methods can also be used to increase the computational efficiency of HMM methods [18]. The phased information from the heuristic approach, which can be obtained with a very limited computational burden, can be used to supply the HMM with an accurate, relevant and relatively small set of reference haplotypes. This set of reference haplotypes can reduce the computational demand of the HMM permitting it to handle large populations better. Similar ideas are used in minimac, which achieves very fast computation by using pre-phased data [19].

In this study, we present a combined method for genotype imputation that takes advantage of the accuracy and speed provided by heuristic methods and the robustness of HMM methods. In particular, the method improves the heuristic method of AlphaImpute [3] with the HMM implemented in MaCH [13], and has been released as the version v1.5 of AlphaImpute. Performance of the combined method using real and simulated data is compared with results obtained separately from heuristic and HMM methods.

Methods

The imputation method presented in this study is a combined heuristic and HMM method to impute genotypes of all individuals in a population for all markers. The method incorporates the heuristic imputation approach of AlphaImpute and the HMM method of MaCH. AlphaImpute implements the heuristic method explained in Hickey et al. [3] and MaCH implements the HMM explained in Li et al. [13]. All three methods are briefly described in the following sections.

Heuristic method of AlphaImpute

AlphaImpute combines: (1) basic rules of Mendelian inheritance; (2) segregation analysis; (3) long-range phasing; and (4) haplotype library imputation in order to phase and impute genotype data of all individuals in a population [3]. The program iterates across these four sets of actions multiple times to accumulate information and determine the haplotype that each individual carries at each position along the genome.

The basic rules of Mendelian inheritance and the segregation analysis are used in conjunction with all pedigree and genotype information to derive phase for as many alleles as possible under the assumption that each locus is inherited independently of its neighbours at this step.

Long-range phasing and haplotype library imputation, which are both implemented in the AlphaPhase software and described in full detail in Hickey et al. [4], are used to derive the haplotypes that are carried by the individuals that are genotyped at high-density. Both long-range phasing and haplotype library imputation work by dividing the genome into genome regions, referred to as cores, and resolving the haplotypes within the cores for the individuals concerned. Cores of different lengths are used in several runs to phase each locus as part of overlapping cores and to facilitate the identification of phasing errors. These phasing steps generate a library of haplotypes for each core that are used later.

Missing alleles are then imputed by matching haplotypes that are obtained during the long-range phasing to alleles that are imputed and phased by the basic rules of Mendelian inheritance and by the segregation analysis. All haplotypes stored in the haplotype libraries are considered candidates of the true haplotype of the proband (i.e., the individual being phased) for each core of each phasing round. Alleles that are imputed and phased by the basic rules of Mendelian inheritance and by the segregation analysis are compared to corresponding alleles in each of the haplotypes in the library and haplotypes that are consistent with the alleles of a proband are retained as candidate haplotypes. This is repeated for each core. For a given marker position, individual alleles are imputed such that all remaining haplotypes across all of the cores that span this marker are in agreement. To impute from parental haplotypes this process is also repeated with a restriction that the haplotypes retained in the haplotype library comprise only those haplotypes that are carried by the parents. Libraries are updated with any new haplotype found. The matching process is iterated a defined number of times and at the end of each iteration each chromosome of each individual is traversed in each direction to detect recombination locations and to model the imputation of alleles in the regions of uncertainty that are adjacent to these recombinations as a weighted average of the two gametes carried by the relevant parent [20]. At the end of the final iteration, the segregation analysis is repeated and used to fill in alleles that remain unimputed.

Hidden Markov model algorithm of MaCH

In its diploid form, MaCH implements a HMM that characterises the unphased genotypes, G, as a mosaic of pairs of haplotypes taken from a set of template haplotypes,

H [13]. The mosaic of pairs of haplotypes represents the hidden sequence of states. At a locus, $i = 1, \ldots, M$, a hidden state is represented as $S_i = (h_i, k_i)$, where h_i corresponds to the haplotype on the first gamete, and k_i corresponds to the haplotype on the second gamete. The total number of possible states is $|H|^2$, corresponding to all the combinations of pairs of haplotypes, $S = \{(h, k) | h, k \in H\}$, where h and k are two haplotypes of *H*. The objective is to deduce the hidden sequence of states that best fits the data, i.e., to estimate the best pair of haplotypes that explain the unphased observed genotypes. This information in turn enables imputation at the unobserved loci.

In the HHM implemented in MaCH:

1. the prior probability is defined by assuming that every state is equally likely at the first marker,
2. the probability of transition from one state to another is given as a function of $m - 1$ crossover parameters, θ_i, which reflect the recombination rates, and
3. the probability of observing a genotype at each locus given a particular state is defined as a function of the *m* error parameters, ε_i, which indicate the effect of genotype errors and mutations.

MaCH uses a Monte-Carlo procedure to estimate the best pair of haplotypes for each individual [13]. Initially, the program samples a pair of haplotypes that is compatible with the observed genotype data. Alleles in the initial pair of haplotypes are phased randomly for heterozygous markers and sampled according to population allele frequencies for markers with missing data. Each individual is then updated with two new haplotypes that are sampled from the template haplotypes. This step involves calculating likelihoods of the hidden states, which is solved using the computationally efficient Baum's forward–backward algorithm. To limit computational complexity, the set of template haplotypes is represented by a random subset of all the current haplotype estimates.

The updating process is repeated an arbitrary number of iterations over the whole population. Each iteration involves an update of the model parameters (mutation and recombination rates). The crossovers and the errors are set to 0.01 at first, and are re-estimated at each iteration. For this purpose, the algorithm stores the number and position of recombinations, and the number of times the inferred genotype agrees with the observed one. The new estimates of the parameters improve the model and, thus, the haplotype sampling. The final pair of haplotypes for each individual is the consensus pair that minimises the total proportion of switches in haplotypes (i.e., switch error) when compared to the haplotypes sampled at each round.

MaCH also implements a haploid version of the model described above to improve the model when phase data is available. For phased individuals, the Baum's forward–backward algorithm is used twice to independently sample the two haplotypes from the template haplotypes.

The computational complexity of analysing a single individual in MaCH scales as $O(M \times |H|^2)$, where M is the number of markers and $|H|$ is the number of template haplotypes. The complexity of the forward–backward algorithm is $O(M \times |S|^2)$, where $|S|$ is the number of hidden states. Therefore, the diploid HMM of the MaCH algorithm scales as $O(M \times |S|^2) = O(M \times |H|^4)$, thus the number of template haplotypes is usually limited to reduce computational load. However, it is possible to further reduce the computational requirements of the algorithm to $O(M \times |H|^2)$ by taking advantage of regular patterns within the transition probability matrix.

Motivation for an algorithm that combines AlphaImpute and a HMM

AlphaImpute is computationally feasible for large datasets and is highly accurate for most of the genome of most individuals in typical livestock populations, but it has some weaknesses. For example, an individual that is to have its high-density genotypes imputed is genotyped at low-density, its sire and maternal grandsire are both genotyped at high-density and there is sufficient high-density information on their relatives to enable their genotype data to be completely and accurately phased. However, no genotype information is available on the dam and the maternal granddam. Thus, this data would enable all of the gametes the individual inherited from its sire to be accurately imputed, with the exception of the few small regions that are adjacent to recombination events. The imputation of the alleles on the gamete inherited from the individual's dam is more complex. While the portion of this gamete that derives from the maternal grandsire would be imputed with a relatively high level of accuracy, the lack of information available to impute the portion of this gamete that derives from the maternal granddam makes its imputation more difficult. These situations are common in livestock populations, where genotyping strategies often involve only genotyping male ancestors at high-density.

Another common situation is that the genotyping strategy may involve genotyping all parents of selection candidates at high-density and the selection candidates themselves at low-density. The quality control checks of the genotype data may show that one of the parents of a selection candidate is incorrectly identified in the pedigree records and must be set to missing. Thus, AlphaImpute would be able to accurately impute the gamete that the selection candidate inherited from one of its parents, but would have limited ability to perform imputation of the gamete inherited from the parent that has been set to missing. Also an individual that is part of the population being imputed by AlphaImpute may have unknown pedigree information for both its sire and its dam, due to any number of reasons, while most of the rest of the population has complete or partially complete pedigree information. In all three of these examples, a HMM could impute the segments that AlphaImpute is not able to impute well.

HMM methods are computationally intensive and typically less accurate than AlphaImpute in circumstances in which the pedigree information and population structure allow AlphaImpute to perform well. Much of the computational requirements of a HMM, and imputation errors that result from it, derive from establishing the template haplotypes and estimating the recombination rates between markers. AlphaImpute can accurately resolve haplotypes with computational efficiency.

Based on the different strengths, the two methods could be used to augment each other in several ways. For instance, accurate haplotypes that are resolved with heuristic methods could be fed as template haplotypes to a HMM and thus be used to increase its imputation accuracy and computational efficiency. Template haplotypes could be chosen for a particular genome segment with or without regard to any available pedigree information. A haploid version of the HMM could be used for individuals for which AlphaImpute could perform imputation for one gamete but not the other or not for part of one of the gametes while a diploid version of the same model could be used for individuals for which AlphaImpute could not perform imputation for either gamete or for the same region of both gametes. In addition, in some situations the heuristic rules of AlphaImpute may be able to partially impute some of the alleles on a gamete or segment of a gamete. This would effectively increase marker density and could increase the accuracy of the subsequent imputation by the HMM.

Hybrid algorithm

The method we propose (hybrid method) combines the different components of AlphaImpute with the HMM that underlies MaCH into a single framework. The hybrid method begins by applying the imputation method of AlphaImpute: basic rules of Mendelian inheritance, segregation analysis, long-range phasing, and haplotype library imputation. This gives accurate imputation for all individuals for which enough information is available. It also gives a large haplotype library that includes some whole-chromosome phased gametes. The HMM is then applied to impute the remaining missing alleles of any individual that are not imputed by AlphaImpute. The haplotype library is used to sample the template haplotypes used in the HMM. The haplotype update is run through several iterations so the parameters of the model converge. The solutions of the first iterations are disregarded as burn-in. Allele dosages and the final imputation are calculated using the remaining iterations. Finally, haplotypes are constructed from the most frequent alleles for each locus, and allele probabilities are assessed as the average across runs.

Within the hybrid method we devised three training modes and two imputation modes for the HMM in order to suit particular situations. When applying the method to a dataset, a user can use any combination of these modes:

Training mode 1 Use the high-density genotypes in the diploid HMM to estimate the model parameters;

Training mode 2 Sample a set of template haplotypes from a haplotype library and use them in the haploid HMM to estimate the model parameters;

Training mode 3 Use both of the data types and versions of the HMM from training modes 1 and 2 jointly while ensuring that genomic data for any individual enters the training set only once, with the haplotype information taking precedence;

Imputation mode 1 Use the haploid HMM to impute segments of a gamete that have not been imputed by AlphaImpute;

Imputation mode 2 Use the diploid HMM to impute segments of both gametes that have not been imputed by AlphaImpute.

To determine which training mode should be used for a given dataset, we defined some heuristic rules. When the user chooses not to use the heuristic method of AlphaImpute, Training mode 1 is used. When the number of phased gametes from AlphaImpute is above a user specified threshold Training mode 2 is used by default. When the number of phased gametes from AlphaImpute is below the user specified threshold Training mode 3 is used by default. Training mode 3 could also be used in datasets with peculiar properties, such as in some F_1 crossbred datasets in which AlphaImpute has sufficient information to resolve the haplotypes in one of the parental breeds but not in the other.

To determine which imputation mode should be used for a given segment of a gamete or pairs of individuals' gametes, the proportions of alleles imputed for the individual by AlphaImpute are examined. When the number of imputed alleles of a gamete from AlphaImpute is above a user-specific threshold Imputation mode 1 is used. When the number of imputed alleles of both individuals' gametes from AlphaImpute is above the user-specific threshold Imputation mode 2 is used.

The hybrid algorithm is designed to work for biallelic markers. Alleles are coded as 0, 1 or 9, where 9 is a missing allele; allele probabilities range from 0 to 1. Genotypes are determined as the sum of the allele codes, and allele dosages are assessed as the sum of the allele codes weighted by the allele probabilities. Therefore, genotypes are coded as 0, 1, 2 or 9, where 0 is the reference homozygote, 1 is the heterozygote, 2 is the alternative homozygote, and 9 is a missing genotype. Allele dosages range continuously from 0 to 2.

Datasets

Performance of the hybrid method was tested using both a real pig dataset [courtesy of The Pig Improvement Company (PIC)] and simulated data.

Real data

In the real pig dataset, genotype information was spread sparsely across multiple generations. The pedigree consisted of 6473 animals, including 3213 genotyped at high-density with the Illumina PorcineSNP60 BeadChip. Some animals had multiple generations of high-density genotyped ancestors available, while others had no, or very few, ancestors genotyped at high-density. This population came from a single PIC breeding line and therefore, animals were moderately to highly related to each other as is typical in most livestock breeding programs.

The genotyped animals were divided into training and testing sets. The testing set comprised the 509 most recently born animals genotyped at high-density. We used a single chromosome with 3129 quality-controlled high-density markers. To explore the effect of different genotyping strategies, animals in the testing set had different number of low-density markers selected from high-density, which were used to impute the remaining markers. We used 15, 30, 300, 600, or 2000 low-density markers on this chromosome, which were selected at random. These numbers are roughly equivalent to 300, 600, 6000, 12,000 and 40,000 markers per 20 chromosome genome. The training set comprised the remaining 2704 animals genotyped at high-density for all markers.

To explore the effect of having genotype data for the ancestors, animals in the testing set were grouped into six categories. These categories represented patterns of relationship between the animals in the population and their most recent high-density ancestors. The categories were: both parents genotyped (Both); sire and maternal grandsire genotyped (SireMGS); dam and paternal grandsire genotyped (DamPGS); sire genotyped (Sire); dam genotyped (Dam); and other relatives genotyped (Other).

Simulated data

Two sets of data were also simulated. For the first set, the pedigree of the real pig dataset was used and genotype data were simulated for different high and low-density panels. For the second set, a five-generation pedigree was simulated by mating 25 sires with 500 dams to produce 1000 progeny per generation. Genotype data were simulated for high and low-density panels. To explore the effect of imputation in the absence of pedigree data, another pedigree was created by randomly removing the sire and dam links for 500 individuals in the last generation of the simulated pedigree. These individuals without pedigree information were treated as unrelated individuals and the remaining pedigree information was used for imputing other animals and for resolving haplotypes of the training set. The parent's genotypes were retained in the training set.

Data for five replicates encompassing different genotyping strategies were simulated. The simulation of the genotype data required the following three steps:

1. Generate whole-genome sequence data,
2. Generate the marker genotypes, and
3. Mask the genotype information for markers that are not in the low-density panels.

Sequence data was generated using the Markovian Coalescent Simulator (MaCS) [21] and AlphaDrop [22] for 1000 base haplotypes for each of 30 chromosomes. Chromosomes were each 100 cM long, comprised 10^8 base pairs and were simulated using a per site mutation rate of 2.5×10^{-8}, a per site recombination rate of 1.0×10^{-8} and an effective population size that varied over time in accordance with estimates for the Holstein cattle population [23]. The population size was set to 100 in the final generation of the coalescent simulation, to 1256 at 1000 years ago, to 4350 at 10,000 years ago, and to 43,500 at 100,000 years ago, with linear changes in between these time-points. The resulting sequence had approximately 1.7 million segregating sites in total.

Chromosomes of individuals in the first generation were sampled from the 1000 simulated base haplotypes and those in the following generations were sampled from the chromosomes of their parents involving recombination. Crossovers occurred with 1% probability per cM and were uniformly distributed along the chromosomes.

For each chromosome, high-density panels were created by randomly selecting 2000 (H2k) and 10,000 (H10k) of the segregating sites. These numbers are roughly equivalent to 60,000 and 300,000 markers per 30-chromosome genome.

To explore the effect of genotyping strategies, low-density panels were simulated by masking the genotype information of some individuals in the pedigree. For the simulated dataset using the real pig pedigree, we masked the genotypes of the same 509 animals included in the testing set of real pig dataset. For the dataset with the simulated pedigree, we masked the genotypes of the 1000 individuals in the last generation. The genotype information was masked by selecting 15 (L15), 30 (L30), 300 (L300), 600 (L600), and 2000 (L2k) markers at random and removing the genotype information of the remaining markers from the H2k and H10k high-density panels. Low-density panels had densities equivalent to 500, 1000, 9000, 18,000 and 60,000 markers per 30 chromosome genome. A summary of the high and low-density panels is in Table 1.

Comparison

The hybrid method was compared to the AlphaImpute and MaCH imputation methods using simulated and real data. The comparison was made in terms of computation time and imputation accuracy for different imputation strategies. Final results for the comparison of the simulated data are presented as the mean of five replicates.

The computation time was measured as the CPU time that the three methods used to run the simulated data. The CPU time is the total amount of time the CPU spent executing instructions. If the software were run in serial mode, the CPU time would be comparable to the real time the software takes to run. However, AlphaImpute and the hybrid method allow parallelization of some calculations. For instance, the phasing runs of the long-range phasing step of AlphaImpute and the hybrid method can be run simultaneously. Also the updating process of an individual's haplotypes of the hybrid method is independent for each animal and can be run in parallel.

The imputation accuracy was computed as the average of animal-wise Pearson correlations between the true and imputed allele dosages [24]. Explicitly this involved calculating the correlation between the true and imputed allele dosages for each animal and averaging these correlations across animals in a category. For the simulated data that were generated using the simulated pedigree, the average was computed over the 1000 individuals in the last generation when the pedigree information was available, and over the unrelated individuals without pedigree information otherwise. For the real data and the simulated data using the real pig pedigree, averages were computed over the 509 animals in the testing set but according to the six categories of animals grouped by their pattern of relationship to their most recent densely genotyped ancestors. For instance, the averaged imputation accuracy of the 'Both' category was computed among the 46 animals within that category. A summary of the number of animals per category is in Table 2.

The imputation accuracy was also computed as the average of marker-wise Pearson correlations between the true and imputed allele dosages. Averages were computed

Table 2 Number of animals per category

Category	Count
Both	46
SireMGS	63
DamPGS	21
Sire	36
Dam	19
Other	324
Total	509

Number of animals per category based on the relationship to their most recent densely genotyped ancestors: both parents genotyped (Both); sire and maternal grandsire (SireMGS); dam and paternal grandsire (DamPGS); sire only (Sire); dam only (Dam); and other relatives (Other)

Table 1 Description of the simulated marker panels

Panel code	Panel design	Number of markers per chromosome	Number of markers across the genome
H10k	High-density	10,000	300,000
H2k	High-density	2000	60,000
L2k	80%	2000	60,000
L600[a]	94/70%	600	18,000
L300[a]	97/85%	300	9000
L30[a]	99.7/98.3%	30	900
L15[a]	99.8/99.2%	15	450

Low-density panels were simulated by randomly masking the genotype information of the H10k and H2k high-density panels. 15, 30, 300, 600, and 2000 markers per chromosome were selected at random from both H10k and H2k high-density panels, to simulate densities equivalent to 450, 900, 9000, 18,000 and 60,000 markers per 30 chromosome genome

[a] Values correspond to percentage of markers masked from the H10k and H2k high-density panels, respectively

among genotypes categorized into intervals of minor allele frequencies: [0.0, 0.025], [0.025, 0.05], [0.05, 0.075], [0.075, 0.10], [0.10, 0.15], [0.15, 0.20], [0.20, 0.25], [0.25, 0.30], [0.30, 0.35], [0.35, 0.40], [0.40, 0.45], and [0.45, 0.50].

For the simulated data, several imputation strategies using different combinations of the number of the template haplotypes and the number of iterations of the HMM parameters were set. The number of template haplotypes ranged from 100 to 300, with increments of 50 haplotypes. The number of iterations ranged from 20 to 50 with increments of 10, with 5 burn-in iterations.

Other parameters for both the heuristic and the HMM were set by default and are described in the following section.

Parameter settings

For the heuristic methods, a default set of 10 core lengths ranging from 500 to 9000 high-density markers was used for the various iterations of the long-range phasing and haplotype library imputation process. The set of cores was chosen to minimize both the computational intensity of the phasing processes and to maximize the number of times an allele was phased as a part of cores spanning different markers. The number of processors available was set to 20 in order to run simultaneously all the phasing runs of the long-range phasing step of the hybrid method and AlphaImpute. The number 20 comes from two times for each of the 10 core lengths, as described in the original publication of AlphaImpute [3]. The number of iterations of the heuristic rules was set by default to 5 for both the hybrid method and AlphaImpute.

For the HMM, the thresholds that control the training and imputation modes of the hybrid method and the number of processors available were set by default. A threshold of 50% was chosen for the training mode. Thus, if more than 50% of the individuals were phased by the previous heuristic step, their phased gametes were used to populate the template haplotypes in the HMM. An individual is considered to be phased if 99% or more of its markers have been phased. A threshold of 90% was chosen for the imputation mode. Thus, if more than 90% of the markers of an individual were imputed by the previous heuristic step, then the haploid version of the HMM was used for that individual, otherwise the diploid version was used. The number of processors available for the HMM was set to 8. This is arbitrary and depends on the number of processors that are available. The number of burn-in iterations was set to 5.

Results

We compared the computation time and imputation accuracy, of the hybrid method, AlphaImpute and MaCH. AlphaImpute had the fastest computation time,

followed by the hybrid method and then MaCH. The hybrid method was the most accurate and its accuracy increased with the number of template haplotypes and with the marker density of the low-density panel. Removing pedigree information reduced imputation accuracy for the hybrid method and AlphaImpute, but the hybrid method remained the most accurate. The imputation accuracy of all three methods decreased with the minor allele frequency and the hybrid method was the most accurate across minor allele frequencies.

Computation time

AlphaImpute always required the same CPU time under the parameter settings considered, whereas the CPU time required by the two HMM increased with the number of template haplotypes and with the number of iterations. AlphaImpute was the fastest, and the hybrid method was faster than MaCH regardless of the number of template haplotypes or the number of iterations.

Figure 1 shows computation times, in CPU hours, of the three imputation methods for the number of template haplotypes and iterations. Real times are reported in Additional file 1: Figure S1. Panels (a) and (b) of Fig. 1 show the computation times for imputing to the H10k high-density panel (H10k) and H2k high-density panel (H2k), respectively. The computation time of AlphaImpute was independent of the number of template haplotypes and the number of iterations, whereas the computation time of the hybrid method and MaCH varied with the number of template haplotypes and iterations. The computation time of the hybrid method increased almost linearly with the number of template haplotypes whereas computation time of MaCH increased quadratically with the number of template haplotypes. This quadratic increase made MaCH the slowest method for large numbers of template haplotypes. The computation time of both the hybrid method and MaCH increased linearly with the number of iterations.

Computation times were measured using an Intel Xeon E5-2630 v3 (2.4 GHz) 16-cores processor based cluster running on a 64-bit Linux (Scientific Linux 7).

Imputation accuracy
Simulated data

In this section, we compare the imputation accuracies of the three methods using different numbers of template haplotypes, varying low-density panels and for imputing with and without pedigree information. The main results can be summarized in five points:

1. When pedigree information was fully available, the hybrid method was comparable to AlphaImpute and better than MaCH.

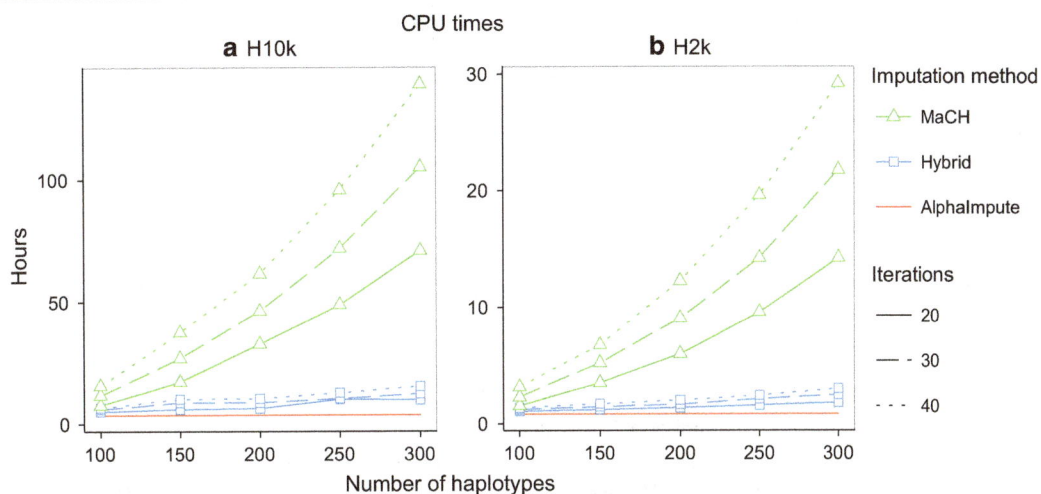

Fig. 1 CPU times. Computation time in CPU hours for imputing to **a** the H10k and **b** the H2k high-density panels. Subfigures show CPU times of the hybrid method (*blue*), AlphaImpute (*red*) and MaCH (*green*) imputation methods for different numbers of template haplotypes and iterations (*different line styles*). AlphaImpute is independent of the number of template haplotypes and iterations, and is shown as a *horizontal line*

2. When pedigree information was not available, the hybrid method was about twice as accurate as AlphaImpute and MaCH for the low-density panels.
3. When pedigree information was not available, the imputation accuracy of the hybrid method and MaCH increased with the number of template haplotypes.
4. The imputation accuracy of the hybrid method increased with the marker density of the low-density panels and was higher than the accuracy of AlphaImpute and MaCH across all the low-density panels.
5. The imputation accuracy of the hybrid method was the highest and remained stable for six categories of individuals depending on which of their immediate ancestor were genotyped at high-density.

When pedigree information was available, the hybrid method was comparable to AlphaImpute and better than MaCH. Figure 2 shows the imputation accuracies of all three methods for different numbers of template haplotypes. Panels (a) and (b) of Fig. 2 show the imputation accuracies for imputing from the L30 low-density panel to the H10k high-density panel (H10k-L30) and from the L30 low-density panel to the H2k high-density panel (H2k-L30), respectively. The imputation accuracies of the hybrid method and AlphaImpute were 0.95 for imputing to both the H10k and H2k high-density panels, whereas the imputation accuracy of MaCH ranged from 0.20 to 0.35. Because MaCH is pedigree-free, we made a separate comparison of its imputation accuracy with those of the hybrid method and AlphaImpute without pedigree, as well.

Figure 2 also shows that without pedigree information, the imputation accuracy of AlphaImpute and MaCH was about 55% (H10k) and 47% (H2k) worse than that of the hybrid method, which increased with the number of template haplotypes. When pedigree information was not used, the imputation accuracy of the hybrid method ranged between 0.50 and 0.65, while imputation accuracy of MaCH ranged from 0.20 to 0.35, depending on the number of template haplotypes. The imputation accuracy of AlphaImpute is independent of the number of template haplotypes and its imputation accuracy was 0.31 and 0.34 for imputing from L30 to H2k and H10k high-density panels. For both high-density panels, the accuracies of AlphaImpute and MaCH were comparable if MaCH used more than 250 (H10k) and more than 200 (H2k) template haplotypes and were about half the accuracy of the hybrid method (>0.50).

Figure 2 shows that the imputation accuracies for imputing to the H10k high-density panel were always slightly less accurate than imputation to the L2k panel. For this reason, we consider only the H10k high-density panel hereinafter.

Next, we compared the imputation accuracies of the hybrid method and MaCH using an imputation of 200 template haplotypes over 20 iterations, which represents a good compromise between time and accuracy (see Additional file 2: Table S1).

The imputation accuracy of the hybrid method increased with increased marker density of the low-density panels and exceeded the accuracy of AlphaImpute and MaCH when pedigree information was missing. Figure 3 shows the imputation accuracies from

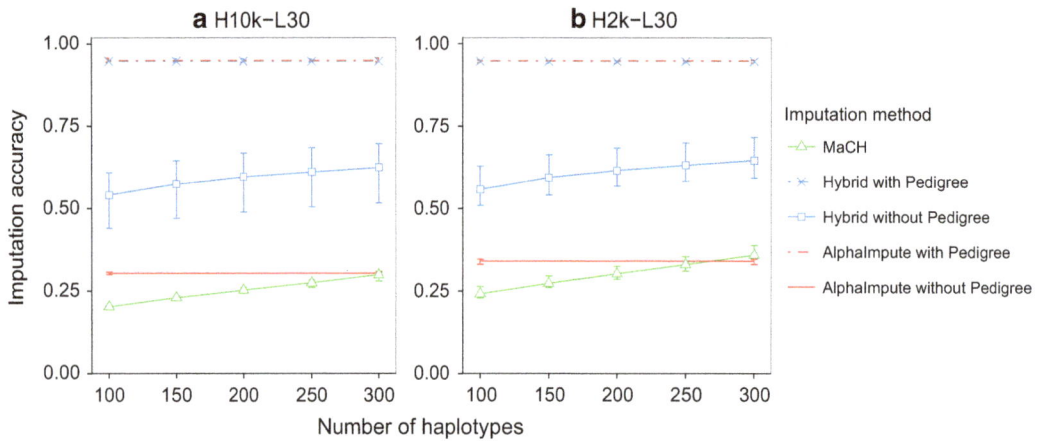

Fig. 2 Imputation accuracy for the number of template haplotypes. Imputation accuracies for different numbers of template haplotypes corresponding to the imputation from the L30 low-density panel to **a** the H10k and **b** the H2k high-density panels. Within each plot the *different lines* show imputation accuracies of the hybrid method (*blue*), AlphaImpute (*red*) and MaCH (*green*) for different numbers of template haplotypes. Imputation was performed with (*dashed*) and without (*solid*) pedigree. The accuracy of AlphaImpute is independent of the number of template haplotypes and iterations, and is shown as a *horizontal line* across haplotypes

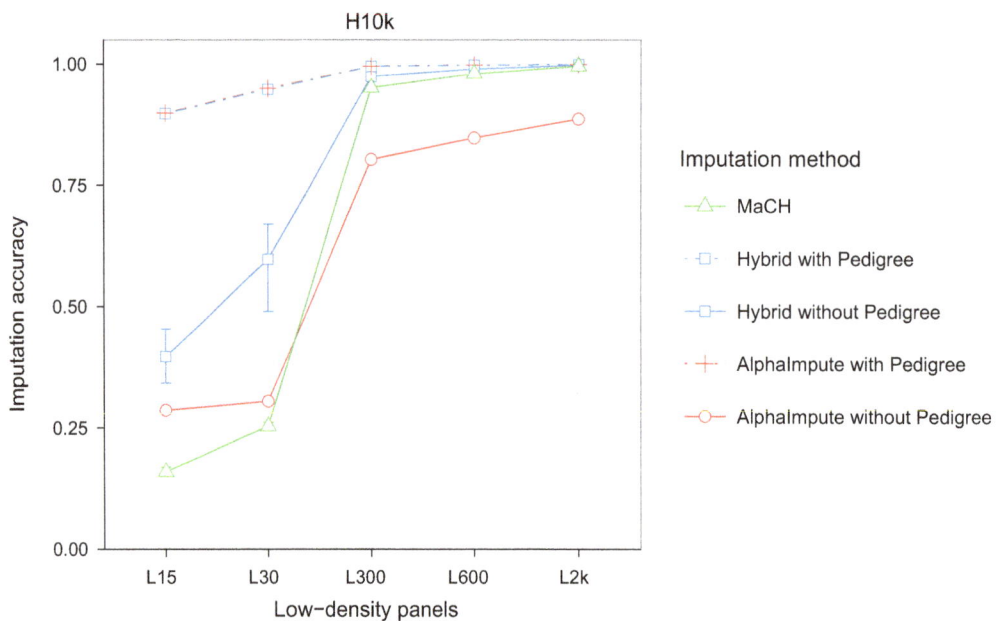

Fig. 3 Imputation accuracies for different low-density panels. The imputation accuracy of the hybrid method (*blue*), AlphaImpute (*red*) and MaCH (*green*) were calculated for genotyping strategies corresponding to imputation from low-density panels to the H10k high-density panel. Imputation was performed with (*dashed*) and without (*solid*) pedigree

the low-density panels to H10k high-density panel. When the pedigree information was not used, the accuracies of all three methods showed a massive jump from the L30 to the L300 low-density panel and more gradual linear changes above that. Without pedigree information, the hybrid method was always the most accurate with imputation accuracies above 0.97 for the three highest

low-density panels (L300, L600 and L2k), MaCH was nearly as accurate and AlphaImpute was 15 to 20% worse. For the lowest low-density panels (L15 and L30), the hybrid method was the most accurate method followed by AlphaImpute and then MaCH when pedigree information was missing. When pedigree information was available, the hybrid method and AlphaImpute were the most accurate with imputation accuracies above 0.99 for the three highest low-density panels and above 0.90 for the lowest low-density panels.

The imputation accuracy of all three methods increased with increased marker density of the low-density panels for different levels of genotype information available on immediate ancestors. Figure 4 shows imputation accuracy of all three methods plotted against the marker density of the low-density panel and the H10k high-density panel for simulated data and for six categories of individuals labeled according to which immediate ancestor was genotyped at high-density. Identical results for the H2k high-density panel are provided in Additional file 3: Figure S2. The imputation accuracies of all the three methods were similar and stable (>0.97) across varying levels of genotype information available on immediate ancestors for the three highest low-density panels. For the L15 low-density panel, the imputation accuracy of the hybrid and AlphaImpute decreased up to 9.5% and that of MaCH up to 66% with respect the L300 low-density panel.

However, this reduction in imputation accuracy varied across different categories.

Reducing the amount of genotype information available on immediate ancestors affected the imputation accuracy. Panels (a)–(f) of Fig. 4 show imputation accuracies of all the three methods for the categories where both parents (Both), sire and maternal grandsire (SireMGS), dam and paternal grandsire (DamPGS), sire only (Sire), dam only (Dam), and other relatives (Other) were genotyped at high-density, respectively. The imputation accuracy of the hybrid and AlphaImpute was higher than 0.90 across all the categories for the two lowest low-density panels. The imputation accuracy of MaCH decreased to less than 0.73 (L15) when at least one of the parents was genotyped at high-density and dropped to 0.33 (L15) when none of the parents were genotyped at high-density (Other).

Real data

This section shows a comparison of the imputation accuracies of all three methods when real data was used (see

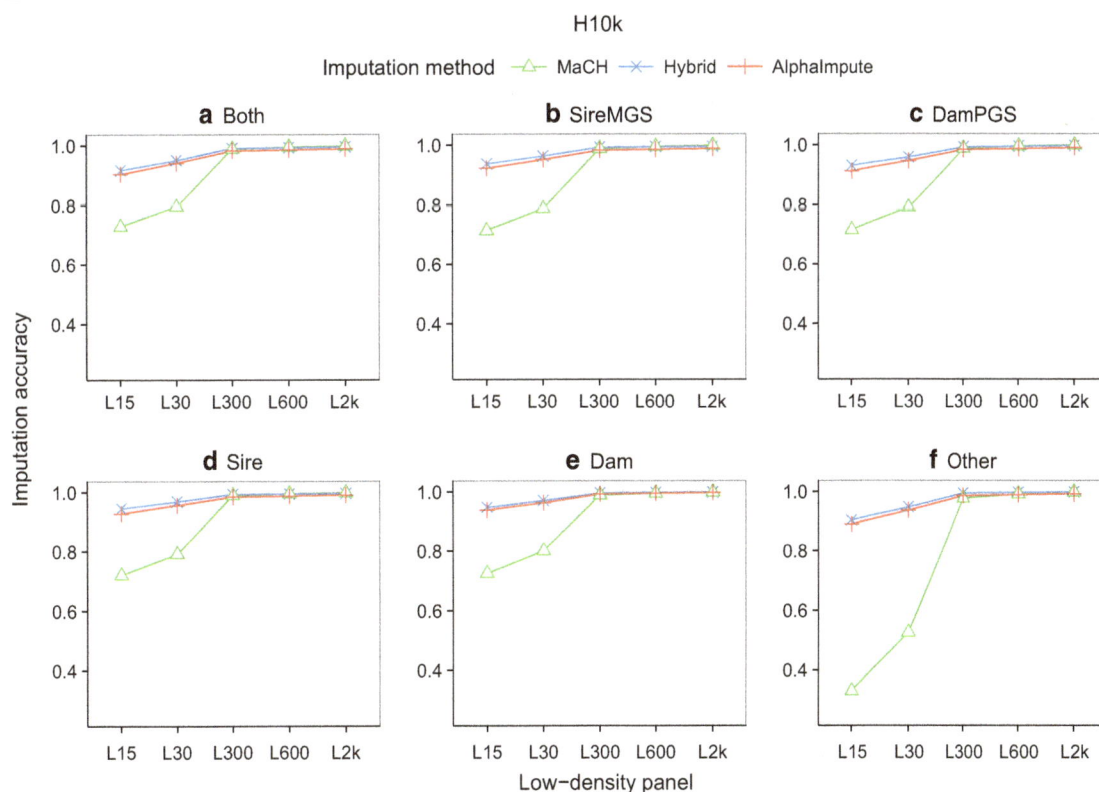

Fig. 4 Imputation accuracy in simulated data for low-density panels across categories of genotype information available on immediate ancestors. Imputation accuracies of the hybrid method (*blue*), AlphaImpute (*red*) and MaCH (*green*) for low-density panels and for six categories of animals based on which immediate ancestor was genotyped at high-density: both parents genotyped (Both); sire and maternal grandsire (SireMGS); dam and paternal grandsire (DamPGS); sire only (Sire); dam only (Dam); and other relatives (Other). The categories depend on the relationship of the animals to their most recent densely genotyped ancestors

Additional file 4: Table S2). For the comparison of the hybrid method and MaCH, we considered 200 template haplotypes and 20 estimation iterations as parameters of the HMM. The comparison provided two main results regarding the low-density panels and the genotype information available on immediate ancestors.

The imputation accuracy of all three methods increased with increased marker density of the low-density panels for different categories of available genotype information on immediate ancestors. Figure 5 shows the imputation accuracy of all three methods plotted against the marker density of the low-density panel for real data and for animals that differ based on which close ancestor was densely genotyped. The imputation accuracies of the three methods were similar (>0.90) for the three highest low-density panels and for all the categories. For the lowest low-density panels, the hybrid method and AlphaImpute performed marginally better than MaCH (>0.80) when at least one parent was genotyped at high-density.

Reducing the amount of genotype information of both parents decreased the imputation accuracy. Panels (a)–(f) of Fig. 5 show imputation accuracies of all three methods for the categories where both parents (Both),

sire and maternal grandsire (SireMGS), dam and paternal grandsire (DamPGS), sire only (Sire), dam only (Dam), and other relatives (Other) were genotyped at high-density, respectively. The imputation accuracies of the three methods were above 0.95 for the highest low-density panels and slightly decreased when none of the parents was genotyped at high-density (Other). For low-density panels with the lowest marker density, the imputation accuracies decreased from ~0.8 across all categories to ~0.6 for the hybrid method and AlphaImpute, and to less than 0.4 for MaCH when none of the parents were genotyped at high-density (Other).

Minor allele frequency

In this section, we compare the imputation accuracies of the three imputation methods against the minor allele frequency in simulated data. Two hundred template haplotypes and 20 iterations were used as parameters of the HMM. The results showed:

1. the imputation accuracy of the three methods increased with increases in the minor allele frequency,

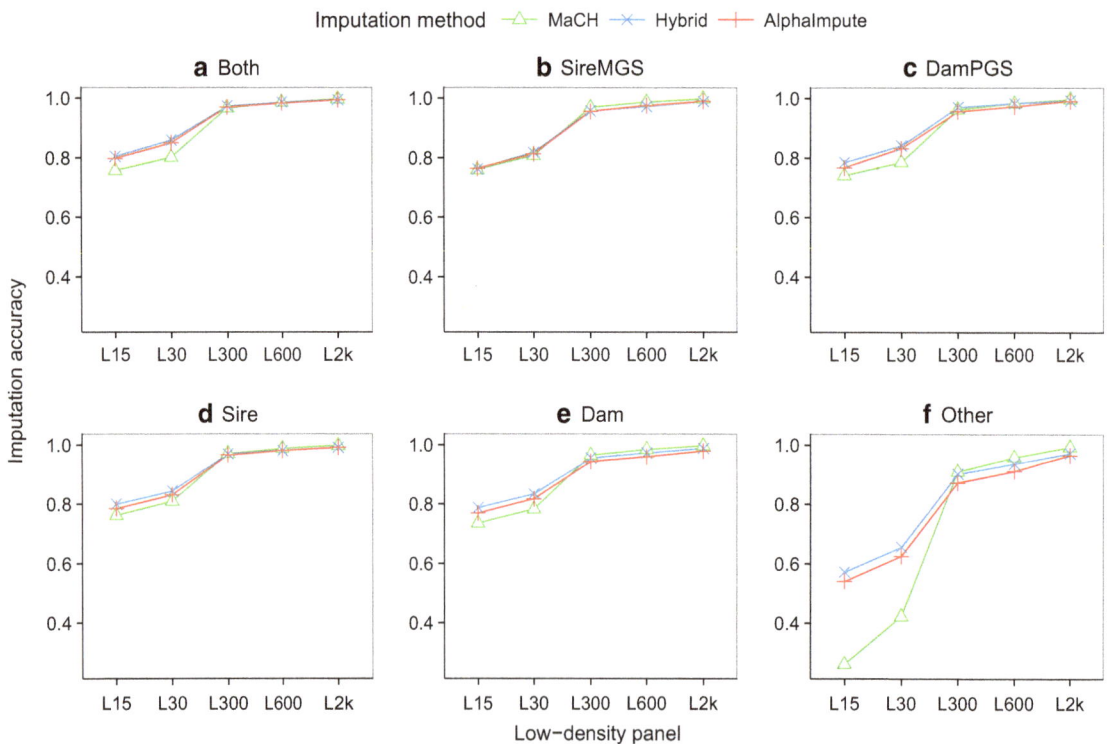

Fig. 5 Imputation accuracy in real data for low-density panels across categories of genotype information that is available on immediate ancestors. Imputation accuracies of the hybrid method (*blue*), AlphaImpute (*red*) and MaCH (*green*) for low-density panels and for six categories of animals based on which immediate ancestor was genotyped at high-density: both parents genotyped (Both); sire and maternal grandsire (SireMGS); dam and paternal grandsire (DamPGS); sire only (Sire); dam only (Dam); and other relatives (Other)

2. the imputation accuracy of the three methods increased as the genotype information of the low-density panels increased, and

3. the hybrid method always had the greatest accuracy when pedigree information was missing.

Imputation accuracy in relation to minor allele frequency

The hybrid method was always the most accurate imputation method across the full spectrum of minor allele frequencies when pedigree information was missing. Figure 6 shows the imputation accuracy of MaCH and those of the hybrid method and AlphaImpute with and without pedigree information for different minor allele frequencies. Figure 6 shows the imputation accuracies for imputing from the L30 low-density panel to the H10k high-density panel (H10k-L30). Similar results for the H2k high-density panel and the other low-density panels are provided in Additional file 5: Figure S3 and Additional file 6: Figure S4.

The hybrid method and AlphaImpute were the most accurate and their imputation accuracy remained above 0.94 across all minor allele frequencies when the pedigree was available. Figure 6 also shows that when the pedigree information was removed, the imputation accuracy

of the three methods decreased with decreases in the minor allele frequency. In this case, the hybrid method was the most accurate and its accuracy increased gradually as the minor allele frequency increased. Its accuracy was always above 0.60 except for minor allele frequencies that were lower than 0.05 for which the accuracy dropped below 0.50. In the absence of pedigree information, AlphaImpute was more accurate than MaCH except for rare alleles with minor allele frequencies higher than 0.1. However, even at the highest minor allele frequencies MaCH never exceeded the accuracy attained by the hybrid method at the lowest minor allele frequencies.

Figure 7 shows the imputation accuracy of the hybrid method and AlphaImpute when pedigree information was missing for imputation from a low-density panel to the H10k panel as a function of the marker density of the low-density panels with minor allele frequency as a parameter. The imputation accuracy of the three methods increased with increased marker density of the low-density panels and, for any given method and marker density, accuracy increased with allele frequency. The hybrid method was always the most accurate. From moderate to high genotype densities of the low-density panels (L300, L600 and L2k), the accuracy of the hybrid method

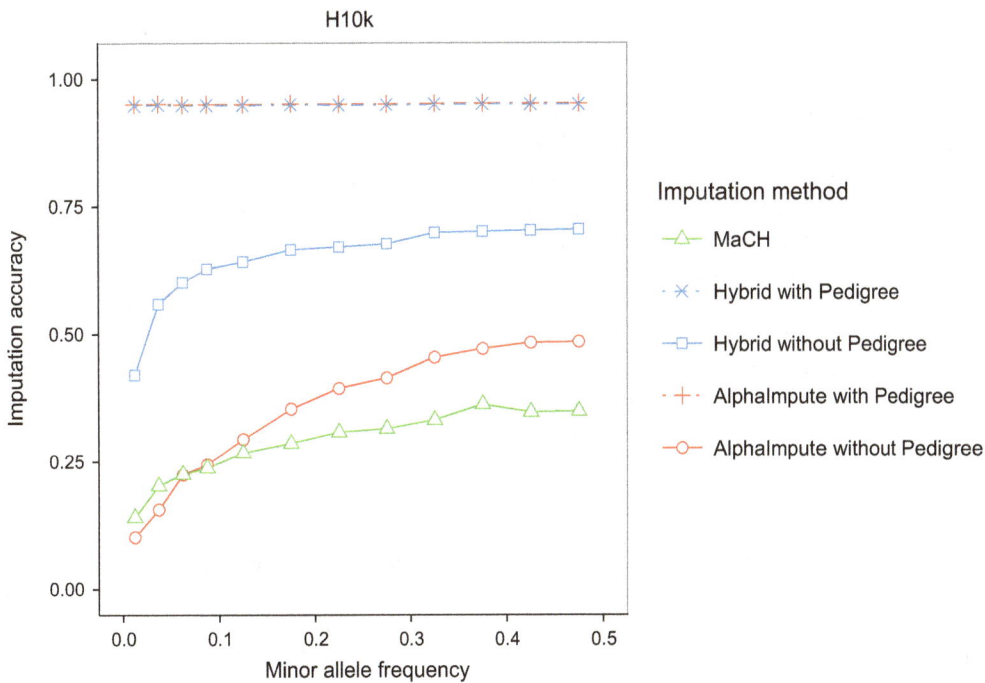

Fig. 6 Imputation accuracies for different minor allele frequency values. Imputation accuracies for different minor allele frequency values of the marker being imputed for imputation from the L30 low-density panel to the H10k. The figure shows the imputation accuracy of the hybrid method (*blue*) and AlphaImpute (*red*) and MaCH (*green*). Imputation with the hybrid method and AlphaImpute was performed with (*dotted*) and without (*solid*) pedigree. MaCH is pedigree-free. The imputation accuracies were computed among genotypes categorized into groups of allele frequencies in the following intervals: [0.0, 0.025], [0.025, 0.05], [0.05, 0.075], [0.075, 0.10], [0.10, 0.15], [0.15, 0.20], [0.20, 0.25], [0.25, 0.30], [0.30, 0.35], [0.35, 0.40], [0.40, 0.45], and [0.45, 0.50]

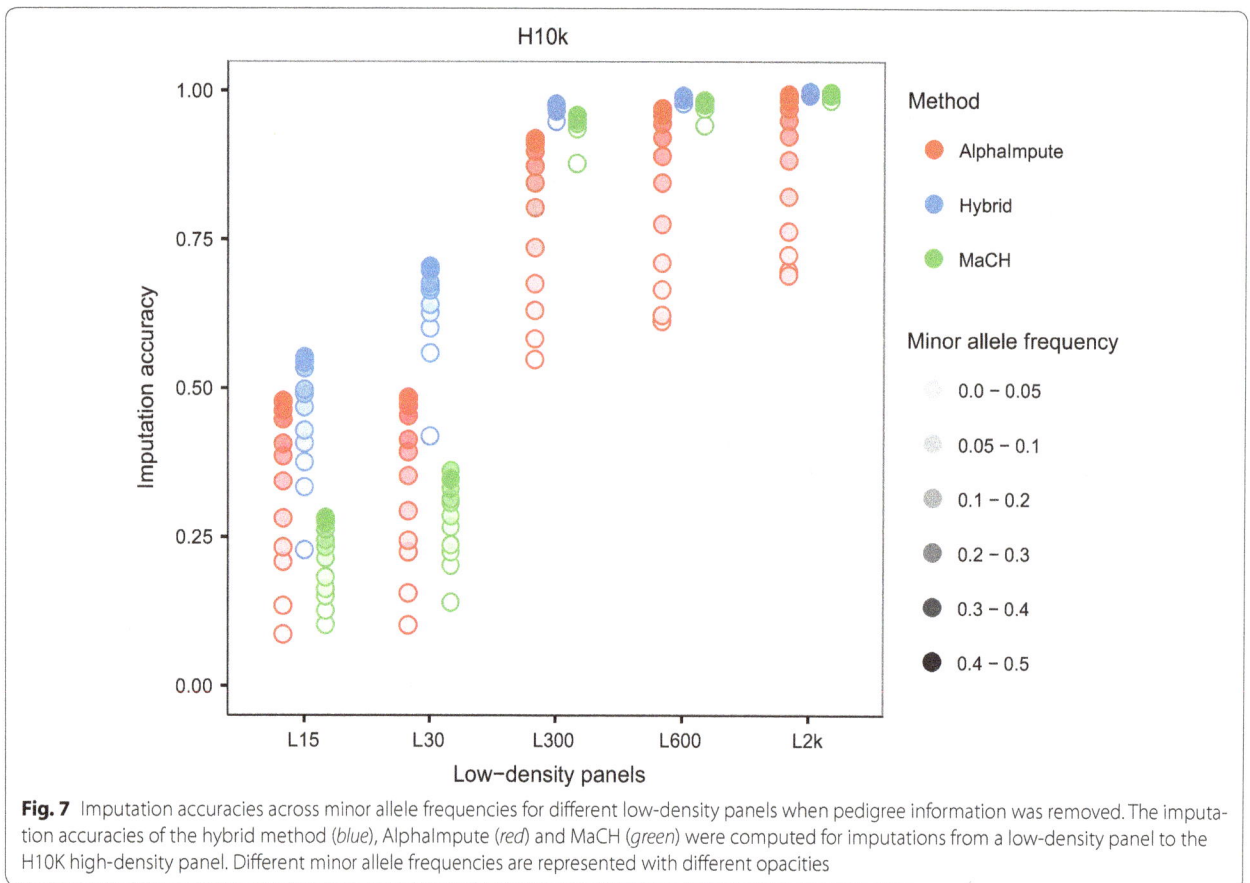

Fig. 7 Imputation accuracies across minor allele frequencies for different low-density panels when pedigree information was removed. The imputation accuracies of the hybrid method (*blue*), AlphaImpute (*red*) and MaCH (*green*) were computed for imputations from a low-density panel to the H10K high-density panel. Different minor allele frequencies are represented with different opacities

was above 0.95 for all allele frequencies in the absence of pedigree information. MaCH came second and had imputation accuracies above 0.85. AlphaImpute was the least accurate imputation method and was the most sensitive to minor allele frequency with accuracies below 0.75 for low values of minor allele frequencies when pedigree information was missing.

For the lowest marker density of the low-density panels (L15 and L30), the imputation accuracy of all three methods was substantially lower and more sensitive to allele frequency. The hybrid method was the most accurate for all values of minor allele frequency. In the absence of pedigree information, AlphaImpute was more accurate than MaCH for higher values of minor allele frequency, and slightly worse for very low values.

Discussion

The results show that among those tested, the hybrid imputation method is the most accurate method at different genotype densities with or without pedigree information. It is also faster than MaCH and only slightly slower than AlphaImpute. These results raise four points for discussion:

1. What are the advantages of using the hybrid method versus using a faster and nearly as accurate imputation method like AlphaImpute?
2. How does the computation time of the hybrid method benefit from the combination of heuristic and probabilistic approaches?
3. What is the best way to measure imputation accuracy of genotypes with rare alleles?
4. How does the hybrid method benefit from the combination of heuristic and probabilistic approaches to impute genotypes with rare alleles?

The advantage of using the hybrid method although it is slower than AlphaImpute is its higher accuracy when pedigree information is missing. In addition, AlphaImpute fails to impute genotypes that do not meet the imputation heuristic rules even when pedigree information is fully available. This causes some genotypes and segments of haplotypes to remain unimputed after the imputation. The HMM implemented in the hybrid method overcomes this limitation and calls the genotype of all markers for all individuals in the population.

Imputation methods usually comprise a phasing step that resolves the haplotypes of individuals genotyped at high-density, and an imputation step that identifies which combinations of these haplotypes are carried by the individuals genotyped at low-density. The hybrid method creates a reference set of extremely accurate haplotypes by applying a long-range phasing and haplotype library heuristic imputation algorithm, AlphaPhase [4]. The advantage of AlphaPhase is that the phasing algorithm can use the genotype information of surrogate parents (i.e., individuals that share a haplotype with the proband and that do not have any opposing homozygous genotypes with the proband). This makes it unnecessary to genotype multiple generations of ancestors at high-density in order to phase genotypes. The hybrid method then uses these well-phased haplotypes to impute the missing genotypes in a combined heuristic and probabilistic approach.

In the imputation step, the hybrid method accurately imputes genotypes by identifying the exact pair of haplotypes for single markers or groups of consecutive markers with basic heuristic pedigree-based phasing rules [3]. However, the heuristic rules are not sufficient to model the recombination of those haplotypes. Failures are particularly pronounced for individuals that are genotyped at very low-density because it is very likely that recombination occurs at markers for which genotype information cannot be inferred, leading to haplotype switch errors. To avoid haplotype switch errors, the hybrid method uses the HMM of MaCH [13] to estimate the recombination rates between markers and the most likely position for recombination in a given haplotype.

The accurate pre-phasing helps to reduce the computation time of the hybrid method without loss in imputation accuracy by decreasing the number of iterations and the number of template haplotypes. This makes the hybrid method faster than similar HMM methods such as MaCH. Unlike MaCH, the hybrid method takes advantage of the pedigree information to pre-phase and unambiguously impute genotypes. Pre-phasing genotype data provides very accurate haplotypes to be used by the HMM. If the number of well-phased haplotypes is high, the faster haploid version of the HMM can be used in most cases instead of the slower diploid version. In fact, our results show how the computation time of the hybrid method scales almost linearly with the number of template haplotypes, whereas the computation time of MaCH scales quadratically. Moreover, seeding a probabilistic method with well-phased haplotypes means that the model parameters are estimated with fewer iterations. Also, the hybrid method is more robust than MaCH in terms of the number of template haplotypes especially when pedigree information is available (see

Additional file 7: Figure S5, Additional file 8: Figure S6). By speeding up the HMM via integration into the hybrid method, it can be applied to larger datasets and its own inherent accuracy can be higher since a user can use more template haplotypes for a fixed amount of computation time. More template haplotypes make the HMM more accurate.

Correct quantification of imputation accuracy of genotypes with rare alleles is especially important for sequence data for which there are many alleles with low frequencies, and the marker-wise correlation between the true genotypes and allele dosages is the best way to measure this imputation accuracy. Imputation accuracy computed in this way measures how much is gained by a sophisticated imputation method in contrast to naïve procedures of imputing two times the allele frequency or the most frequent genotype. The correlation between true genotypes and genotype dosages is a better estimator of imputation accuracy than imputation error rates, particularly for the imputation of genotypes with very rare alleles (<0.1). Imputation error rate, computed as the percentage of genotypes that are imputed incorrectly is strongly affected by the minor allele frequency [24, 25]. When the minor allele frequency is very low, missing genotypes are almost certain to be homozygous for the common allele and naïve imputation methods yield an imputation error rate of almost 0%. The prior uncertainty increases with minor allele frequency, and thus imputation error rate increases as the minor allele frequency increases [25]. In contrast, the correlation coefficient between true genotypes and the allele dosages is an unbiased estimator of the imputation accuracy [24]. The Pearson correlation coefficient (used to calculate imputation accuracy in this paper) assumes that the two variables are normally distributed and requires that the mean and variance be standardized for both true genotypes and genotype dosages in order to calculate the correlation across individuals for a single locus. Thus, true genotypes and dosages were normalized by the mean and standard deviation of the true values. As a consequence, loci with a low minor allele frequency, for which genotypes are difficult to impute, are given more importance.

Marker-wise imputation accuracies of the hybrid method show that the combination of heuristic methods with a HMM helps to impute genotypes with rare alleles. Our results demonstrate that probabilistic methods based on linkage disequilibrium such as the HMM in MaCH are less accurate than heuristic methods based on linkage such as AlphaImpute, which agrees with previous studies [13, 24, 25]. However, the hybrid method was more accurate than both AlphaImpute and MaCH across minor allele frequencies and across genotype densities especially when pedigree information was missing (see Additional file 5: Figure S3). A possible explanation for this is that rare alleles become more frequent in the template haplotypes used by the HMM because heuristic methods are able to impute some of these rare alleles without ambiguity.

Conclusions

The hybrid heuristic and probabilistic imputation method proposed in this paper imputes accurate genotypes quickly for large populations and large numbers of markers even when limited pedigree information is available. It is usually more accurate and never significantly less accurate than a purely heuristic method and is faster than a standard probabilistic method.

Additional files

Additional file 1: Figure S1. Real computation time in hours for imputing to **a** the H10k and **b** the H2k high-density panels. Subfigures show real times of the hybrid method (*blue*), AlphaImpute (*red*) and MaCH (*green*) imputation methods for different numbers of template haplotypes and iterations (*different line styles*). AlphaImpute is independent of the number of template haplotypes and iterations, and is shown as a *horizontal line*.

Additional file 2: Table S1. Summary of imputation accuracies. Imputation accuracy for different imputation from L15, L30, L300, L600 and L2k low-density panels to H2k and H10k high-density panels. Imputation accuracies of the hybrid method and AlphaImpute were computed for imputation performed with and without pedigree. MaCH is always pedigree-free. Imputation accuracies of the hybrid method and MaCH correspond to a parameter setting that is equal to 200 template haplotypes and 20 iterations.

Additional file 3: Figure S2. Imputation accuracy in simulated data for different low-density panels across categories of genotype information available on immediate ancestors. Imputation accuracies of the hybrid method (*blue*), AlphaImpute (*red*) and MaCH (*green*) for different low-density panels and for six categories of animals according to which of their immediate ancestor are genotyped at high-density: both parents genotyped (Both); sire and maternal grandsire (SireMGS); dam and paternal grandsire (DamPGS); sire only (Sire); dam only (Dam); and other relatives (Other). The categories depend on the relationship of the animals to their most recent densely genotyped ancestors.

Additional file 4: Table S2. Summary of imputation accuracies for the real data. Imputation accuracy calculated as the correlation between the true genotypes and the genotype dosages across different categories of animals in the testing set. Animals in the testing set were grouped into six different categories according to which of their immediate ancestors are genotyped at high-density: both parents genotyped (Both); sire and maternal grandsire (SireMGS); dam and paternal grandsire (DamPGS); sire only (Sire); dam only (Dam); and other relatives (Other). Imputation accuracy of the hybrid method and MaCH correspond to a parameter setting that is equal to 200 template haplotypes and 20 iterations.

Additional file 5: Figure S3. Imputation accuracies according to minor allele frequency when pedigree information was removed. Imputation accuracies according to minor allele frequency of the marker being imputed when pedigree information was removed for 500 of the 1000 individuals in the last generation. The imputation accuracy of the hybrid method (*blue*), AlphaImpute (*red*) and MaCH (*green*) are plotted in different subfigures corresponding to the imputation strategies from the L15, L30, L300, L600 and L2k low-density panels to the H10k (**a–e**) and H2k (**f–i**) high-density panels. The imputation accuracies were computed among genotypes categorized into groups of allele frequencies in the

following intervals: [0.0, 0.025], [0.025, 0.05], [0.05, 0.075], [0.075, 0.10], [0.10, 0.15], [0.15, 0.20], [0.20, 0.25], [0.25, 0.30], [0.30, 0.35], [0.35, 0.40], [0.40, 0.45], and [0.45, 0.50].

Additional file 6: Figure S4. Imputation accuracies according to minor allele frequency of the marker being imputed when the pedigree information was fully available. The imputation accuracy of the hybrid method (*blue*), AlphaImpute (*red*) and MaCH (*green*) are plotted in different subfigures corresponding to the imputation strategies from the L15, L30, L300, L600 and L2k low-density panels to the H10k (**a–e**) and H2k (**f–i**) high-density panels. The imputation accuracies were computed among genotypes categorized into groups of allele frequencies in the following intervals: [0.0, 0.025], [0.025, 0.05], [0.05, 0.075], [0.075, 0.10], [0.10, 0.15], [0.15, 0.20], [0.20, 0.25], [0.25, 0.30], [0.30, 0.35], [0.35, 0.40], [0.40, 0.45], and [0.45, 0.50].

Additional file 7: Figure S5. Imputation accuracies of all three methods for different numbers of template haplotypes when pedigree information was removed. The different subfigures correspond to the imputation strategies from the L15, L30, L300, L600 and L2k low-density panels to the H10k (**a–e**) and H2k (**f–i**) high-density panels. Each subfigure plots imputation accuracies of the hybrid method (*blue*), AlphaImpute (*red*) and MaCH (*green*) for different numbers of template haplotypes. AlphaImpute is independent of number of template haplotypes and iterations and is shown as a *horizontal line* across haplotypes.

Additional file 8: Figure S6. Imputation accuracies of all three methods for different numbers of template haplotypes when pedigree information was fully available. The different subfigures correspond to the imputation strategies from the L15, L30, L300, L600 and L2k low-density panels to the H10k (**a–e**) and H2k (**f–i**) high-density panels. Each subfigure plots imputation accuracies of the hybrid method (*blue*), AlphaImpute (*red*) and MaCH (*green*) for different number of template haplotypes. AlphaImpute is independent of the number of template haplotypes and iterations and is shown as a *horizontal line* across haplotypes.

Authors' contributions

JMH and RA conceived the original ideas, designed the algorithm, and wrote the first version of the program with input from CN, GG and DM. RA performed the analysis and wrote the first draft of the paper. All authors read and approved the final manuscript.

Author details

[1] The Roslin Institute and Royal (Dick) School of Veterinary Studies, The University of Edinburgh, Easter Bush Research Centre, Midlothian EH25 9RG, Scotland, UK. [2] Division of Scientific Computing, Department of Information Technology, Science for Life Laboratory, Uppsala University, Lägerhyddsvägen 2, Box 337, 751 05 Uppsala, Sweden.

Acknowledgements

The authors acknowledge the financial support from the BBSRC ISPG to The Roslin Institute "BB/J004235/1", from Genus PLC and from Grant Ns. "BB/M009254/1", "BB/L020726/1", "BB/N004736/1", "BB/N004728/1", "BB/L020467/1", "BB/N006178/1" and Medical Research Council (MRC) Grant No. "MR/M000370/1". This work has made use of the resources provided by the Edinburgh Compute and Data Facility (ECDF) (http://www.ecdf.ed.ac.uk/). The authors thank Dr. Susan Cleveland (WI, USA) and Dr. Andrew Derrington (Scotland, UK) for assistance in refining the manuscript.

Competing interests

The authors declare that they have no competing interests.

References

1.   Goddard ME. The use of high density genotyping in animal health. In: Pinard M-H, Gay C, Pastoret P-P, Dodet B, editors. Developments in

biologicals. Basel: KARGER; 2008. p. 383–9. http://www.karger.com/doi/10.1159/000317189. Accessed 1 Sept 2016.

2. Habier D, Fernando RL, Dekkers JCM. Genomic selection using low-density marker panels. Genetics. 2009;182:343–53.

3. Hickey JM, Kinghorn BP, Tier B, van der Werf JH, Cleveland MA. A phasing and imputation method for pedigreed populations that results in a single-stage genomic evaluation. Genet Sel Evol. 2012;44:9.

4. Hickey JM, Kinghorn BP, Tier B, Wilson JF, Dunstan N, van der Werf JH. A combined long-range phasing and long haplotype imputation method to impute phase for SNP genotypes. Genet Sel Evol. 2011;43:12.

5. VanRaden PM, Null DJ, Sargolzaei M, Wiggans GR, Tooker ME, Cole JB, et al. Genomic imputation and evaluation using high-density Holstein genotypes. J Dairy Sci. 2013;96:668–78.

6. Sargolzaei M, Chesnais JP, Schenkel FS. A new approach for efficient genotype imputation using information from relatives. BMC Genomics. 2014;15:478.

7. Rabiner L. A tutorial on hidden Markov models and selected applications in speech recognition. Proc IEEE. 1989;77:257–86.

8. Stephens M, Smith NJ, Donnelly P. A new statistical method for haplotype reconstruction from population data. Am J Hum Genet. 2001;68:978–89.

9. Scheet P, Stephens M. A fast and flexible statistical model for large-scale population genotype data: applications to inferring missing genotypes and haplotypic phase. Am J Hum Genet. 2006;78:629–44.

10. Browning SR, Browning BL. Rapid and accurate haplotype phasing and missing-data inference for whole-genome association studies by use of localized haplotype clustering. Am J Hum Genet. 2007;81:1084–97.

11. Delaneau O, Marchini J, Zagury J-F. A linear complexity phasing method for thousands of genomes. Nat Methods. 2012;9:179–81.

12. Howie BN, Donnelly P, Marchini J. A flexible and accurate genotype imputation method for the next generation of genome-wide association studies. PLoS Genet. 2009;5:e1000529.

13. Li Y, Willer CJ, Ding J, Scheet P, Abecasis GR. MaCH: using sequence and genotype data to estimate haplotypes and unobserved genotypes. Genet Epidemiol. 2010;34:816–34.

14. Abecasis GR, Cherny SS, Cookson WO, Cardon LR. Merlin—rapid analysis of dense genetic maps using sparse gene flow trees. Nat Genet. 2002;30:97–101.

15. Nettelblad C, Holmgren S, Crooks L, Carlborg Ö. cnF2freq: efficient determination of genotype and haplotype probabilities in outbred populations using Markov models. In: Rajasekaran S, editor. Bioinformatics and computational biology. Berlin: Springer; 2009. p. 307–19. doi:10.1007/978-3-642-00727-9_29.

16. O'Connell J, Gurdasani D, Delaneau O, Pirastu N, Ulivi S, Cocca M, et al. A general approach for haplotype phasing across the full spectrum of relatedness. PLoS Genet. 2014;10:e1004234.

17. Nettelblad C. Breakdown of methods for phasing and imputation in the presence of double genotype sharing. PLoS One. 2013;8:e60354.

18. Druet T, Georges M. A hidden Markov model combining linkage and linkage disequilibrium information for haplotype reconstruction and quantitative trait locus fine mapping. Genetics. 2010;184:789–98.

19. Howie B, Fuchsberger C, Stephens M, Marchini J, Abecasis GR. Fast and accurate genotype imputation in genome-wide association studies through pre-phasing. Nat Genet. 2012;44:955–9.

20. Hickey JM, Kranis A. Extending long-range phasing and haplotype library imputation methods to impute genotypes on sex chromosomes. Genet Sel Evol. 2013;45:10.

21. Chen GK, Marjoram P, Wall JD. Fast and flexible simulation of DNA sequence data. Genome Res. 2009;19:136–42.

22. Hickey JM, Gorjanc G. Simulated data for genomic selection and genome-wide association studies using a combination of coalescent and gene drop methods. G3 Genes Genomes Genet. 2012;2:425–7.

23. Villa-Angulo R, Matukumalli LK, Gill CA, Choi J, Van Tassell CP, Grefenstette JJ. High-resolution haplotype block structure in the cattle genome. BMC Genet. 2009;10:19.

24. Calus MPL, Bouwman AC, Hickey JM, Veerkamp RF, Mulder HA. Evaluation of measures of correctness of genotype imputation in the context of genomic prediction: a review of livestock applications. Animal. 2014;8:1743–53.

25. Hickey JM, Crossa J, Babu R, de los Campos G. Factors Affecting the accuracy of genotype imputation in populations from several maize breeding programs. Crop Sci. 2012;52:654–63.

10

Genetic analysis of teat number in pigs reveals some developmental pathways independent of vertebra number and several loci which only affect a specific side

Gary A. Rohrer[*] and Dan J. Nonneman

Abstract

Background: Number of functional teats is an important trait in commercial swine production. As litter size increases, the number of teats must also increase to supply nutrition to all piglets. Therefore, a genome-wide association analysis was conducted to identify genomic regions that affect this trait in a commercial swine population. Genotypic data from the Illumina Porcine SNP60v1 BeadChip were available for 2951 animals with total teat number (TTN) records. A subset of these animals (n = 1828) had number of teats on each side recorded. From this information, the following traits were derived: number of teats on the left (LTN) and right side (RTN), maximum number of teats on a side (MAX), difference between LTN and RTN (L − R) and absolute value of L − R (DIF). Bayes C option of GENSEL (version 4.61) and 1-Mb windows were implemented. Identified regions that explained more than 1.5% of the genomic variation were tested in a larger group of animals (n = 5453) to estimate additive genetic effects.

Results: Marker heritabilities were highest for TTN (0.233), intermediate for individual side counts (0.088 to 0.115) and virtually nil for difference traits (0.002 for L − R and 0.006 for DIF). Each copy of the *VRTN* mutant allele increased teat count by 0.35 (TTN), 0.16 (LTN and RTN) and 0.19 (MAX). 15, 18, 13 and 18 one-Mb windows were detected that explained more than 1.0% of the genomic variation for TTN, LTN, RTN, and MAX, respectively. These regions cumulatively accounted for over 50% of the genomic variation of LTN, RTN and MAX, but only 30% of that of TTN. *Sus scrofa* chromosome SSC10:52 Mb was associated with all four count traits, while SSC10:60 and SSC14:54 Mb were associated with three count traits. Thirty-three SNPs accounted for nearly 39% of the additive genetic variation in the validation dataset. No effect of piglet sex or percentage of males in litter was detected, but birth weight was positively correlated with TTN.

Conclusions: Teat number is a heritable trait and use of genetic markers would expedite selection progress. Exploiting genetic variation associated with teat counts on each side would enhance selection focused on total teat counts. These results confirm QTL on SSC4, seven and ten and identify a novel QTL on SSC14.

Background

Genetic selection for increased litter size in pigs has resulted in many sows giving birth to more live piglets than they are capable of nursing. The competition for teats leads to increased pre-weaning mortality due to crushing and starvation [1]. Therefore, selection on teat number has begun to ensure that sows can nurture all of their piglets [2]. Number of piglets born in the largest 25% of litters in purebred Danish Large White and Landrace exceeded 18 in sows born in 2009 [3], which indicates that the number of piglets born was larger than the number of teats for a substantial proportion of litters. Number of teats in pigs is a variable and heritable trait. Number of teats differs between breeds, for example [4–6], and is moderately heritable [7–10]. Numerous

*Correspondence: gary.rohrer@ars.usda.gov
U.S. Meat Animal Research Center, USDA, Agricultural Research Service, Clay Center, NE, USA

genome scans have been conducted for number of teats in pigs (QTLdb; http://www.animalgenome.org/cgi-bin/QTLdb/SS/index), yet to date, few causative genes (or variants) have been discovered.

Most early studies used crosses between Meishan and occidental F_2 swine populations and detected the largest QTL on either *Sus scrofa* chromosome SSC1 or 7 [6, 11, 12]. These QTL on SSC1 and seven coincided with QTL for vertebra number or carcass length, which led to the hypothesis that vertebra and teat number were controlled by common genes [6, 13]. Putative causative genetic variations for vertebra number in the *NR6A1* gene [14] on SSC1 and the *VRTN* gene [15] on SSC7 have been associated with variation in teat number in Meishan × occidental cross populations [6, 12]. *VRTN* has also been associated with teat number in commercial swine populations [13, 16, 17].

While the presence of mammary glands is a defining character of species in the class Mammalia, location and number of mammary glands across species are quite variable [18]. Mammary glands commonly exhibit bilateral symmetry [19] and variation in number of functional mammary glands within a species is relatively low. Among the farmed artiodactyl species, only pigs have thoracic/pectoral and abdominal mammary glands, in addition to the inguinal mammary glands that are present in all artiodactyls. Mice are a common model mammalian species, yet they lack abdominal mammary glands and male pups do not have any visible teats at all. A greater understanding of mammary gland development is necessary to fully exploit the genetic variation present in pigs.

In early embryonic development of mammals, three separate streaks of multilayered surface ectoderm will form a mammary line that spans from the axilla to the inguen (groin) of the embryo. Mammary line cells will either group together or regress and eventually form mammary rudiments, which can later develop into functional mammary glands. As the gland continues to form, milk canals and nipples develop, completing the process. The developmental process in mice suggests that each pair of mammary glands develops at its own pace and may be regulated by different mechanisms [19, 20]. While male mouse embryos develop mammary rudiments, these structures typically regress prior to birth [19].

Studies have shown that spontaneous events and genetic mutations can result in bilateral asymmetry of mammary development in mice. Fernández et al. [9] speculated that the observed fluctuating asymmetry in the number of nipples in pigs may be caused by disruption of co-adaptive gene complexes, which results in developmental instability. Fluctuating asymmetry has been studied in numerous species and it is often

associated with increased stress or disease during critical development time periods.

Therefore, to increase our knowledge on the genetic factors that regulate mammary gland development in pigs, we conducted genome-wide association studies (GWAS) for various measurements of teat number in a composite population of commercial pigs. Individual counts of number of teats on each side were collected to evaluate bilateral symmetry and to determine if selection on total number of teats was the most effective measurement to record. Single nucleotide polymorphisms (SNPs) that were highlighted in the GWAS were then evaluated in an expanded population which contained germplasm from additional lines of commercial pigs. The results presented will be useful to enhance selection for increased lactation capacity as well as identify potential candidate genes that affect mammary gland development.

Methods
Data collection
Description of the population
The population of pigs used for this study was a ½ Landrace–¼ Duroc–¼ Yorkshire composite population that was created in 2001 and maintained as a closed population through 2010 as previously described [21]. Animals born in 2011 were from dams of this population and sired by Landrace boars from industry suppliers, while animals born in 2012 were sired by Yorkshire sires from industry sources. All pigs produced were processed at 1 day of age, when the number of teats was recorded and tail docked and stored for DNA extraction as part of the standard operating procedure which has been approved by the USMARC IACUC committee. Animals born from May 2008 to August 2009 had number of teats recorded for left and right sides.

Genotypic data
Extraction of DNA from tail tissue was done using the Wizard® genomic DNA purification kit for genomic DNA purification according to the manufacturer's protocols (Promega Corp., Madison, WI, USA). Approximately 75 ng of genomic DNA was then used in the reactions for the Illumina genotyping platforms. Assays using the Illumina Porcine SNP60 BeadChip v1 were done at USMARC and the chips were scanned at USDA-ARS-BARC Bovine Functional Genomics Laboratory, while all other genotyping analyses were done at GeneSeek (Lincoln, NE, USA). In total, 2951 Landrace–Duroc–Yorkshire pigs were genotyped with the Illumina Porcine SNP60v1 BeadChip and used for the GWAS. In addition, 2502 animals from either the closed Landrace–Duroc–Yorkshire population (n = 1275) or animals sired by commercial boars (n = 1227) were genotyped using one

of the other three Illumina-based genotyping platforms (Illumina Porcine SNP60v2 BeadChip; NeoGen Porcine GGP and NeoGen Porcine GGPHD) and were included in the validation phase analyses.

Data analysis
Model development
The appropriate statistical model for genome-wide association was determined based on analyses that were conducted using all animals that had recorded teat counts for left and right sides (n = 6472). Evaluated phenotypes were total (TTN), left (LTN) and right (RTN) teat number. In addition, maximum number of teats on one side (MAX), left minus right side teat number (L − R) and the absolute value of L − R were also analyzed (DIF). An animal model, which fit sex and contemporary group as fixed effects, percentage of males born in the litter as a covariate and litter as a random effect, was initially run as the full model using WOMBAT [22]. Reduced models in which one effect was eliminated were run and the residual and phenotypic variances were estimated, and then these were compared with the estimated variances from the full model. The order in which effects were eliminated was based on the predicted effects and was as follows: percentage of males in the litter, sex of the animal, and lastly the random effect of litter. This procedure was conducted on all six phenotypes studied.

Genome-wide association analyses
The dataset for GWAS included only animals with teat count records (total or left and right side data) and that were genotyped with the Illumina Porcine SNP60v1 BeadChip (n = 2951) using GENSEL v4.61R (http://bigs.ansci.iastate.edu). However, only 1828 animals had data for individual side counts. BayesC π was initially run to estimate variance components and π for the final genome-wide association analyses for which a minimum of 4000 iterations were conducted after removing 100 iterations for burn-in. Posterior estimates of π were evaluated to determine if the estimate of π had converged. For some traits, it was necessary to run more iterations to obtain a stable estimate of π. Genome-wide associations were conducted running BayesC with a prior as determined in the BayesC π runs. A total of 41,000 iterations were performed with the first 1000 discarded for posterior summaries. A 1-Mb window approach was conducted as described by Rohrer et al. [23]. Therefore, SNPs were required to have a unique position in the current swine genome (Build 10.2; [24]) resulting in 41,148 SNPs included in the final analyses. A fixed effect for contemporary group was included for all traits. A covariate for number of copies of the B allele of SNP NV090 [15], to account for the effect of *VRTN* alleles, was fitted for

TTN, LTN, RTN and MAX. Genotypes for SNP NV090 were predicted as previously described [16]. SNP NV090 was selected because it is located 6 kb upstream of the transcriptional start site for *VRTN* (based on GenBank Accession AB554652), gave very reliable genotypes in our lab and was found to be in complete linkage disequilibrium with the 291-bp insertion into the intron of *VRTN* (NV123) [23], which may actually be the causative mutation. Windows that explained more than 1% of the genomic variation are presented.

SNP validation
The SNP that explained the highest proportion of genomic variation for each 1-Mb window explaining more than 1.5% of the genomic variation for any count trait (TTN, LTN, RTN or MAX) was identified (see Additional file 1: Table S1) and used for this evaluation. All animals from the USMARC herd with SNP genotypes that were obtained from any of the four Illumina-based platforms were included (n = 5453). This included an additional 1275 animals from the closed Landrace–Duroc–Yorkshire population and 1227 animals sired by commercial boars. For SNPs that were not genotyped on an animal, genotypes were imputed using FImpute [25] based on information on flanking SNPs and at least three generations of pedigree data. All SNPs were fitted simultaneously by including a covariate for number of copies of the B allele. When two SNPs had a linkage disequilibrium coefficient exceeding 0.8, one of the SNPs was eliminated from the model. The only phenotypic data available for all animals was TTN. An animal model was fit using at least three generations of pedigree data and including fixed effects for contemporary group and breed of sire as well as a covariate for the effect of *VRTN* alleles (also imputed for this dataset using FImpute). Additive genetic variance was estimated in this dataset both with and without fitted SNPs to determine the percentage of genetic variation represented by the effects of SNPs. Candidate genes were selected by manually inspecting genes that were positioned within QTL regions based on the UCSC Genome Browser Gateway (www.genome.ucsc.edu/cgi-bin/hgGateway) using Sscrofa 10.2 genome build.

Results
Descriptive statistics and estimates of heritabilities obtained from the analysis of 6472 animals with data recorded for TTN, LTN, RTN, MAX, L − R and DIF are in Table 1. The fixed effect of sex as well as the regression coefficient for percentage of males in a litter did not affect estimates of residual or phenotypic variance for any trait analyzed and were removed from the model. The random effect of litter accounted for only ~2.5% of the phenotypic

Table 1 Descriptive statistics and estimates of variance components for the population of animals (n = 6472) used for statistical model development

Trait	Residual variance	Additive genetic variance	Phenotypic variance	Heritability (SE)	Mean	Range
TTN	0.592	0.578	1.171	0.494 (0.038)	14.73	8 to 21
LTN	0.242	0.149	0.390	0.381 (0.037)	7.32	5 to 13
RTN	0.319	0.137	0.456	0.301 (0.035)	7.41	2 to 12
MAX	0.234	0.154	0.389	0.397 (0.037)	7.59	6 to 13
L − R	0.525	0.000	0.525	0.000 (0.003)	−0.08	−4 to 7
DIF	0.324	0.002	0.326	0.006 (0.005)	0.45	0 to 7

LTN left side teat number, *MAX* maximum teat number of a side, *RTN* right side teat number, *TTN* total teat number, *L − R* difference between LTN and RTN, and *DIF* absolute value of L − R

variation for TTN, LTN, RTN and MAX and heritabilities of 0.41, 0.32, .023 and 0.29 were estimated, respectively. When the random effect of litter was removed from the model for the four count traits, phenotypic variance increased by approximately 3% while the estimate of additive genetic variation increased by an average of 21% and residual variance decreased by an average of 6%. Therefore, it was concluded that the random effect of litter was absorbing some of the additive genetic variation and removing this term seemed appropriate. When the random effect of litter was removed, estimated heritabilities were highest for TTN (0.49), intermediate for MAX (0.40), LTN (0.38) and RTN (0.30) and nil for L − R and DIF.

Genome-wide association study

Summary statistics from the GENSEL analyses are in Table 2. Genotypic data from the Illumina Porcine SNP60v1 BeadChip [26] were available for 2951 animals with TTN records, of which a subset (n = 1828) had number of teats recorded on each side. Genotypic data for SNPs that had a unique location on the *S. scrofa* build 10.2 [24], a minor allele frequency higher than 0.05

and a call rate higher than 80% were considered, which resulted in 41,148 SNPs after editing. SNP heritabilities (i.e. genomic heritabilities) were highest for TTN (0.233), intermediate for individual side counts (0.088 to 0.115) and virtually nil for difference traits (0.002 for L − R and 0.006 for DIF).

The number of 1-Mb windows that explained more than 1% of the genomic variation detected was 15 for TTN, 18 for LTN, 13 for RTN and 18 for MAX (Fig. 1). These regions cumulatively accounted for over 50% of the genomic variation in LTN, RTN and MAX, while they only explained 30% of the genomic variation in TTN. Ten 1-Mb windows were associated with more than one trait. Most notable was the chromosome SSC10:52 Mb which was associated with all four count traits, while SSC10:60 and SSC14:54 Mb were associated with three of the four count traits. The 1-Mb windows that explained more than 1% of the genomic variation are in Fig. 1. The estimated additive effects of *VRTN* were equal to 0.35 for TTN, 0.16 for LTN, 0.19 for RTN and 0.17 for MAX. Information on all 1-Mb windows that explained more than 0.40% of the genomic variation and the SNPs that are associated with the most variation are in Additional file 1: Table S1.

Table 2 Summary statistics from GENSEL genome-wide association analysis of animals genotyped with the Illumina Porcine SNP60v1 BeadChip (n = 2951)

Trait	Genomic variance (GV)	Phenotypic variance	Genomic heritability	Number of 1-Mb windows >1% GV	Percent of GV explained by 1-Mb windows[a]	Mean	Range
TTN	0.221	0.948	0.233	15	30.6	15.8	8 to 20
LTN	0.034	0.356	0.096	18	57.4	7.6	5 to 10
RTN	0.035	0.396	0.088	13	50.6	7.7	2 to 10
MAX	0.040	0.345	0.115	18	54.2	7.9	6 to 10
L − R	0.001	0.528	0.002	4	9.2	−0.05	−3 to 4
DIF	0.001	0.312	0.006	6	26.0	0.47	0 to 4

LTN left side teat number, *MAX* maximum teat number of a side, *RTN* right side teat number, *TTN* total teat number, *L − R* difference between LTN and RTN, and *DIF* absolute value of L − R

[a] Percentage of phenotypic variation explained by markers, as predicted by GENSEL, that was explained by SNPs contained in 1-Mb windows that exceeded the 1% genomic variation threshold

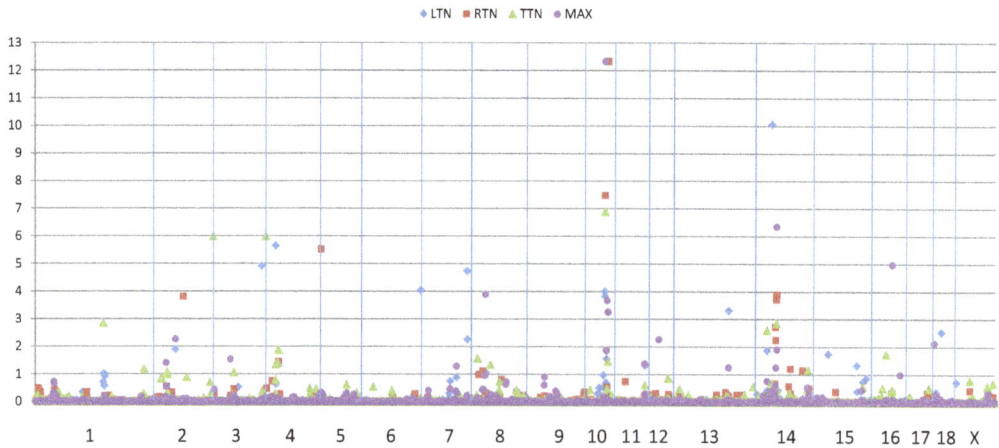

Fig. 1 Manhattan plot of GENSEL genome-wide association analysis for count traits. *Horizontal* axis is the position on the swine genome (Build 10.2) and the *vertical* axis is the percent of genomic variation associated with each 1-Mb window

Several 1-Mb windows that explained more than 1% of the genomic variation for L − R (n = 4) and DIF (n = 6) were detected; however, these results are likely meaningless since the estimated genomic variation was nearly zero (Table 2). A region with a large effect on DIF was found on SSC8 between 95 and 96 Mb and accounted for over 20% of the genomic variation. Cumulatively, the regions that explained more than 1% of the genomic variation for these traits accounted for <0.2% of the observed phenotypic variation.

Thirty-six SNPs (including the SNP in *VRTN*) that were located in 32 unique 1-Mb windows were selected for the validation phase. Three SNPs were removed because they created multi-colinearity among the regression coefficients due to high linkage disequilibrium. All three SNPs were within a 1-Mb window represented by multiple SNPs, which were associated with different phenotypes. In total, the SNPs accounted for ~39% of the additive genetic variation estimated for this dataset. The effect of *VRTN* was 0.335 for TTN. The magnitudes of the estimated additive effects for eight SNPs were between 0.10 and 0.14, 12 SNPs had additive effects that ranged from 0.05 to 0.10 and 12 SNPs had estimated effects that were <0.05 (Table 3).

Discussion

How the development of the mammary gland has evolved is an interesting issue. It was observed by Aristotle more than 2000 years ago (as referenced by Diamond [27]) and shown more recently by Gilbert [28] that natural selection results in a number of teats equal to the maximum litter size expected, which is approximately twice the average litter size. However, selection for increased fecundity in livestock species (primarily swine and sheep)

has resulted in litter sizes that exceed lactational capacity for many litters, thus requiring artificial rearing and/or cross-fostering of young to increase survival. Therefore, in these species, selective pressure needs to be placed on lactational capacity to increase neonatal survival and reduce the cost of production. Understanding the genetic mechanisms that regulate mammary gland development and teat number will contribute to the design of an optimal strategy for selection.

Genetic factors that control teat number

We detected several interesting genomic regions that affect number of teats in commercial-type pigs. Foremost, was the confirmation of the association between SNPs in *VRTN* and number of teats. This association was also found by Ding et al. [6] and Duijvesteijn et al. [13]. While these two groups speculated that teat number and vertebra number are controlled by a similar set of genes, none of the regions reported in Table 3 for TTN were found to be associated with vertebra number in this population [23]. However, two regions for individual side counts did overlap with QTL for vertebra number, i.e. SSC5:0 and SSC12:26 Mb. The region on SSC5:0 Mb which explained 5.53% of the genomic variation for RTN is adjacent to the region that explained 2.22% of the genomic variation for thoracic vertebra number. Two candidate genes in this region include *ceramide kinase* (*CERK*) which produces ceramide-1-phosphate and has a role in cell proliferation and migration [29] and *CELSR1*, a regulator of planar cell polarity [30]. The region on SSC12:26 Mb explained 2.27% of the genomic variation for MAX, 8.59% of the genomic variation for lumbar vertebra number and 4.74% of the genomic variation for thoracolumbar vertebra number. This region is within

Table 3 Results of additive effects of SNPs in the validation population (n = 5453) for TTN with positions based on *S. scrofa* build 10.2

Chr.	Position	SNP name	Additive effect	Previous association	Previous references
1	181,741,697	ASGA0005093	0.116	TTN[a]	
2	59,489,740	INRA0008845	−0.073	LTN, MAX	
2	81,675,738	ALGA0014021	−0.031	RTN	
3	49,321,597	H3GA0009450	0.059	MAX	[6, 11, 13]
3	134,660,431	MARC0090699	0.069	LTN	[11, 13]
4	25,899,175	DRGA0004616	−0.103	LTN	[11, 40]
4	33,780,262	ALGA0024379	0.031	RTN, TTN	[11, 40]
5	252,858	H3GA0017369	−0.061	RTN	[34]
6	157,649,704	M1GA0009139	−0.071	LTN	[11, 50]
7	103,208,408	*VRTN*/NV090	0.335	LTN, MAX, RTN, TTN	[6, 12, 13, 16, 17, 50]
7	124,146,658	MARC0073407	−0.039	LTN	[11]
8	16,445,414	ALGA0046611	0.077	TTN	[12, 51]
8	37,492,529	ALGA0047617	−0.127	LTN, MAX	[13]
10	51,681,377	DIAS0002581	−0.034	LTN	[13, 33–35, 43]
10	52,456,152	H3GA0030271	−0.062	LTN, MAX, RTN, TTN	[13, 33–35, 43]
10	52,679,135	MARC0018399	0.137	LTN, MAX, RTN, TTN	[13, 33–35, 43]
10	56,365,810	ASGA0103067	0.027	MAX	[11, 13, 33, 34]
10	58,071,987	ASGA0048302	0.029	MAX	[11, 13, 33, 34]
10	60,511,977	ASGA0048404	−0.044	MAX, RTN, TTN	[11, 33, 34]
12	26,420,000	ALGA0065784	−0.038	MAX	[11, 33, 34]
13	146,433,577	H3GA0037388	−0.123	LTN, MAX	
14	29,780,586	M1GA0018459	0.123	LTN, TTN	
14	41,043,761	ASGA0062848	0.106	LTN	
14	51,173,806	ASGA0063286	0.017	RTN	
14	52,942,907	ALGA0077532	0.016	RTN	
14	53,370,377	ASGA0063370	−0.062	MAX, RTN	
14	54,744,215	H3GA0040220	0.093	MAX, RTN, TTN	
14	54,791,585	ASGA0063388	NA[a]	MAX, RTN, TTN	
14	54,867,498	MARC0059175	NA[a]	MAX, RTN, TTN	
14	55,003,669	ASGA0063395	0.125	MAX, RTN	
14	55,429,701	ASGA0063406	NA	MAX, RTN	
15	37,205,130	ASGA0069274	−0.030	LTN	[50]
16	35,144,613	ALGA0090150	0.068	TTN	[11, 52]
16	50,977,092	MARC0028125	0.083	MAX	[11, 52]
18	4,400,270	ASGA0095800	−0.063	MAX	[13]
18	22,976,763	ALGA0097407	0.039	LTN	[53]

LTN left side teat number, *MAX* maximum teat number of a side, *RTN* right side teat number, *TTN* total teat number

[a] This SNP was not analyzed due to high linkage disequilibrium with another SNP included in the analysis

[b] Trait which originally exceeded the 1.4% genomic variation

the *COL1A1* gene. Mutations in *COL1A1* cause osteogenesis imperfecta leading to reduced bone mass and increased fracture. Two other potential candidate genes are the homeobox proteins *DLX3* and *DLX4* at 26.2 Mb on SSC12 but, to our knowledge, no role of these genes in mammary gland development has been described. *DLX3* induces the degradation of p63 [31] a transcription factor that is necessary for epidermal–mesenchymal interactions during embryonic development. Mice that

lack p63 have no mammary glands [32]. In spite of the large contributions of these regions in the GWAS, their additive effects in the validation phase of this study were extremely low (0.061 and 0.038, respectively).

The region with the most consistent and largest effects on all teat count traits (SSC10:52) was also found by Duijvesteijn et al. [13] in Large White pigs as well as in several studies that used Meishan by occidental F_2 populations [33–35]. This region contains the candidate genes

MPP7 and *FRMD4A*. *FRMD4A* resides within a copy number variation (CNV) region [36] and both *FRMD4A* and *MPP7* regulate the polarization of epithelial cells [37, 38]. The region on SSC10:60 Mb was reported by Guo et al. [11] who suggested *PLXDC2* as a possible candidate gene, which encodes a transmembrane receptor for the neurotrophic factor PEDF [39]. Hirooka et al. [33] and Rodríguez et al. [34] reported broad QTL intervals that spanned all of the 1-Mb windows on SSC10 reported in the current study.

The region on SSC4:25 Mb explained a high percentage of the genomic variation for LTN and had an additive effect on TTN of 0.103 in the validation population. Guo et al. [11] and Tortereau et al. [40] reported that this region segregated in Meishan cross populations. A potential candidate gene is *TRPS1*, which encodes a transcriptional repressor that regulates epithelial-mesenchymal transition [41] is required for morphogenesis during embryonic mammary gland development [42].

The region on SSC14:51–55 Mb has not been associated with teat number in pigs based on QTLdb and no obvious positional candidate genes were identified. This region had a large effect on RTN and MAX and the association with TTN was validated in the larger population. The SNP with the greatest estimated effect on TTN in the validation data was located at 55.00 Mb on SSC15 only 70 kb from the *T-box 1 transcription factor* gene (54.93 Mb). T-box transcription factors are critically important for normal tissue and organ development in the embryo. GWAS results for this region revealed a broad peak that spans several Mb. Several of the SNPs tested in the validation phase were in high linkage disequilibrium which made it difficult to directly pinpoint which SNP had the largest effect. Thus, multiple causative genes/variants are possible.

Other novel QTL regions with estimated additive effects exceeding 0.10 were on SSC1:181, SSC8:37, SSC13:146, SSC14:29 and SSC14:41 Mb. Unfortunately, identification of obvious candidate genes was unsuccessful. Although the SSC1:181 Mb region includes two genes for multiple epidermal growth factors (*MEGF6* and *MEGF11*) and the SSC14:41 Mb region contains two genes that are involved in the modulation of the NOTCH signaling pathway (*DTX1* and *RITA1*), these genes have not been shown to affect mammary gland development.

An unexpected finding is that among the eight regions discussed above, all with additive effects >0.1 in the validation population, five (SSC4:25, SCC8:37, SCC10:52, SCC13:146, and SCC14:55 Mb) were identified to have CNV segregating within the studied population [36]. Additional research is necessary to validate these CNV, determine their inheritance and test for association with teat counts.

Mammary gland development in the pig

The concept that additional mammary glands arise from somite division in the developing embryo is supported by the effects of the *VRTN* gene that were observed on vertebra number and teat count [6, 12, 16, 17, 23] and of the *NR6A1* gene on SSC1 [6]. However, since genetic variation within *NR6A1* is only observed in crosses between Asian and European breeds, we were not able to evaluate this result. In spite of these two co-localizations of vertebra number and teat count in the proximity of *VRTN* and *NR6A1* as well as the speculation mentioned in Duijvesteijn et al. [13], none of the regions which exceeded 1% of the genomic variation for TTN (Fig. 1) were associated with vertebra number in this population [23]. The only region in the validation phase associated with vertebra number [23] was SSC12:26 Mb, which it had an extremely low additive effect (0.038). The region on SSC5:0 Mb was near a QTL for thoracic vertebra number (SSC5:1 Mb), which harbors *WNT7B* as candidate gene. However, there are clearly additional genetic factors that affect teat count since the USMARC Meishan population averages a 2.6 greater TTN [43] while having 1.5 fewer ribs and 2.0 fewer thoracolumbar vertebra than the Landrace–Duroc–Yorkshire population used in this GWAS.

Final teat number is likely a composite trait for which the underlying genetic model begins with somite division, followed either by proliferation of the mammary buds and/or regression of milk buds which results in teat number at birth. If genetic variation exists for all segments of the mammary gland development, then more progress may be possible if selection is applied to the component traits. This was the hypothesis on which was based the study of the maximum number of teats on one side since this would be the best estimate of an animal's true genetic potential for the initial phase of somite proliferation. Measures of difference (L − R and DIF) may reflect regression of mammary buds. Based on the lack of genomic variation detected for L − R and DIF, these traits appear to be controlled by non-genetic factors. These factors must have a role during gestation, but the estimated effects of litter variation were zero for both L − R and DIF in the model development phase of this study. Similarly, Borchers et al. [8] and Fernández et al. [9] found virtually no effects of litter on L − R as well as only a minor effect of litter on count traits. While environmental factors such as stress and disease have been associated with fluctuating asymmetry in mammals [44], the factors that regulate asymmetrical mammary development in pigs is still unknown.

Few studies have actually evaluated left and right teat count values in pigs. While approximately 60% of pigs have the same number of teats on each side [8, 10] and the current study, a range of −3 to +3 for L − R was reported by Borchers et al. [8] and Fernández et al. [9]. A much wider range was observed in the current study (−4 to +7). Based on a limited number of studies, it appears that, in pig, teat number on the left side is a more heritable than teat number on the right side. Borchers et al. [8] reported a higher heritability for LTN than for RTN (0.20 vs. 0.18, respectively) although this difference was not statistically significant. Similarly, the current study found a 10% increase in heritability for LTN versus RTN in the model development dataset (Table 1) and in the subset of these animals used for GWAS (Table 2). Ding et al. [6] reported ten QTL for LTN and only seven QTL for RTN (30% fewer RTN than LTN) and 18 versus 13 (28% fewer RTN than LTN) 1-Mb regions explained more than 1% of the genomic variation in this study. The observation of identical trends and nearly identical magnitudes of differences is compelling. Furthermore, a region on SSC6:136 Mb that was reported by Ding et al. [6] to only affect LTN coincides with our results. In most studies [6, 9, 10] and the current study, the mean RTN was slightly larger than the mean LTN; however, this trend was not found by Borchers et al. [8]. Polythelia and polymastia were reported to occur more frequently on the right side in both humans [45] and mice [46] while missing mammary glands are most frequently observed on the left side in mice [46], thus the larger number of teats on the right side concurs with these phenomena. The current study indicates that MAX has an even stronger genetic component than either LTN or RTN. As this study is the first to report the genetic analysis of MAX, more studies are needed before this result can be confirmed because the differences in the estimated heritabilities were not statistically significant.

Other factors that affect mammary gland development
Studies of embryonic development in mice have shown that mammary glands of male fetuses regress between day 13.5 and 15.5 of gestation due to circulating androgens, such that after birth male pups do not possess mammary gland tissue [47]. Based on this hypothesis, Drickamer et al. [48] studied the effect of the percentage of males in a litter on teat number in pigs and found that the mean teat number decreased as the percentage of male pigs increased. Some studies in pigs have included a fixed effect for sex of the pig when analyzing teat number but either they did not indicate if the effect was significant or they did not present estimates of the effect [8, 13]. Willham and Whatley [10] reported similar number of teats for male versus female piglets. Our data did not show any impact of the sex of the piglet (mean TTN of 14.73 for both male and female piglets) or of the percentage of males in a litter (regression coefficients were <0.001 for all analyses). In fact, a query on three

populations of commercial-type pigs born at USMARC (representing sampling industry animals in the 1993, 2000 and 2010–2015) resulted in virtually identical mean numbers of teats among male and female piglets (14.64 vs. 14.65; respectively) among more than 110,000 piglets.

Borchers et al. [8] reported evidence that teat count was correlated with birth weight. In the USMARC populations studied here, this trend was also present (Fig. 2) in all three commercial-type populations that are maintained since the early 1990s as well as in a population created in 1980. Surprisingly, TTN was not associated with number of piglets born or vertebra numbers. Since mammary gland and teat development is already evident by 28 days of gestation, the causative factor(s) for the correlation between birth weight and TTN must occur in early stages of development. Based on the estimated litter variance for all analyzed traits, the factor(s) are not common to all fetuses within a litter. An improved understanding of these factors may enable increasing both teat number and birth weights in commercial swine.

Selection for increased number of teats
Estimated heritabilities indicate that selection would be successful for increased teat number. This was documented in sheep by Alexander Graham Bell [49] who observed that some sheep were born with up to eight teats versus the normal 2. Selection pressure for teat number in swine has typically relied on a minimum threshold (independent culling level) where pigs that were above this threshold were selection candidates. Thresholds of either 12 or 14 were often implemented. A more dramatic increase in teat number could be obtained if more rigorous selection was applied.

Estimated heritabilities for teat number often range from 0.2 to 0.4 [8]. The estimated heritability presented in Table 1 (0.49) was higher than the genomic heritability from GWAS (0.31) and the estimated heritability in the validation phase when no SNPs were fit (0.37). A contributing factor to these differences was the animals included in the analyses. During the development of the model, all animals born were included. However, for all other analyses, only genotyped animals were included. Most of the genotyped animals were females that had been retained for breeding so a minimum of 12 teats were required. As shown by the differences between minimum and maximum values in Tables 1 and 2, less phenotypic variation was present in the genotyped animals and the genetic variation estimated in these animals was lower. Arakawa [50] found a lower heritability based on SNPs (0.34) than that estimated from pedigree information (0.43), which is similar to the current study.

If final teat number is the result of the number of mammary buds initially developed and a maintenance (or

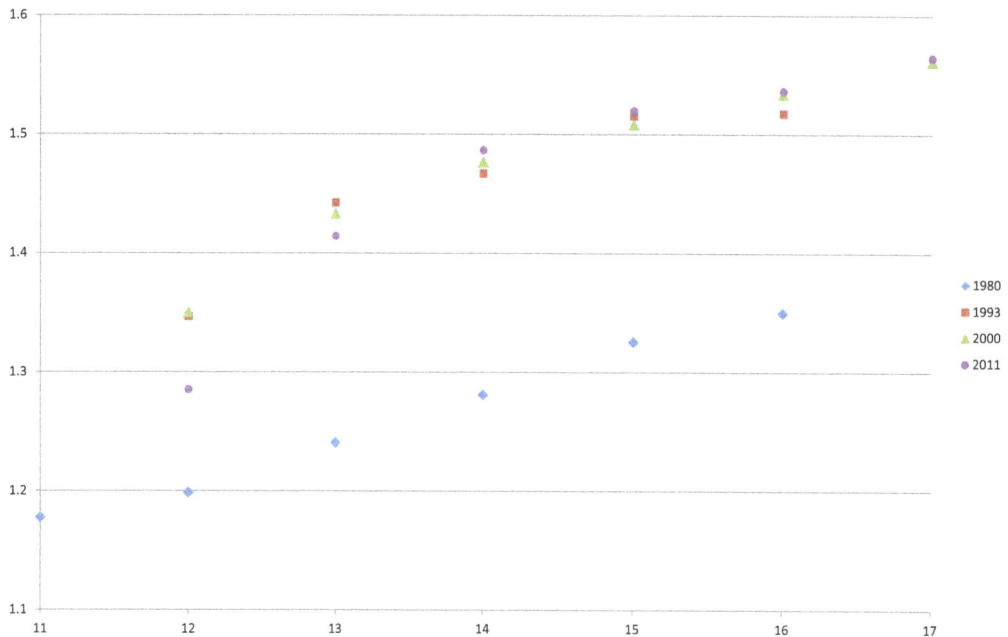

Fig. 2 Average birth weight (kg) for piglets born by total number of teats from four different populations at USMARC derived from commercial genetics in 1980, 1993, 2000 and 2011. Each class has at least 800 animals represented in the mean value. No values for the 11 teat number class for 1993, 2000 or 2011 and the 17 teat number class for 1980 or 1993 are provided due to too few pigs

regression) component, which acts randomly relative to the side of the developing organism, then MAX would be a better indicator of the initial number of mammary buds than RTN or LTN. L − R or DIF could predict the maintenance or regression component; however, genetic variation for these traits was nil, thus selection based on these measures would likely be ineffective. Among the traits analyzed here, TTN is the trait with the highest heritability and the most phenotypic variation and since it is the most important trait for swine production, selective pressure to increase lactational capacity in commercial sows should focus on TTN. Based on the association between birth weight and teat number, this would result in a serendipitous increase of piglet birth weight if this association is due to genetics.

Conclusions

Selection to increase the number of teats is possible in pigs and use of genetic markers should expedite progress. Since individual side counts of teats is not commonly recorded, exploitation of loci that independently control RTN or LTN would be most effective with genetic markers. These results validate the effect of *VRTN* on teat number as well as the QTL located on SSC4:25 and SSC10:52 Mb, and we identified an important novel region on SSC14:51–55 Mb, which needs to be further studied. The most heritable trait that also possessed the

most phenotypic variation was TTN. Therefore, without SNPs, single-trait traditional selection for TTN would yield the greatest gains; however, a selection index including TTN and MAX might slightly improve selection response. In this study, factors such as sex of the pig or of other pigs in the litter had no effect on TTN and there was no common environmental effect associated with fluctuating asymmetry for teat counts. The observed correlation between birth weight and teat number was unexpected and deserves further investigation. While it is clear that teat number can be increased in pigs, whether an increase in number of teats will result in increased total milk production still needs to be addressed.

Additional file

Additional file 1: Table S1. Information on 1-Mb windows that explained more than 0.4% of the genomic variation as determined by GENSEL. Description: A listing of the information from all 1-Mb windows (defined in SSC and Mb) that exceeded 0.4% of the genomic variation as determined by GENSEL. The number of SNPs within each window (#SNPs), percentage of genomic variation explained by the SNPs on average across all post-burn-in samples (%Var), the frequency at which a 1-Mb window explained more than the average amount of genomic variation (p > Average), the position of the first SNP (map_pos0) and last SNP (map_posn) in the window. The last eight columns pertain to the SNPs within the 1-Mb window with the largest estimated effect: SNP name, location in build 10.2, effect size, standard error of the estimate, frequency of SNP retention in each sample, allele frequency of the B allele in the population, T-test for the effect and simple P-value corresponding to the T-test.

Authors' contributions
GAR designed the study, collected phenotypic and genotypic data and conducted analyses; DJN performed functional analyses of genomic regions; both authors contributed to writing the manuscript. Both authors read and approved the final manuscript.

Acknowledgements
The authors would like to acknowledge Dr. JM Veltmaat for stimulating discussions pertaining to mammary gland development, Dr. JF Schneider for conducting preliminary analyses, Ms. K Simmerman for expert technical support, USDA-ARS-BARC BFGL lab for reading BeadChips, USMARC Swine operations staff for animal care and Ms. L. Parnell for manuscript preparation. Mention of trade names or commercial products in this publication is solely for the purpose of providing specific information and does not imply recommendation or endorsement by the U.S. Department of Agriculture. The U.S. Department of Agriculture (USDA) prohibits discrimination in all its programs and activities on the basis of race, color, national origin, age, disability, and where applicable, sex, marital status, familial status, parental status, religion, sexual orientation, genetic information, political beliefs, reprisal, or because all or part of an individual's income is derived from any public assistance program. (Not all prohibited bases apply to all programs.) Persons with disabilities who require alternative means for communication of program information (Braille, large print, audiotape, etc.) should contact USDA's TARGET Center at (202) 720-2600 (voice and TDD). To file a complaint of discrimination, write to USDA, Director, Office of Civil Rights, 1400 Independence Avenue, S.W., Washington, DC 20250-9410, or call (800) 795-3272 (voice) or (202) 720-6382 (TDD). USDA is an equal opportunity provider and employer.

Competing interests
Both authors declare they have no competing interests.

Funding
Funding for this project was supplied by USDA CRIS #3040-31000-094-00D.

References
1. Andersen IL, Nævdal E, Bøe KE. Maternal investment, sibling competition, and offspring survival with increasing litter size and parity in pigs (*Sus scrofa*). Behav Ecol Sociobiol. 2011;65:1159–67.
2. Merks JWM, Mathur PK, Knol EF. New phenotypes for new breeding goals in pigs. Animal. 2012;6:535–43.
3. Nielsen B, Su G, Lund MS, Madsen P. Selection for increased number of piglets at d 5 after farrowing has increased litter size and reduced piglet mortality. J Anim Sci. 2013;91:2575–82.
4. Cassady JP, Young LD, Leymaster KA. Heterosis and recombination effects on pig reproductive traits. J Anim Sci. 2002;80:2303–15.
5. Kim JS, Jin DI, Lee JH, Son DS, Lee SH, Yi YJ, et al. Effects of teat number on litter size in gilts. Anim Reprod Sci. 2005;90:111–6.
6. Ding N, Guo Y, Knorr C, Ma J, Mao H, Lan L, et al. Genome-wide QTL mapping for three traits related to teat number in a White Duroc × Erhualian pig resource population. BMC Genet. 2009;10:6.
7. Pumfrey RA, Johnson RK, Cunningham PJ, Zimmerman DR. Inheritance of teat number and its relationship to maternal traits in swine. J Anim Sci. 1980;50:1057–60.
8. Borchers N, Reinsch N, Kalm E. Teat number, hairiness and set of ears in a Piétrain cross: variation and effects on performance traits. Arch Tierz Dummerstorf. 2002;45:465–80.
9. Fernández A, Toro M, Rodríguez C, Silió L. Heterosis and epistasis for teat number and fluctuating asymmetry in crosses between Jiaxing and Iberian pigs. Heredity. 2004;93:222–7.
10. Willham RL, Whatley JA Jr. Genetic variation in nipple number in swine. J Anim Breed Genet. 1962;78:350–63.
11. Guo YM, Lee GJ, Archibald AL, Haley CS. Quantitative trait loci for production traits in pigs: a combined analysis of two Meishan × Large White populations. Anim Genet. 2008;39:486–95.
12. Sato S, Atsuji K, Saito N, Okitsu M, Sato S, Komatsuda A, et al. Identification of quantitative trait loci affecting corpora lutea and number of teats in a Meishan × Duroc F2 resource population. J Anim Sci. 2006;84:2895–901.
13. Duijvesteijn N, Veltmaat JM, Knol EF, Harlizius B. High-resolution association mapping of number of teats in pigs reveals regions controlling vertebral development. BMC Genomics. 2014;15:542.
14. Mikawa S, Morozumi T, Shimanuki S-I, Hayashi T, Uenishi H, Domukai M, et al. Fine mapping of a swine quantitative trait locus for number of vertebrae and analysis of an orphan nuclear receptor, germ cell nuclear factor (NR6A1). Genome Res. 2007;17:586–93.
15. Mikawa S, Sato S, Nii M, Morozumi T, Yoshioka G, Imaeda N, et al. Identification of a second gene associated with variation in vertebral number in domestic pigs. BMC Genet. 2011;12:5.
16. Lopes MS, Bastiaansen JWM, Harlizius B, Knol EF, Bovenhuis H. A genome-wide association study reveals dominance effects on number of teats in pigs. PLoS One. 2014;9:e105867.
17. Verardo LL, Silva FF, Lopes MS, Madsen O, Bastiaansen JWM, Knol EF, et al. Revealing new candidate genes for reproductive traits in pigs: combining Bayesian GWAS and functional pathways. Genet Sel Evol. 2016;48:9.
18. Veltmaat JM, Ramsdell AF, Sterneck E. Positional variations in mammary gland development and cancer. J Mammary Gland Biol Neoplasia. 2013;18:179–88.
19. Propper AY, Howard BA, Veltmaat JM. Prenatal morphogenesis of mammary glands in mouse and rabbit. J Mammary Gland Biol Neoplasia. 2013;18:93–104.
20. Veltmaat JM, Relaix F, Le LT, Kratochwil K, Sala FG, van Veelen W, et al. Gli3-mediated somitic Fgf10 expression gradients are required for the induction and patterning of mammary epithelium along the embryonic axes. Development. 2006;133:2325–35.
21. Schneider JF, Rempel LA, Rohrer GA. Genome-wide association study of swine farrowing traits. Part I: genetic and genomic parameter estimates. J Anim Sci. 2012;90:3353–9.
22. Meyer K. WOMBAT: a tool for mixed model analyses in quantitative genetics by restricted maximum likelihood (REML). J Zhejiang Univ Sci B. 2007;8:815–21.
23. Rohrer GA, Nonneman DJ, Wiedmann RT, Schneider JF. A study of vertebra number in pigs confirms the association of vertnin and reveals additional QTL. BMC Genet. 2015;16:129.
24. Groenen MAM, Archibald AL, Uenishi H, Tuggle CK, Takeuchi Y, Rothschild MF, et al. Analyses of pig genomes provide insight into porcine demography and evolution. Nature. 2012;491:393–8.
25. Sargolzaei M, Chesnais JP, Schenkel FS. A new approach for efficient genotype imputation using information from relatives. BMC Genomics. 2014;15:478.
26. Ramos AM, Crooijmans RPMA, Affara NA, Amaral AJ, Archibald AL, Beever JE, et al. Design of a high density SNP genotyping assay in the pig using SNPs identified and characterized by next generation sequencing technology. PLoS One. 2009;4:e6524.
27. Diamond JM. Evolutionary adaptations. Aristotle's theory of mammalian teat number is confirmed. Nature. 1987;325:200.
28. Gilbert AN. Mammary number and litter size in Rodentia: the "one-half rule". Proc Natl Acad Sci USA. 1986;83:4828–30.
29. Hoeferlin LA, Wijesinghe DS, Chalfant CE. The role of ceramide-1-phosphate in biological functions. Handb Exp Pharmacol. 2013;215:153–66.
30. Tissir F, Goffinet AM. Atypical cadherins Celsr1-3 and planar cell polarity in vertebrates. Prog Mol Biol Transl Sci. 2013;116:193–214.

31. Di Costanzo A, Festa L, Duverger O, Vivo M, Guerrini L, La Mantia G, et al. Homeodomain protein Dlx3 induces phosphorylation-dependent p63 degradation. Cell Cycle. 2009;8:1185–95.

32. Mills AA, Zheng B, Wang X-J, Vogel H, Roop DR, Bradley A. p63 is a p53 homologue required for limb and epidermal morphogenesis. Nature. 1999;398:708–13.

33. Hirooka H, de Koning DJ, Harlizius B, van Arendonk JAM, Rattink AP, Groenen MAM, et al. A whole-genome scan for quantitative trait loci affecting teat number in pigs. J Anim Sci. 2001;79:2320–6.

34. Rodríguez C, Tomás A, Alves E, Ramirez O, Arqué M, Muñoz G, et al. QTL mapping for teat number in an Iberian-by-Meishan pig intercross. Anim Genet. 2005;36:490–6.

35. Dragos-Wendrich M, Moser G, Bartenschlager H, Reiner G, Geldermann H. Linkage and QTL mapping for Sus scrofa chromosome 10. J Anim Breed Genet. 2003;120:82–8.

36. Wiedmann RT, Nonneman DJ, Rohrer GA. Genome-wide copy number variations using SNP genotyping in a mixed breed swine population. PLoS One. 2015;10:e0133529.

37. Ikenouchi J, Umeda M. FRMD4A regulates epithelial polarity by connecting Arf6 activation with the PAR complex. Proc Natl Acad Sci USA. 2010;107:748–53.

38. Stucke VM, Timmerman E, Vandekerckhove J, Gevaert K, Hall A. The MAGUK protein MPP7 binds to the polarity protein hDlg1 and facilitates epithelial tight junction formation. Mol Biol Cell. 2007;18:1744–55.

39. Cheng G, Zhong M, Kawaguchi R, Kassai M, Al-Ubaidi M, Deng J, et al. Identification of PLXDC1 and PLXDC2 as the transmembrane receptors for the multifunctional factor PEDF. Elife. 2014;3:e05401.

40. Tortereau F, Gilbert H, Heuven HCM, Bidanel JP, Groenen MAM, Riquet J. Combining two Meishan F2 crosses improves the detection of QTL on pig chromosomes 2, 4 and 6. Genet Sel Evol. 2010;42:42.

41. Huang JZ, Chen M, Zeng M, Xu SH, Zou FY, Chen D, et al. Down-regulation of TRPS1 stimulates epithelial–mesenchymal transition and metastasis through repression of FOXA1. J Pathol. 2016;239:186–96.

42. Boras-Granic K, Chang H, Grosschedl R, Hamel PA. Lef1 is required for the transition of Wnt signaling from mesenchymal to epithelial cells in the mouse embryonic mammary gland. Dev Biol. 2006;295:219–31.

43. Rohrer GA. Identification of quantitative trait loci affecting birth characters and accumulation of backfat and weight in a Meishan–White composite resource population. J Anim Sci. 2000;78:2547–53.

44. Møller AP. A review of developmental instability, parasitism and disease: infection, genetics and evolution. Infect Genet Evol. 2006;6:133–40.

45. Schmidt H. Supernumerary nipples: prevalence, size, sex and side predilection—a prospective clinical study. Eur J Pediatr. 1998;157:821–3.

46. Little CC, McDonald H. Abnormalities of the mammae in the house mouse. J Hered. 1945;36:285–8.

47. Veltmaat JM, Mailleux AA, Thiery JP, Bellusci S. Mouse embryonic mammogenesis as a model for the molecular regulation of pattern formation. Differentiation. 2003;71:1–17.

48. Drickamer LC, Rosenthal TL, Arthur RD. Factors affecting the number of teats in pigs. J Reprod Fertil. 1999;115:97–100.

49. Castle WE. The genetics of multi-nippled sheep: an analysis of the sheep-breeding experiments of Dr. and Mrs. Alexander Graham Bell at Beinn Bhreagh, N. S. J Hered. 1924;15:75–85.

50. Arakawa A, Okumura N, Taniguchi M, Hayashi T, Hirose K, Fukawa K, et al. Genome-wide association QTL mapping for teat number in a purebred population of Duroc pigs. Anim Genet. 2015;46:571–5.

51. Cassady JP, Johnson RK, Pomp D, Rohrer GA, Van Vleck LD, Spiegel EK, et al. Identification of quantitative trait loci affecting reproduction in pigs. J Anim Sci. 2001;79:623–33.

52. Bidanel JP, Rosendo A, Iannuccelli N, Riquet J, Gilbert H, Caritez JC, et al. Detection of quantitative trait loci for teat number and female reproductive traits in Meishan × Large White F2 pigs. Animal. 2008;2:813–20.

53. Hernandez SC, Finlayson HA, Ashworth CJ, Haley CS, Archibald AL. A genome-wide linkage analysis for reproductive traits in F2 Large White × Meishan cross gilts. Anim Genet. 2014;45:191–7.

11

Systematic genotyping of groups of cows to improve genomic estimated breeding values of selection candidates

Laura Plieschke[1]* [ID], Christian Edel[1], Eduardo C. G. Pimentel[1], Reiner Emmerling[1], Jörn Bennewitz[2] and Kay-Uwe Götz[1]

Abstract

Background: Extending the reference set for genomic predictions in dairy cattle by adding large numbers of cows with genotypes and phenotypes has been proposed as a means to increase reliability of selection decisions for candidates.

Methods: In this study, we explored the potential of increasing the reliability of breeding values of young selection candidates by genotyping a fixed number of first-crop daughters of each sire from one or two generations in a balanced and regular system of genotyping. Using stochastic simulation, we developed a basic population scenario that mimics the situation in dual-purpose Fleckvieh cattle with respect to important key parameters. Starting with a reference set consisting of only genotyped bulls, we extended this reference set by including increasing numbers of daughter genotypes and phenotypes. We studied the effects on model-derived reliabilities, validation reliabilities and unbiasedness of predicted values for selection candidates. We also illustrate and discuss the effects of a selected sample and an unbalanced sampling of daughters. Furthermore, we quantified the role of selection with respect to the influence on validation reliabilities and contrasted these to model-derived reliabilities.

Results: In the most extended design, with 200 daughters per sire genotyped from two generations, single nucleotide polymorphism (SNP) effects were estimated from a reference set of 420,000 cows and 4200 bulls. For this design, the validation reliabilities for candidates reached 80 % or more, thereby exceeding the reliabilities that were achieved in traditional progeny-testing designs for a trait with moderate to high heritability. We demonstrate that even a moderate number of 25 genotyped daughters per sire will lead to considerable improvement in the reliability of predicted breeding values for selection candidates. Our results illustrate that the strategy applied to sample females for genotyping has a large impact on the benefits that can be achieved.

Background

Genomic selection and genomic breeding value estimation were implemented in several cattle breeding programs in the last few years. Since the introduction of this methodology, there has been a constant attempt to further improve it and to increase the reliabilities of genomic breeding values. One key factor is the size of the reference set [1, 2]. Nowadays, there are several international organizations that promote the exchange of genotypes on a regular basis to enlarge reference sets and to improve the quality of genomic predictions of the participating countries. In dual-purpose Fleckvieh (FV) cattle, genomic selection was implemented in 2011 and genetic evaluation centers in Germany and Austria cooperate in a joint genetic and genomic evaluation that uses a common genotype pool [3]. Currently, the reference set for FV includes approximately 9000 bulls with phenotypic measures on most traits.

Several studies have reported that sharing genotypes within breeds results in large benefits for the reliability of

*Correspondence: Laura.Plieschke@lfl.bayern.de
[1] Bavarian State Research Center for Agriculture, Institute of Animal Breeding, Prof.-Dürrwaechter-Platz 1, 85586 Poing-Grub, Germany
Full list of author information is available at the end of the article

genomic predictions e.g. [4–6]. However, most opportunities to increase the genotype pool by exchanging genotypes have been exploited and, in most cases, the growth of reference sets within breeds is restricted to the yearly increase in number of genomically preselected young bulls receiving daughter proofs. As a consequence, fewer bulls are progeny-tested than in pre-genomic selection programs [7, 8] and the proportion of old bulls increases over time. Since the reliability of genomic predictions also depends on the degree of relationship between reference and predicted animals [9], this 'aging' of the reference set may lead to decreased reliabilities. As a demonstration of that effect, Cooper et al. [10], for example, excluded subsets of old bulls and found that older bulls in the reference set had only a minimal impact on the reliability of the genomic breeding values of predicted animals. In addition, preselection of young reference bulls may influence the quality of genomic predictions. Schaeffer [11] predicted a situation where considerable bias was introduced on genomic evaluations by strong preselection [12–14] of young bulls based on their genomic breeding values.

Another possibility to increase the size of the reference set is to use information from genotyped and phenotyped females, which can have a beneficial influence on the quality of genomic predictions. Thomasen et al. [15] found that, by adding female genotypes in the reference set, more genetic gain with a lower rate of inbreeding can be achieved compared to a breeding scheme where the reference set grows only from the addition of newly progeny-tested bulls. Pryce et al. [8] showed that by adding 10,000 cows to a reference set of 3000 Holstein bulls, the reliability of genomic predictions of 437 young bulls in the validation set was improved by 4 to 8 %. Calus et al. [16] also combined cows and bulls in a single reference set and found that the highest validation accuracies were achieved with the combined dataset compared to scenarios with a reference set that included only cows or only bulls. Furthermore, since usually cows are not strongly preselected, inclusion of their genotypes and phenotypes may also contribute to reduce the biasing effects of preselection as pointed out by Schaeffer [11]. Last but not least, genotyping cows might be especially important for creating reference sets for so-called new traits or expensive-to-measure traits [7, 17, 18] and, most likely, will be the basis of new and useful management tools for farmers [8].

If female genotypes are to be included in a genomic system, one of the key questions is which cows should be genotyped. Pryce et al. [8], Wiggans et al. [19] and Dassonneville et al. [20] discussed preferential treatment as a potential problem related to the inclusion of bulls' dams into the reference set. Dassonneville et al. [20] found that

the inclusion of records on elite cows resulted in overestimation of genomic enhanced breeding values for all animals. Thus, even if genotypes are available for elite cows as a consequence of using genomic predictions for the selection of bulls' dams, in the end, they should not be part of the reference set.

In a preliminary study [21], we performed a deterministic simulation based on nuclear pedigrees extracted from the German-Austrian FV population and showed that there is a benefit from including genotyped cows into the reference set. We quantified the effects of this inclusion on the reliability of genomic breeding values of young selection candidates and found marginal to considerable gains in reliability (between 1 and 40 %) depending on the scenario. However, we were not able to quantify the effects of selection on the results and we could not quantify the cumulative effects at the population level. Therefore, in this study, we examined the following three main effects by means of a stochastic simulation: (1) effects of selection on validation reliability, (2) effects of genotyping randomly selected cows on the accuracy of prediction, and (3) effects of some alternative strategies for sampling the genotyped daughters.

Methods
Simulation

We used the open access software QMSim [22] to run a simulation with five repetitions. Our aim was to simulate a population that resembled the German-Austrian dual-purpose Fleckvieh cattle population for several key characteristics (e.g. linkage disequilibrium (LD) structure, allele frequencies and effective population size).

Simulation of the population

QMSim first simulated a so-called historical population, which consisted of 2000 unrelated animals with a balanced sex ratio. These animals were randomly mated for 2500 generations. To create a sufficiently strong LD structure as observed in FV, a bottleneck was introduced after 2500 generations by reducing the number of breeding animals to 150 for one generation, which corresponds approximately to the effective population size in FV i.e. 160 based on the observed LD structure [23]. This estimate is quite close to that based on pedigree data [24]. After this bottleneck, population size was increased within one generation again to 31,500 animals (30,000 dams and 1500 sires), which represented the founder animals (generation 0) of the so-called 'recent' or pedigreed population. The recent population was propagated for another 10 generations. In each generation of the recent population, 15,000 female and 15,000 male offspring were generated by mating 30,000 dams and 1500 breeding sires. Generations overlapped and in each generation 30 % of the dams and 70 % of the sires were replaced. These two replacement parameters were quite similar to the situation observed in the real FV population. Breeding animals were selected based on their estimated breeding value (EBV) which was calculated within QMSim with a reliability of 0.6. This was done to mimic a genomic selection program where dams are selected based on a combination of pedigree information and own performance and sires are selected on their genomic breeding value.

Males of generation 5 to 10 were genotyped (Table 1). Sires belonging to generations 5 to 8 (n = 4200) were assigned to the reference set. The remaining animals of generations 9 and 10 were used as validation set for forward prediction. Note that whereas sires in generation 9 (n = 1050) were young bulls that were selected by QMSim based on a genomic breeding value but without daughter performances, the animals of generation 10 (n = 15,000) were unselected candidates. The validation animals were further characterized by the status of their sire i.e. a reference animal or not. Figure 1 gives an overview of the structure of the simulation.

Simulation of the genome

We simulated 30 chromosomes, each 100 cM long. On each chromosome, 1660 single nucleotide polymorphisms (SNPs) and 30 quantitative trait loci (QTL) were evenly distributed (49,800 SNPs and 900 QTL in total). After routine checks [3, 25], nearly 38,000 valid SNPs and approximately 700 QTL that were still segregating in the reference set (both numbers slightly varying between replicates of the simulation) were available. The routine checks were as follows: (1) SNPs that deviated from Hardy–Weinberg equilibrium (HWE) with a p-value less than 10^{-5} and (2) SNPs with a minor allele frequency (MAF) lower than 0.02 were excluded from the dataset. We assumed a sex-linked trait and a single

observation for each female with a heritability set to 0.4. The polygenic nature of the trait was ensured by the relatively large number of QTL and their effects were drawn from a uniform distribution (option 'uniform' from QMSim) to prevent the occurrence of a few isolated large QTL effects. With a uniform distribution, the mean of the effects is related to the variance and, thus, the range of the QTL effects is limited. We performed a couple of tests with QMSim and the results confirmed our assumptions (data not shown).

Simulation of the daughter sets

In the main part of the simulation, we generated 200 daughters for each of the reference bulls of generations 7 and 8 (which represented a total of 420,000 additional female genotypes and phenotypes). Due to memory requirements and some limitations of the QMSim software, we did not simulate the daughter genotypes with QMSim directly. Instead, based on the known haplotypes (SNPs and QTL) of the reference bulls of these two generations, we simulated different male gametes by recombination and randomly mated them with gametes of potential dams of the same cohort (excluding sisters, daughters and dams) that was simulated by applying the same strategy. Assuming a Poisson distribution for cross-overs, recombination was simulated by generating on average one random cross-over per Morgan for each chromosome. Using the observed QTL status of each daughter and the known (true) QTL effects from the QMSim simulation, we calculated the true breeding value (TBV) for each daughter.

Phenotypes

We generated yield deviations (YD, [26]) for daughters using the TBV and a random residual. Depending on the design investigated, these daughter phenotypes were used to calculate daughter yield deviations (DYD, [26])

Table 1 Assignment of animals to the reference or validation set in the different scenarios

Generation	Number of individuals			Explanation
	Base scenario	Extended scenarios step 1	Extended scenarios step 2	
5	1050	1050	1050	Reference set
6	1050	1050	1050	
7	1050	1050	1050 + daughters	
8	1050	1050 + daughters	1050 + daughters	
9	1050	1050	1050	Validation set
10a	4516	4516	4516	
10b	10,484	10,484	10,484	

Validation animals were further divided according to the status of the corresponding sire (member of the reference set or not), resulting in three validation groups.
Sires of validation animals in generations 9 and 10a were part of the reference set and sires of validation animals in generation 10b were not part of the reference set.
First, daughters of the sires of generation 8 were added to the reference set (step 1) and then daughters of the sires of generation 7 were also added (step 2)

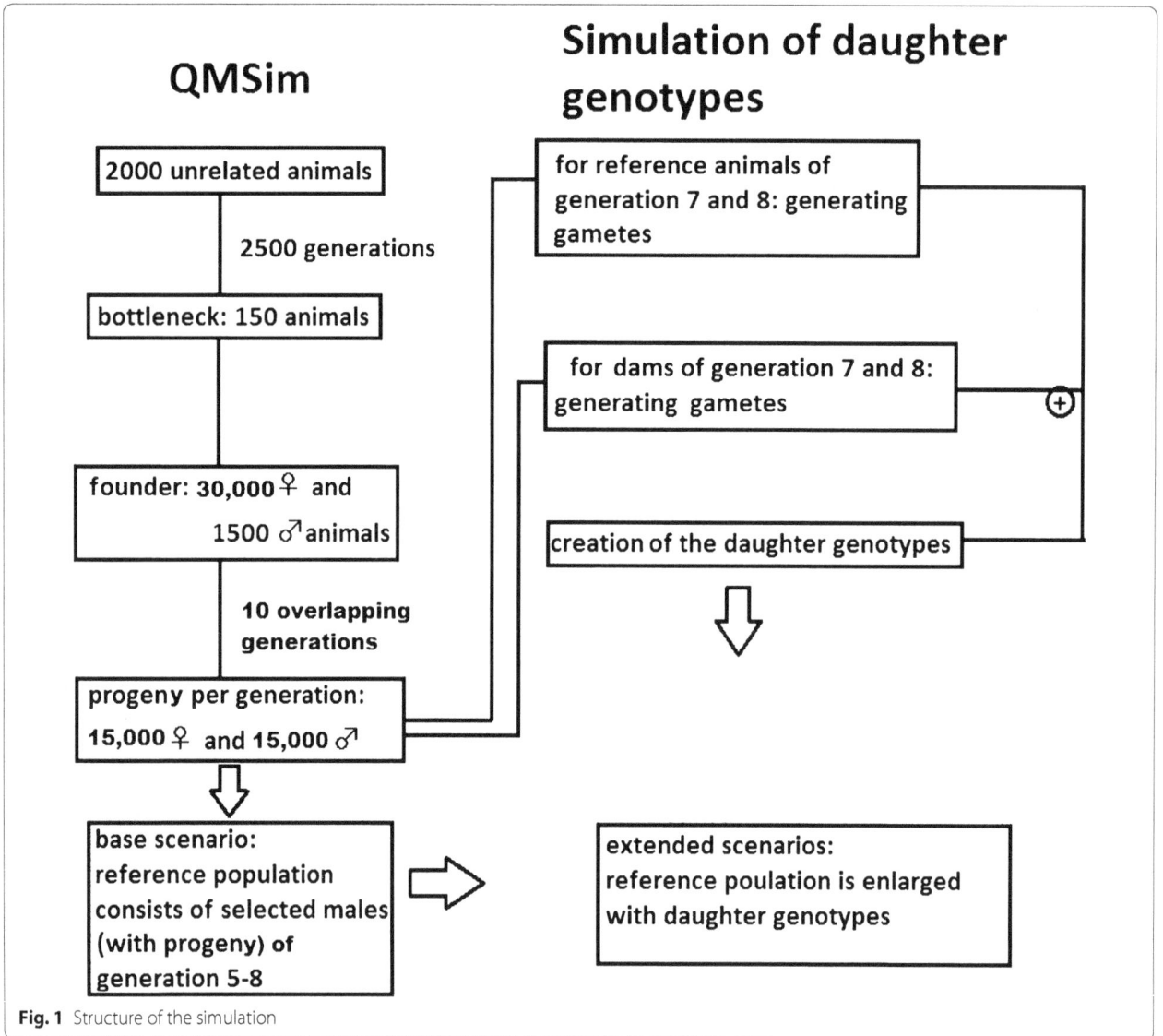

Fig. 1 Structure of the simulation

of the corresponding bull or were directly included in the reference set. In this way, YD of the reference daughters were automatically omitted from the daughter yield deviation (DYD) of the sire and double-counting in the extended scenarios was avoided. To account for different variances of the YD and DYD, phenotypes were weighted with the equivalent number of own performances (EOP, [27]) calculated as

$$EOP = \lambda \frac{R^2_{phen}}{1-R^2_{phen}},$$

where $\lambda = \frac{\sigma^2_e}{\sigma^2_a}$ with σ^2_a being the additive genetic variance and σ^2_e the residual variance and R^2_{phen} the reliability of the DYD or YD.

Designs

In a more general analysis, we investigated the effects of selection on validation reliability and model-derived reliability parameters. To be able to identify these selection effects, we repeated the basic scenario using the same parameters for QMSim except that we replaced directional selection on EBV by random selection.

In the main part of the simulation, we included large numbers of genotyped cows into the reference set. The general sampling strategy was to genotype a random sample of fixed size of phenotyped daughters of each artificial insemination (AI) bull in defined cohorts. We investigated 10 different scenarios: one base scenario and nine extended scenarios. In the base scenario, the reference set consisted only of sires of generations 5 to 8. For the extended scenarios, an increasing number of the generated female genotypes and phenotypes were integrated into the reference set. Tables 1 and 2 give an overview of the different scenarios.

To assess how robust the benefits are with respect to our general sampling strategy, we changed the composition of the sample of scenario −/50 (Table 2). Instead of including a random sample of daughters as was done in scenario −/50, we selected the best 50 daughters of each sire for scenario −/50$_s$ (selection was done on YD). In the scenario −/25$_r$25$_s$, we selected 25 daughters at random and combined them with the 25 best remaining daughters of the corresponding sire. Finally, we also ran one unbalanced scenario (−/50$_{ub}$) with different numbers of daughters per sire to test the effect of moderate unbalancedness but the overall number of genotyped females was kept the same as in scenario −/50. This was done by randomly selecting five daughters for 330 sires, 50 daughters for 621 sires and all 200 daughters for 99 sires. The different numbers of the daughter sets per sire were chosen arbitrarily but we ensured that the total number of genotyped females was maintained and that each sire was represented by at least some daughters. Moreover, random assignment of the different numbers of daughters to the sires was also conducted.

Table 2 Scenarios with corresponding number of animals and composition of the reference set

Scenario	Reference set	
	Number of sires	Number of daughters
Base	4200	0
−/25	4200	26,250
−/50	4200	52,500
−/100	4200	105,000
−/200	4200	210,000
50/50	4200	105,000
100/100	4200	210,000
200/200	4200	420,000
−/50$_s$	4200	52,500
−/25$_r$25$_s$	4200	52,500
−/50$_{ub}$	4200	52,500

The names of the extended scenarios are derived from the number of daughters per sire which are included in the reference set and the sire's generation. The number before the slash in the scenario's name is the number of daughters per progeny-tested bull of generation 7 (i.e. step 2 of the extended scenarios) and the number after the slash is the number of daughters per progeny-tested bull of generation 8 (i.e. step 1 of the extended scenarios). The −/50$_s$ is a scenario in which the best daughters were selected to be genotyped, −/25$_r$25$_s$ is a scenario in which 25 random daughters per sire and the 25 best daughters per sire were selected and genotyped and −/50$_{ub}$ is a scenario in which an unbalanced number of daughters for all sires was selected

Genomic prediction

Due to the large number of genotypes, we used a SNP-best linear unbiased prediction (BLUP) model [28] to calculate direct genomic values (DGV) and reliabilities. The model equation is as follows:

$$\mathbf{y} = \mathbf{Xb} + \mathbf{Mg} + \mathbf{e},$$

and the corresponding mixed model equations are:

$$\begin{pmatrix} \mathbf{X'R^{-1}X} & \mathbf{X'R^{-1}M} \\ \mathbf{M'R^{-1}X} & \mathbf{M'R^{-1}M} + \mathbf{I}/\sigma_g^2 \end{pmatrix} \begin{pmatrix} \hat{\mathbf{b}} \\ \hat{\mathbf{g}} \end{pmatrix} = \begin{pmatrix} \mathbf{X'R^{-1}y} \\ \mathbf{M'R^{-1}y} \end{pmatrix}$$

where

$$\sigma_g^2 = \frac{\sigma_a^2}{\sum_{j=1}^m (2p_j q_j)},$$

and $\mathbf{y}$ is the vector of observations (here DYD or YD), $\mathbf{b}$ the vector of fixed effects (in our simulation only an overall mean), $\mathbf{g}$ is the vector of random marker effects, and $\mathbf{e}$ the vector of residual effects. Matrix $\mathbf{X}$ is a design matrix which links the observations to the respective fixed effects and $\mathbf{M}$ is the centered coefficient matrix of marker genotypes and p_j and q_j are base allele frequencies of marker j estimated for generation 0 [29]. Centering was done by subtracting two times the base allele frequency estimate from the corresponding column of $\mathbf{M}$. Matrix $\mathbf{R}$ is a diagonal matrix with σ_e^2/w_i on the diagonal, where w_i is the EOP of the i-th observation and matrix $\mathbf{I}$ is an identity matrix of order m (number of markers).

DGV are calculated as:

$$\text{DGV} = \hat{\mathbf{b}} + \mathbf{M}\hat{\mathbf{g}},$$

and the corresponding predicted error variances (pev) are calculated as:

$$\text{pev(DGV)} = \mathbf{M}^* \mathbf{C}_s^{-1} \mathbf{M}^{*\prime},$$

where $\mathbf{M}^*$ is matrix $\mathbf{M}$ extended with a column of ones, and $\mathbf{C}_s^{-1}$ is the inverse of the left hand side of the SNP-BLUP-MME (mixed model equation). The inclusion of the overall mean in the calculation of the pev can be questioned and may lead to slightly higher theoretical reliabilities. We empirically compared results including and omitting the overall mean and found differences that were smaller than the rounding precision of the results. Moreover, because the overall mean is included in each scenario, its impact on the contrasts between scenarios can be ignored.

The reliability of the DGV of the i-th animal can then be calculated as:

$$R_i^2 = 1 - \frac{\text{diag(pev(DGV))}_i}{\text{diag}(\mathbf{G})_i \sigma_a^2},$$

where diag(pev(DGV))_i is the i-th diagonal element of the pev(DGV) and $\text{diag}(\mathbf{G})_i$ the i-th diagonal element of the genomic relationship matrix (**G**) which is 1 plus the genomic inbreeding coefficient. Matrix **G** is defined as follows:

$$\mathbf{G} = \frac{\mathbf{MM'}}{\sum_{j=1}^{m}(2p_j q_j)}.$$

In addition, we calculated a weighted regression of TBV on DGV for validation animals. We used the model fit of this regression as a measure of validation reliability (ρ^2) and the slope (b) as a measure of the bias that describes the inflation of estimates [30].

To quantify the effect of incomplete LD between SNPs and QTL on the difference between model-derived theoretical reliabilities and validation reliabilities, we included an analysis where we extended the marker genotype coefficient matrix **M** by QTL genotypes. We present the results in the context of the comparison between designs with directional and with random selection (ρ^2_{QTL}).

Results

Comparison of the simulated dataset with the Fleckvieh population

Comparison of the extent of LD between the simulated dataset and the real Fleckvieh dataset [31], revealed a good agreement with slightly higher values of the linkage parameter r^2 [32] for the simulated data at shorter distances. The average distance between a QTL and the nearest SNP in the simulated data was 60 kb. Allele frequencies for the simulated dataset were more evenly distributed than those for the real FV data, for which a slight shift to lower allele frequencies was observed. These results are illustrated in Figure S1 [see Additional file 1: Figure S1] and Figure S2 [see Additional file 2: Figure S2].

Simulation

For ease of interpretation, we separated the presentation of results for generation 9 from those for generation 10, in order to highlight the fact that generation 9 represents a group of individuals that are already pre-selected on an

EBV including Mendelian sampling information in the course of the simulation process. This selection does have an effect on validation statistics [30]. In contrast, generation 10 is strictly unselected. Results for generation 10 were further divided according to the status of the sire (member of the reference group or not). A more detailed categorization of the results for these two generations is provided in Tables S1 and S2 [see Additional file 3: Tables S1 and S2]. There was a general tendency for scenarios with the same number of genotyped females (scenario −/100 compared to scenario 50/50 and scenario −/200 compared to scenario 100/100) showing nearly identical results. For the sake of clarity, we do not present results for the redundant scenarios. All the results shown are averages over five repetitions of the simulation. Standard errors of the results presented in the main body of the paper were less than 1.3 % for validation reliabilities (except for one scenario i.e. $-/25_r25_s$ where standard errors were between 3.2 and 4.1 %) and less than 0.02 for regression slopes.

General effects of selection

Table 3 shows model-derived reliabilities (R^2) and validation reliabilities (ρ^2) for a scenario with directional selection and a scenario with random selection. Model-derived reliabilities were slightly higher for the scenario with directional selection than for the scenario with random selection, which indicated that, with directional selection, the pattern of family sizes differs and results in a more informative structure for validation animals. Comparing R^2 with ρ^2 for randomly selected populations, we found slightly lower validation reliabilities when only SNPs were considered. When QTL were included in the SNP panel, validation reliabilities (ρ^2_{QTL}) were slightly higher than R^2. In the scenario with directional selection the validation reliabilities for generation 10 were lower than with random selection (40 to 51 and 33 to 40 %, respectively). When, in addition, the validation sample was selected on information that included Mendelian sampling information as in generation 9, the decrease

Table 3 Model-derived reliabilities (R^2) and validation reliabilities (ρ^2) in the base scenario with directional and random selection

Validation set	Sire status	Number of individuals	Base scenario					
			Random selection			Directional selection		
			R^2	ρ^2	ρ^2_{QTL}	R^2	ρ^2	ρ^2_{QTL}
9	Reference	1050	54	51	59	58	26	32
10a	Reference	4516	54	51	58	58	40	48
10b	Not reference	10,484	48	40	49	48	33	41

Validation animals were divided according to whether their sire was in the reference set or not. For the purpose of illustration (and only here), we included results of analyses in which the segregating QTL were included in the SNP panel used for estimation and prediction (ρ^2_{QTL})

in validation reliabilities was even more pronounced (26 to 51 %). Selection on parent average (PA) in the validation group did not result in inflated predictions (slope estimates ranged from 0.93 to 0.99 for generation 10a).

Effects of genotyped daughters

Table 4 presents validation reliabilities for the three validation groups for the basic scenario and five extended scenarios. Using results for group 10a as a starting point, it can generally be stated that introducing an increasing number of genotyped daughters into the reference set clearly had a positive impact on the validation reliability. Beginning with scenario −/100, validation reliabilities reached values of 70 % and more. If the sire of a validation animal was not a member of the reference set (generation 10b), the overall validation reliability was reduced, but the general trend observed was the same. As expected, the effect of the contribution of a missing sire to the overall reliability decreased as information increased. When the validation group itself was selected (generation 9), the validation reliabilities for all scenarios were lower than for the other validation groups. Again, the impact of this decrease was more pronounced when the number of cows in the reference set was smaller.

Effects of the composition of the daughter samples

Table 5 illustrates some aspects of the composition of the sample of daughters that were chosen for genotyping. Starting with values for R^2, ρ^2 and b for scenario −/50 as

a reference point, we found a lower validation reliability and a noticeable increase in inflation of genomic predictions when a selected daughter group was genotyped (scenario −/50$_s$), even if selection was based on the criterion of moderate reliability as in this case. Comparing the base scenario (Table 4) to scenario −/50$_s$ (Table 5), the benefit from adding 52,500 genotyped daughters was small with respect to validation reliability. The negative effect of this preselection can be partially compensated by a combination of directly and randomly selected daughters (scenario −/25$_r$25$_s$, Table 5), but nevertheless the results were lower than those for a scenario where only 25 randomly selected daughters per sire were included (scenario −/25, see Table 4). A moderately unbalanced scenario (scenario −/50$_{ub}$, Table 5), however, had no detectable effect on reliabilities or regression slopes.

Discussion

In this study, we show that even small groups of daughters per sire can have large beneficial effects on model-derived reliabilities as well as validation reliabilities. A straightforward strategy to achieve these beneficial effects is to genotype a balanced random sample of daughters per sire. With respect to the structure of the validation sample, the results for generation 10 represent the ideal validation sample because it comprises the complete male offspring of the previous generation. In the following discussion, we refer to the results for validation group 10a unless otherwise indicated.

Table 4 Validation reliability (ρ^2) for six different scenarios

Validation set	Sire status	Number of individuals	ρ^2					
			Base	−/25	−/50	−/100	100/100	200/200
9	Reference	1050	26	44	53	62	72	80
10a	Reference	4516	40	56	65	73	80	86
10b	Not reference	10,484	32	51	60	69	77	84

Validation animals were divided according to whether their sire was in the reference set or not

Table 5 Model-derived reliabilities (R^2 were virtually equal across all scenarios), validation reliability (ρ^2) and regression slopes of the −/50 scenario and the three additional scenarios

Scenarios				−/50		−/50$_s$		−/25$_r$25$_s$ [a]		−/50$_{ub}$	
Validation set	Sire status	Number of individuals	R^2	ρ^2	b	ρ^2	b	ρ^2	b	ρ^2	b
9	Reference	1050	81	53	0.82	35	0.60	40	0.98	53	0.79
10a	Reference	4516	81	65	0.95	42	0.76	48	1.22	65	0.95
10b	Not reference	10,484	76	60	0.92	37	0.70	44	1.14	60	0.91

Validation animals were divided according to whether their sire was in the reference set or not

[a] Higher standard error compared to the other scenarios

Effects of selection

This section is included to illustrate the general effects of selection on validation statistics and to clarify the extent to which the results obtained can be explained by the fact that our population is under selection. The results (Table 3) are in good agreement with expectations and results found by other authors [33–35]. Surprisingly, at first, model-derived theoretical reliabilities were slightly higher for the scenario with directional selection than for the scenario with random selection. However, by analyzing family structures, we found that with directional selection the pattern of family sizes differed, resulting in a more informative structure for validation animals in scenarios with directional selection (results not shown). Model-derived theoretical reliabilities and validation reliabilities show relatively good agreement for the scenario with random selection. The slightly lower values for validation reliabilities are presumably a consequence of the fact that the LD between SNPs and QTL is not perfect and consequently some parts of the additive-genetic variance are not captured by SNPs [36]. However, by simply adding the QTL to the model, we found that validation reliabilities were slightly higher than model-derived theoretical reliabilities. In this case also, the theoretical model is only an approximation of the underlying true model.

The lower values for validation reliabilities under directional selection must be considered as a consequence of selection in the parental generation [33]. When the validation sample itself was selected on a criterion that included Mendelian sampling information, as was the case in generation 9, the decrease in validation reliabilities was even more pronounced. These results are in agreement with previous studies about the effects of selection on theoretical and validation reliabilities [35, 37].

Size and structure of the daughter samples

We tested different scenarios for which increasing numbers of genotyped and phenotyped daughters per sire were included in the reference set. By genotyping 25 daughters per sire from a single generation (corresponding to an overall number of 26,250 genotyped females, Table 2), the validation reliability was considerably improved, from 40 % in the base scenario to 56 % (Table 4, scenario –/25). As the number of daughters increased, the validation reliability showed a nearly linear increase. If we assume that proofs from progeny-testing typically show a validation reliability of about 70 % [38], this threshold is reached in scenario –/100 for validation group 10a and in scenario 100/100 for all other validation groups. With the largest number of genotyped daughters in scenario 200/200 (corresponding to a total of 420,000 genotyped females in the reference set), all validation

groups reached reliabilities of 80 % or more. This indicates that large numbers of (unselected) females in the reference set can largely compensate for unfavorable effects such as selection in the parental generation or the effect of a sire for which daughter proofs are not available. As already mentioned, we did not find any relevant differences between scenarios with equal total numbers of females (e.g. scenarios 50/50 and –/100). The similarity between the results of these scenarios is interesting. We expected that a scenario with daughters from two generations such as scenario 50/50 would lead to (slightly) higher validation reliabilities than scenario –/100 because with overlapping generations a larger number of sires would have genotyped daughters in scenario 50/50 and therefore more haplotypes would have been sampled. However, it seems that the existing diversity of haplotypes is already sufficiently covered when genotyping only one generation. In addition, beneficial effects can be reduced by an additional round of meiosis [21]. This implies that a large fraction of the benefits can be already generated in the first generation of a genotyping strategy that considers randomly selected females. Other studies found increases in validation reliabilities when including cows in the reference set but the reported increases were generally much lower e.g. [8, 16, 39]. We see several reasons for such differences. The most obvious one is certainly the larger number of cows that were assumed to be genotyped and phenotyped. Pryce et al. [8] and Koivula et al. [39] added approximately 10,000 genotyped cows to the reference set and Calus et al. [16] only ~1600 first lactation heifers. Other reasons might be related to key parameters such as the reliability of the phenotype [36], effective population size or the LD structure. Moreover, all studies mentioned above used real data that can be differently influenced by selection.

The concept that we propose here is based on genotyping and phenotyping a random sample of (preferably) first-crop daughters of each sire from a generation. We examined how deviations from this design would influence results. Comparison of the results of scenario –/50 (random daughter sample, Table 4) and scenario –/50$_s$ (selected daughter sample, Table 5), showed that with scenario –/50$_s$ the beneficial effect of an additional pool of 52,500 genotypes in the reference set on validation reliability is almost null when compared to the base scenario. Even worse, preselection of daughters caused an increase in inflation as indicated by the low regression slopes (Table 5). One possible explanation is that reference animals that are selected based on their within-family deviation lead to biased family means and also to biased estimates of the deviations from the family mean. Schaeffer [11] argued that the animal model might become obsolete due to the fact that, in the future, only preselected young bulls will become reference animals. The consequence of this preselection would be that the phenotyped sons of a sire would not represent a random sample of all sons of this sire. Schaeffer [11] expected a relevant increase in inflation as a consequence of this development and given our results this expectation might be at least partly justified. Although not explicitly covered here, it seems likely that the integration of elite cows in the reference set will result in an even stronger bias, because elite cows are not only selected, they frequently receive also preferential treatment so that even their phenotypes are biased. Studies of Wiggans et al. [19] and Dassonneville et al. [20] dealt with the consequences of preferential treatment and provide further evidence of its biasing effects.

The negative result of scenario $-/50_s$ can be only partly removed by a combination of selected and unselected daughters (scenario $-/25_r25_s$; same number of daughters, Table 5). This result indicates that the combination of selected and unselected data cannot yield precise and unbiased estimates. Moreover, the results of scenario $-/25_r25_s$ are lower than those of scenario $-/25$ (Table 4), which indicates that it might be relevant to exclude the genotypes of (pre-)selected daughters from the reference set if this information is available. This kind of monitoring presents an additional challenge especially to single-step genomic BLUP, in which putting a restriction on the reference set is not conceptually intended, an important aspect that was already emphasized by other authors [40].

Another factor with a strong impact on the validation results is the heritability of the trait. In a pilot study [21], we found that for traits with medium to high heritabilities ($h^2 = 0.35$), 100 genotyped daughters per bull increased the marginal reliability [41] by up to 17 % (depending on the scenario) whereas in situations with very low heritabilities ($h^2 = 0.05$), the same number of daughters increased the reliability by up to 4 % only. Our study was limited to a trait with a heritability of 0.4 to investigate several other questions. However, it may be expected that with a lower heritability, less substantial improvements would be found.

In the literature, there are other strategies for genotyping cows. Jiménez-Montero et al. [42] found higher reliabilities when cows selected from both extremes of the distribution of phenotypes were genotyped instead of the best ones or a random sample. We hypothesize that such a strategy would be better suited for traits for which only a few QTL with large effects segregate. Such traits are not common in dairy cattle [43] and therefore we focused our study on a trait with polygenic characteristics, for which no advantage of genotyping extreme animals is expected. Moreover, such a sampling strategy would require trait-specific daughter samples, which is an obstacle for practical implementation. In Calus et al. [16], cow genotypes

of entire herds are integrated in the reference set. This strategy could indeed ensure the representativeness of the cow sample if some precautions are taken. We found no disadvantages with moderate unbalancedness in scenario $-/50_{ub}$ in which we ensured that each bull was at least represented by a sample of five daughters. Further investigations on this subject are necessary to clarify which degree of unbalancedness can be tolerated before the accuracy of prediction deteriorates.

In real world breeding programs, it is reasonable to assume that there is a limited interest for the farmers to genotype randomly selected cows and to keep all of them for an unbiased performance recording. Thus, for practical implementation, it would be necessary to find a solution to finance the genotyping costs and to keep track of the cows sampled for the reference set. However, this independent financing solution, once established as a component of the breeding program, might be the only way to ensure a neutral, unselected daughter sample in the long term.

The simple balanced genotyping designs proposed here led to very stable improvements as indicated by the small standard errors of reliabilities and slopes. The only exception was for scenario $-/25_r25_s$, which showed more variation in the results. This indicates that some sampling designs are more robust than others with respect to the improvements that can be achieved.

Conclusions

Extending the reference set by adding a large number of cows with genotypes and phenotypes increases the reliability of breeding values of young selection candidates and may overcome the deterioration of validation reliabilities that are caused by intense preselection of young bulls. We showed the benefits from genotyping a random sample of (first-crop) daughters of all sires from one or two generations. It is possible to obtain reliabilities for selection candidates that are as high as, or even higher than, the reliabilities that have been formerly observed for young progeny-tested bulls. We found that the benefits that can be achieved are sensitive to the strategy used to sample females for genotyping.

Additional files

Additional file 1: Figure S1. LD-structure of the real Fleckvieh population (r2_Fleckvieh data, [30]) and of the simulated population (r2_simulated data) according to distance between SNPs in kb.

Additional file 2: Figure S2. Distribution of the allele frequencies. (A) Simulated data, approximately 38,000 segregating SNPs; (B) Real data on Fleckvieh cattle, approximately 41,000 segregating SNPs.

Additional file 3: Tables S1 and S2. More detailed results on validation reliabilities (ρ^2) (**Table S1**) and on the model-derived theoretical reliabilities (R^2) (**Table S2**) for six scenarios. Description: Validation animals were divided according to the generation of their sire.

Authors' contributions

LP performed the analysis and drafted the manuscript. LP, CE, ECGP, RE and KUG designed the study. LP and CE developed methods. CE, ECGP, RE, JB and KUG revised the manuscript. All authors read and approved the final manuscript.

Author details

[1] Bavarian State Research Center for Agriculture, Institute of Animal Breeding, Prof.-Dürrwaechter-Platz 1, 85586 Poing-Grub, Germany. [2] Institute of Animal Science, University Hohenheim, Garbenstraße 17, 70599 Stuttgart, Germany.

Acknowledgements

We gratefully acknowledge the Arbeitsgemeinschaft Süddeutscher Rinderzucht- und Besamungsorganisationen e.V. for their financial support within the research cooperation "Zukunftswege". Furthermore, we wish to thank the editors J van der Werf and H Hayes as well as two unknown reviewers for their helpful suggestions to improve the final manuscript.

Competing interests

The authors declare that they have no competing interests.

References

1. Daetwyler HD, Villanueva B, Woolliams JA. Accuracy of predicting the genetic risk of disease using a genome-wide approach. PLoS One. 2008;3:e3395.
2. Goddard ME. Genomic selection: prediction of accuracy and maximisation of long term response. Genetica. 2009;136:245–57.
3. Edel C, Schwarzenbacher H, Hamann H, Neuner S, Emmerling R, Götz KU. The German-Austrian genomic evaluation system for Fleckvieh (Simmental) cattle. Interbull Bull. 2011;44:152–6.
4. Schenkel FS, Sargolzaei M, Kistemaker G, Jansen GB, Sullivan P, Van Doormaal BJ, et al. Reliability of genomic evaluation of Holstein cattle in Canada. Interbull Bull. 2009;39:51–7.
5. Lund MS, de Roos SP, de Vries AG, Druet T, Ducrocq V, Fritz S, et al. A common reference population from four European Holstein populations increases reliability of genomic predictions. Genet Sel Evol. 2011;43:43.
6. Su G, Ma P, Nielsen US, Aamand GP, Wiggans G, Guldbrandtsen B, et al. Sharing reference data and including cows in the reference population improve genomic predictions in Danish Jersey. Animal. 2016;10:1067–75.
7. Buch LH, Kargo M, Berg P, Lassen J, Sorensen C. The value of cows in the reference populations for genomic selection of new functional traits. Animal. 2012;6:880–6.
8. Pryce JE, Hayes BJ, Goddard ME. Genotyping dairy females can improve the reliability of genomic selection for young bulls and heifers and provide farmers with new management tools. In: Proceedings of the 36th ICAR Biennial Session: 16–20 June 2008; Niagara Falls; 2009.
9. Clark SA, Hickey JM, Daetwyler HD, van der Werf JHJ. The importance of information on relatives for the prediction of genomic breeding values and the implications for the makeup of reference data sets in livestock breeding schemes. Genet Sel Evol. 2012;44:4.
10. Cooper TA, Wiggans GR, VanRaden PM. Short communication: analysis of genomic predictor population for Holstein dairy cattle in the United-States—Effects of sex and age. J Dairy Sci. 2015;98:2785–8.
11. Schaeffer, LR. Is the animal model obsolete in dairy cattle? University of Guelph, personal communication to the Animal Genetics Discussion Group (AGDG). 2014.
12. Vitezica ZG, Aguilar I, Misztal I, Legarra A. Bias in genomic predictions for populations under selection. Genet Res (Camb). 2011;93:357–66.
13. Patry C, Ducrocq V. Accounting for genomic pre-selection in national BLUP evaluations in dairy cattle. Genet Sel Evol. 2011;43:30.
14. Patry C, Ducrocq V. Evidence of biases in genetic evaluations due to genomic preselection in dairy cattle. J Dairy Sci. 2011;94:1011–20.
15. Thomasen JR, Sorensen AC, Lund MS, Guldbrandtsen B. Adding cows to the reference population makes a small dairy population competitive. J Dairy Sci. 2014;97:5822–32.
16. Calus MPL, de Haas Y, Veerkamp RF. Combining cow and bull reference populations to increase accuracy of genomic prediction and genome-wide association studies. J Dairy Sci. 2013;96:6703–15.
17. Calus MPL, de Haas Y, Pszczola M, Veerkamp RF. Predicted accuracy of and response to genomic selection for new traits in dairy cattle. Animal. 2013;7:183–91.
18. Egger-Danner C, Cole JB, Pryce JE, Gengler N, Heringstad B, Bradley A, et al. Invited review: overview of new traits and phenotyping strategies in dairy cattle with a focus on functional traits. Animal. 2015;9:191–200.
19. Wiggans GR, Cooper TA, VanRaden PM, Cole JB. Technical note: adjustment of traditional cow evaluations to improve accuracy of genomic predictions. J Dairy Sci. 2011;94:6188–93.
20. Dassonneville R, Baur A, Fritz S, Boichard D, Ducrocq V. Inclusion of cow records on genomic evaluations and impact on bias due to preferential treatment. Genet Sel Evol. 2012;44:40.
21. Edel C, Pimentel ECG, Plieschke L, Emmerling R, Götz KU. Short communication: the effect of genotyping cows to improve the reliability of genomic predictions for selection candidates. J Dairy Sci. 2016;99:1999–2004.
22. Sargolzaei M, Schenkel FS. QMSim: a large-scale genome simulator for livestock. Bioinformatics. 2009;25:680–1.
23. Pausch H, Aigner B, Emmerling R, Edel C, Götz KU, Fries R. Imputation of high-density genotypes in the Fleckvieh cattle population. Genet Sel Evol. 2014;45:3.
24. Bundesanstalt Für Landwirtschaft Und Ernährung. Endbericht: Erfassungsprojekt Erhebung von Populationsdaten tiergenetischer Ressourcen in Deutschland: Tierart Rind. 2010 (http://download.ble.de/07BE001.pdf). Accessed 24 Feb 2016.
25. Ertl J, Edel C, Emmerling R, Pausch H, Fries R, Götz KU. On the limited increase in validation reliability using high-density genotypes in genomic best linear unbiased prediction: observations from Fleckvieh cattle. J Dairy Sci. 2012;97:487–96.
26. VanRaden PM, Wiggans GR. Derivation, calculation, and use of national animal model information. J Dairy Sci. 1991;74:2737–46.
27. Edel C, Emmerling R, Götz KU. Optimized aggregation of phenotypes for MA-BLUP avaluation in German Fleckvieh. Interbull Bull. 2009;40:178–83.
28. Meuwissen THE, Hayes BJ, Goddard ME. Prediction of total genetic value using genome-wide dense marker maps. Genetics. 2001;157:1819–29.
29. Gengler N, Mayeres P, Szydlowski M. A simple method to approximate gene content in large pedigree populations: applications to the myostatin gene in dual-purpose Belgian Blue cattle. Animal. 2007;1:21–8.
30. Mäntysaari EA, Liu Z, VanRaden PM. Interbull validation test for genomic evaluations. Interbull Bull. 2010;41:17–22.
31. Ertl J, Edel C, Neuner S, Emmerling R, Götz K-U. Comparative analysis of linkage disequilibrium in Fleckvieh and Brown Swiss cattle. In: Proceedings of the 63rd annual meeting of the European federation of animal science: 27–21 August 2012; Bratislava; 2012.
32. Hill WG. Estimation of effective population size from data on linkage disequilibrium. Genet Res. 1981;38:209–16.
33. VanRaden PM. Efficient methods to compute genomic predictions. J Dairy Sci. 2008;91:4414–23.
34. Mäntysaari EA, Koivula M. GEBV validation rest revisited. Interbull Bull. 2012;45:1–5.
35. Gorjanc G, Bijma P, Hickey JM. Reliability of pedigree-based and genomic evaluations in selected populations. Genet Sel Evol. 2015;47:65.
36. Hayes BJ, Bowman PJ, Chamberlain AJ, Goddard ME. Invited review: genomic selection in dairy cattle: progress and challenges. J Dairy Sci. 2008;92:433–43.
37. Edel C, Neuner S, Emmerling R, Götz KU. A note on using 'forward prediction' to assess precision and bias of genomic predictions. Interbull Bull. 2012;46:16–9.
38. Powell RL, Norman HD, Sanders AH. Progeny testing and selection intensity for Holstein bulls in different countries. J Dairy Sci. 2003;6:3386–93.
39. Koivula M, Strandén I, Aamand GP, Mäntysaari EA. Effect of cow reference group on validation reliability of genomic evaluation. Animal. 2016;10:1021–6.
40. Liu Z, Goddard ME, Reinhardt F, Reents R. A single-step genomic model with direct estimation of marker effects. J Dairy Sci. 2014;97:5833–50.
41. Harris B, Johnson D. Approximate reliability of genetic evaluations under an animal model. J Dairy Sci. 1998;81:2723–8.
42. Jiménez-Montero JA, González-Recio O, Alenda R. Genotyping strategies for genomic selection in small dairy cattle populations. Animal. 2012;6:1216–24.
43. Daetwyler HD, Pong-Wong R, Villanueva B, Woolliams JA. The impact of genetic architecture on genome-wide evaluation methods. Genetics. 2010;185:1021–31.

12

Evaluation of the accuracy of imputed sequence variant genotypes and their utility for causal variant detection in cattle

Hubert Pausch[1][*] , Iona M. MacLeod[1], Ruedi Fries[2], Reiner Emmerling[3], Phil J. Bowman[1,4], Hans D. Daetwyler[1,4] and Michael E. Goddard[1,5]

Abstract

Background: The availability of dense genotypes and whole-genome sequence variants from various sources offers the opportunity to compile large datasets consisting of tens of thousands of individuals with genotypes at millions of polymorphic sites that may enhance the power of genomic analyses. The imputation of missing genotypes ensures that all individuals have genotypes for a shared set of variants.

Results: We evaluated the accuracy of imputation from dense genotypes to whole-genome sequence variants in 249 Fleckvieh and 450 Holstein cattle using *Minimac* and *FImpute*. The sequence variants of a subset of the animals were reduced to the variants that were included on the Illumina BovineHD genotyping array and subsequently inferred in silico using either within- or multi-breed reference populations. The accuracy of imputation varied considerably across chromosomes and dropped at regions where the bovine genome contains segmental duplications. Depending on the imputation strategy, the correlation between imputed and true genotypes ranged from 0.898 to 0.952. The accuracy of imputation was higher with *Minimac* than *FImpute* particularly for variants with a low minor allele frequency. Using a multi-breed reference population increased the accuracy of imputation, particularly when *FImpute* was used to infer genotypes. When the sequence variants were imputed using *Minimac*, the true genotypes were more correlated to predicted allele dosages than best-guess genotypes. The computing costs to impute 23,256,743 sequence variants in 6958 animals were ten-fold higher with *Minimac* than *FImpute*. Association studies with imputed sequence variants revealed seven quantitative trait loci (QTL) for milk fat percentage. Two causal mutations in the *DGAT1* and *GHR* genes were the most significantly associated variants at two QTL on chromosomes 14 and 20 when *Minimac* was used to infer genotypes.

Conclusions: The population-based imputation of millions of sequence variants in large cohorts is computationally feasible and provides accurate genotypes. However, the accuracy of imputation is low in regions where the genome contains large segmental duplications or the coverage with array-derived single nucleotide polymorphisms is poor. Using a reference population that includes individuals from many breeds increases the accuracy of imputation particularly at low-frequency variants. Considering allele dosages rather than best-guess genotypes as explanatory variables is advantageous to detect causal mutations in association studies with imputed sequence variants.

Background

Several genotyping arrays that comprise a varying number of single nucleotide polymorphisms (SNPs) are routinely used for genome-wide genotyping in cattle. Cows are usually genotyped using low-density genotyping arrays whereas bulls are mostly genotyped at higher density [1]. Moreover, the sequencing of important ancestors of many cattle breeds yielded genotypes at millions of polymorphic sites [2]. Combining the genotype data from various sources into a single large dataset may

*Correspondence: hubert.pausch@ecodev.vic.gov.au
[1] Agriculture Victoria, AgriBio, Centre for AgriBiosciences, Bundoora, VIC 3083, Australia
Full list of author information is available at the end of the article

enhance the power of genome-wide analyses. The imputation of missing genotypes is necessary to ensure that all individuals have genotypes for a shared set of variants. Genotype imputation may also infer dense genotypes in silico for animals that were genotyped at lower density by using animals that were genotyped at a higher density as a reference population [3].

Algorithms that infer missing genotypes apply family- (e.g., [4]) or population-based (e.g., [5–7]) imputation approaches or a combination thereof (e.g., [8–10]). Family-based imputation approaches rely on Mendelian transmission rules in pedigrees to infer missing genotypes. Population-based imputation approaches exploit linkage disequilibrium (LD) between adjacent markers to predict missing genotypes using a probabilistic framework without (explicitly) considering pedigree information [11]. While population-based imputation approaches are accurate, their computing costs are too high to infer genotypes for a large number of animals and markers for routine applications [12, 13]. Methods that apply a combination of family- and population-based imputation approaches exploit shared haplotypes among relatives thereby enabling rapid imputation of genotypes for tens of thousands of individuals and millions of markers in silico [8–10, 14].

The accuracy of imputation from low to higher density depends on the relationships between target and reference animals, genotype density in both panels, allele frequencies of the imputed variants, LD between adjacent markers and algorithms applied to infer missing genotypes (e.g., [12, 15, 16]). These parameters also affect the accuracy of imputation from dense genotypes to sequence variants [2, 17, 18]. However, the accurate imputation of low-frequency variants is critical with sequence data because rare alleles are more frequent in sequence than array-derived variants and the LD between low-frequency and array-derived variants may be low [19, 20].

In cattle, a number of studies with simulated and real sequence data indicated that using imputed sequence variant genotypes may improve genomic predictions and facilitate the detection of causal trait variants because the polymorphisms that underlie phenotypic variation are included in the data [2, 19, 21–23]. However, the benefits of dense marker maps for genome-wide analyses may be compromised when the accuracy of imputation is low [24–26].

In this paper, we evaluate the accuracy of imputing sequence variant genotypes in two cattle breeds using different imputation algorithms and reference populations. We perform association studies between imputed sequence variant genotypes and milk fat percentage in 6958 Fleckvieh bulls and show that the imputation strategy is critical to pinpoint causal mutations.

Methods
Animal ethics statement
No ethical approval was required for this study.

Sequenced animals
We used whole-genome sequence data of 1577 taurine animals that were included in the fifth run of the 1000 bull genomes project [2]. The reads were aligned to the UMD3.1 bovine reference genome using the BWA-MEM algorithm [27, 28]. SNPs, short insertions and deletions were genotyped for all sequenced animals simultaneously using a multi-sample variant calling pipeline that was implemented with the mpileup module of SAMtools [29] and is described in Daetwyler et al. [2]. The variant calling yielded genotypes at 39,721,987 biallelic sites for 1577 animals. We considered genotypes at 22,737,136 autosomal sequence variants that had a minor allele frequency (MAF) higher than 0.5% to build the genomic relationship matrix among the sequenced animals using the plink (version 1.9) software tool [30]. Principal components of the genomic relationship matrix were calculated using the GCTA (version 1.25.3) software tool [31]. Our analyses focused on the Fleckvieh (FV) and Holstein (HOL) breeds because a large number of animals from both breeds were included in the 1000 bull genomes data. Following the inspection of the top principal components, animals for which breed was uncertain were removed leaving 249 FV and 450 HOL cattle (see Additional file 1: Figure S1).

Design of cross-validation scenarios
The imputation from dense genotypes to full sequence variants was evaluated using 15-fold cross-validation in FV and HOL cattle. We considered only sequence variants that were polymorphic in 249 FV or 450 HOL animals. For each fold, the sequenced animals were divided into reference and validation animals. Forty-nine FV and 100 HOL animals, that were a random subset of the sequenced animals, were considered as validation animals (i.e., the same individual may have been included in several validation sets). All remaining sequenced animals from the 1000 bull genomes project were considered as a multi-breed reference population. The within-breed reference populations consisted of 200 FV or 350 HOL animals. The genotypes of the validation animals were reduced to the variants that were included on the Illumina BovineHD genotyping array (HD) to mimic dense genotypes. The masked genotypes of the validation animals were subsequently inferred in silico using full sequence information from the reference animals. The selection of validation animals and subsequent imputation of genotypes was repeated 15 times for chromosomes 1, 5, 10, 15, 20 and 25.

Evaluation of the accuracy of imputation

The overall accuracy of imputation was the mean correlation between in silico imputed and true (sequenced) genotypes ($r_{IMP,SEQ}$) across 15 folds and six chromosomes analysed. Specifically, for each fold and chromosome, we calculated the correlation between matrices of imputed and true genotypes. These values were averaged to obtain the overall $r_{IMP,SEQ}$. In addition, we grouped the imputed sequence variants into 50 classes with regard to their MAF in the reference population. The $r_{IMP,SEQ}$-value for each MAF class was the mean value across six chromosomes and 15 folds. To detect intra-chromosomal variations in the accuracy of imputation, we calculated $r_{IMP,SEQ}$-values for successive 1-Mb segments along each chromosome analysed. Differences between imputation scenarios were tested for significance using two-tailed t-tests.

Imputation methods

The performance of the *FImpute* (version 2.2) [9] and *Minimac3* (version 2.0.1, henceforth referred to as *Minimac*) [7] software tools was evaluated using default parameter settings. The algorithm implemented in *FImpute* uses family and population-based information to infer haplotypes and missing genotypes. A pedigree consisting of 47,012 FV animals tracing back ancestry information to animals born in 1925 was considered when genotypes were inferred using *FImpute*. The pedigree included ancestry information for 138 (out of 250) sequenced FV animals. Pedigree records for sequenced animals from breeds other than FV were also included if available. However, these pedigrees included only first-degree relatives. Genotypes for the imputed sequence variants were coded as 0, 1 and 2 for homozygous, heterozygous and alternative homozygous animals, respectively.

The algorithm implemented in the *Minimac* software tool uses previously phased genotypes, i.e., it takes haplotypes as input for reference and validation animals. Haplotype phases were estimated separately for the reference and validation animals using the phasing algorithm implemented in the *Eagle* (version 2.3) software tool [32]. Both *Eagle* and *Minimac* do not consider pedigree information to infer haplotypes and missing genotypes. *Minimac* provides best-guess genotypes (coded as 0, 1 and 2 for homozygous, heterozygous and alternative homozygous animals, respectively) and allele dosages (continuously distributed values ranging from 0 to 2) for the imputed sequence variants. The $r_{IMP,SEQ}$-values were calculated for best-guess genotypes ($Minimac_{BG}$) and allele dosages ($Minimac_{DOS}$).

Imputation of sequence variants in 6958 Fleckvieh animals

We imputed genotypes for 23,256,743 autosomal sequence variants that were polymorphic in 249 sequenced FV animals for 6958 FV bulls that had (partially imputed) array-derived genotypes at 603,662 autosomal variants (see [33]). Genotypes were imputed with either *FImpute* or *Minimac* using either within- or multi-breed reference populations that consisted either of 249 FV animals or 1577 animals from various bovine breeds (see above). Haplotype phases for reference and target animals were estimated using the *Eagle* software tool (see above). The LD among imputed sequence variants was calculated as the squared correlation between predicted allele dosages.

Accuracy of imputation at two known causal mutations

712 and 902 FV animals that had imputed sequence variant genotypes also had direct genotypes for two mutations in the *DGAT1* (rs109234250 and rs109326954, p.A232K) and *GHR* (rs385640152, p.F279Y) genes that were obtained for previous studies using TaqMan® genotyping assays (Life Technologies) [34–37]. The accuracy of imputation at both sites was the proportion of correctly imputed genotypes.

Association analyses in Fleckvieh cattle

Association tests between 23,256,743 imputed sequence variants and milk fat percentage were carried out in 6958 FV bulls using a variance component-based approach that was implemented in the *EMMAX* software tool and that accounts for population stratification by fitting a genomic relationship matrix as detailed in [20, 38]. The genomic relationship matrix was built using array-derived genotypes of 603,662 (partially imputed) autosomal SNPs [39]. Daughter yield deviations (DYD) for milk fat percentage were the response variables. The genotypes were coded as 0, 1 and 2 for homozygous, heterozygous and alternative homozygous animals, respectively, when sequence variants were imputed using *FImpute*. When the sequence variants were imputed using *Minimac*, we considered both best-guess genotypes and allele dosages as explanatory variables for the association tests. Sequence variants with P values less than 2.1×10^{-9} (5%-Bonferroni-corrected type I error threshold for 23,256,743 tests) were considered as significantly associated.

Computing environment

All computations were performed on the Biosciences Advanced Scientific Computing (BASC) cluster that is located at AgriBio, Centre for AgriBiosciences, Bundoora, VIC 3083. The memory usage and process time required to infer genotypes for 23,256,743 sequence variants in 6958 animals was quantified on 12-core Intel® Xeon® processors rated at 2.93 GHz with 96 GB of random-access memory (RAM).

Results

We evaluated the accuracy of imputation from dense genotypes to sequence variants in 249 FV and 450 HOL animals using sequence data on bovine chromosomes (BTA for *Bos taurus*) BTA1, 5, 10, 15, 20 and 25. The number of polymorphic sites ranged from 413,371 (BTA25) to 1,444,299 (BTA1) and from 383,072 (BTA25) to 1,382,987 (BTA1) in FV and HOL, respectively, indicating that genetic diversity is higher in FV than HOL cattle (Table 1). Variants with a low MAF were more frequent among the sequence than HD variants; between 58.12 and 60.55% of the sequence variants and between 14.27 and 18.55% of the HD variants had a MAF lower than 10%.

Evaluation of the accuracy of imputation

The correlation between imputed and sequenced genotypes ($r_{IMP,SEQ}$) was higher in HOL than FV cattle (Table 2), which likely reflects lower genetic diversity and larger reference population size in HOL. When within-breed reference populations were considered, the $r_{IMP,SEQ}$-values were 0.898, 0.908 and 0.934 for *FImpute*, *Minimac*$_{BG}$ and *Minimac*$_{DOS}$ in FV and 0.912, 0.929 and 0.951 in HOL, respectively. Adding animals from various breeds to the reference panel increased $r_{IMP,SEQ}$

in FV, particularly when *FImpute* was used to impute missing genotypes ($P = 7.5 \times 10^{-11}$, Table 2). The highest accuracy of imputation in FV ($r_{IMP,SEQ} = 0.939$) was achieved using *Minimac*$_{DOS}$ with a multi-breed reference population. In HOL, a multi-breed reference population increased $r_{IMP,SEQ}$ when *FImpute* was used to infer missing genotypes ($P = 4.0 \times 10^{-4}$). However, the use of a multi-breed reference population had little effect on the $r_{IMP,SEQ}$-values in HOL when *Minimac* was used ($P > 0.12$). Using *Minimac*$_{DOS}$ and a within-breed reference population provided the highest accuracy of imputation ($r_{IMP,SEQ} = 0.951$) in HOL. Regardless of the reference population used, the $r_{IMP,SEQ}$-values were higher for *Minimac*$_{DOS}$ than *Minimac*$_{BG}$ in both breeds ($P_{FV} < 2.9 \times 10^{-6}$, $P_{HOL} < 3.9 \times 10^{-6}$).

A decline in the accuracy of imputation was evident for low-frequency variants across all scenarios tested. While $r_{IMP,SEQ}$ was high for variants with a MAF higher than 10%, it was considerably less for low MAF variants (Fig. 1a, b). In FV, the accuracy of imputation was higher for low MAF variants using multi- than within-breed reference populations ($P < 6.6 \times 10^{-6}$, Table 2). The benefit of multi-breed reference panels was less pronounced in HOL and it was only significant ($P = 0.013$) when *FImpute* was used. The $r_{IMP,SEQ}$-values for low-frequency variants (MAF < 0.1) were higher with *Minimac* than *FImpute* ($P_{FV} < 0.005$, $P_{HOL} < 0.002$, Table 2; Fig. 1a, b).

The number of sequence variants that were not polymorphic after imputation in the validation population varied considerably across the imputation scenarios

Table 1 Number of polymorphic sites in Fleckvieh and Holstein cattle

Chr	Chr length (Mb)	Fleckvieh		Holstein	
		SEQ (MAF < 0.1)	700 K (MAF < 0.1)	SEQ (MAF < 0.1)	700 K (MAF < 0.1)
1	158.32	1,444,299 (60.55)	38,009 (18.40)	1,382,987 (59.11)	37,397 (17.01)
5	121.18	1,098,976 (59.64)	28,173 (18.55)	1,073,964 (59.12)	27,667 (15.90)
10	104.30	959,206 (59.13)	25,787 (17.09)	917,799 (59.48)	25,497 (16.12)
15	85.27	866,151 (58.54)	20,617 (17.24)	827,526 (58.12)	20,360 (14.38)
20	71.98	679,738 (59.94)	18,466 (18.18)	649,768 (59.18)	18,264 (18.13)
25	42.85	413,371 (59.08)	11,370 (14.69)	383,072 (58.14)	11,179 (14.27)

Number of sequence (SEQ) and array-derived (Illumina bovineHD, 700 K) variants that were polymorphic in 249 Fleckvieh and 450 Holstein animals. The proportion of variants with a minor allele frequency (MAF) lower than 10% is given in parentheses

Table 2 Cross-validation imputation accuracy in Fleckvieh and Holstein cattle

Breed	Within-breed reference populations			Multi-breed reference populations		
	Fimpute	*Minimac*$_{BG}$	*Minimac*$_{DOS}$	*Fimpute*	*Minimac*$_{BG}$	*Minimac*$_{DOS}$
Fleckvieh	0.898 (0.674)	0.908 (0.735)	0.934 (0.771)	0.921 (0.774)	0.916 (0.812)	0.939 (0.838)
Holstein	0.912 (0.681)	0.929 (0.772)	0.951 (0.798)	0.926 (0.732)	0.928 (0.797)	0.948 (0.820)

The values represent the correlation between true and imputed genotypes. The correlation between true and imputed genotypes for sequence variants with a minor allele frequency lower than 10% is given in parentheses. *BG* best-guess, *DOS* allele dosage

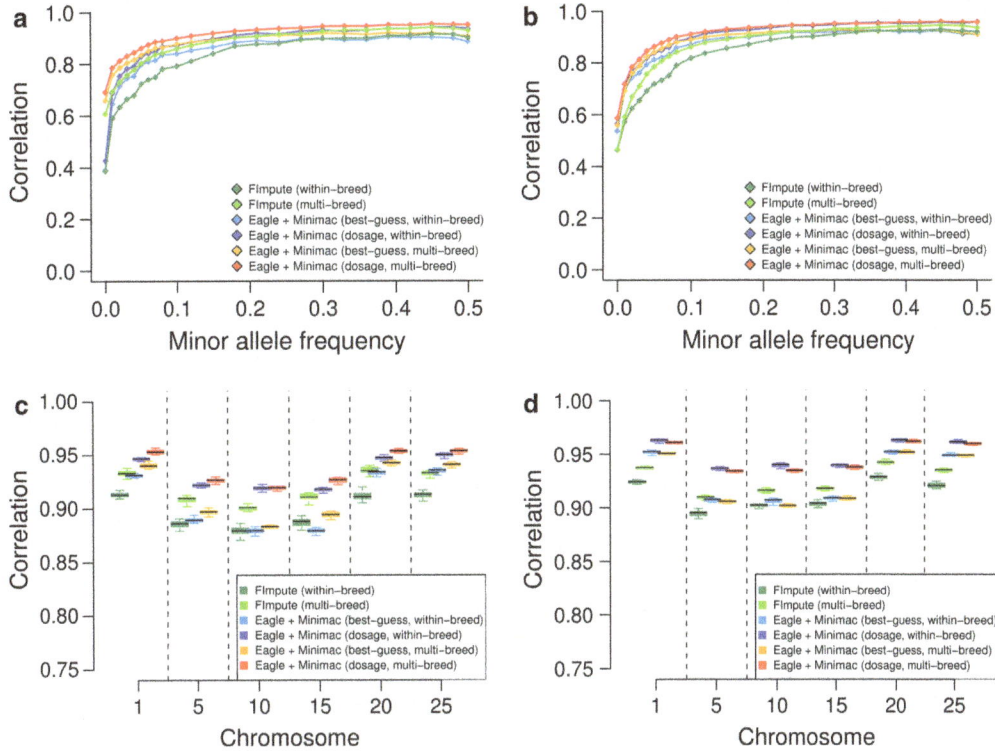

Fig. 1 Accuracy of imputation in Fleckvieh and Holstein cattle. **a**, **b** Correlation between true and imputed genotypes as a function of the frequency of the minor allele in Fleckvieh (**a**) and Holstein (**b**) animals. **c**, **d** The *boxplots* represent the correlation between true and imputed genotypes for sequence variants on six chromosomes in Fleckvieh (**c**) and Holstein (**d**) animals

tested (see Additional file 2: Figure S2). Using $Minimac_{BG}$, between 29.2 and 37.4% of the imputed sequence variants were invariant in the validation populations depending on the breed and reference population considered. The proportion of invariant sites was lower when *FImpute* (between 19.8 and 24.2% of the sequence variants) and $Minimac_{DOS}$ (between 3.6 and 6.7% of the sequence variants) were used for imputation. Variants that had a low MAF in the reference population were more often invariant in the validation population than common variants. When the $r_{IMP,SEQ}$-values were calculated using only sequence variants that were polymorphic in the validation population, $Minimac_{BG}$ provided the most accurate genotypes. However, the accuracy of imputation was also high using $Minimac_{DOS}$ (see Additional file 2: Figure S2c and S2d) which provided genotypes for considerably more sequence variants than $Minimac_{BG}$ and *FImpute*.

The accuracy of imputation varied considerably across chromosomes. The $r_{IMP,SEQ}$-values were high for sequence variants located on BTA1, 20 and 25. However, the accuracy of imputation was lower for sequence variants located on BTA5, 10 and 15 (Fig. 1c, d). This pattern was observed regardless of the reference population and imputation software used. To investigate the reason for

the variable accuracy of imputation across chromosomes, we calculated $r_{IMP,SEQ}$-values, HD SNP coverage and sequence variant density for successive 1-Mb windows along the six chromosomes analysed. Multiple (partially extended) segments with high imputation errors were located on BTA5, 10 and 15 (see Additional file 3: Figure S3) at positions where the bovine genome contains large segmental duplications [40, 41]. These segments were often sparsely covered with HD SNPs but contained a large number of polymorphic sequence variants. Such segments with low imputation quality were not detected along BTA1, 20 and 25, which had an overall higher accuracy of imputation.

Pinpointing causal mutations among imputed sequence variants

We imputed genotypes for 756,135 and 679,738 sequence variants that were located on BTA14 and 20, respectively, in 6958 FV animals using either *FImpute* or *Minimac* and considering either within- or multi-breed reference populations. To evaluate the ability to detect causal mutations, the imputed genotypes were tested for association with milk fat percentage. Milk fat percentage was the target trait because two mutations in the

DGAT1 (rs109326954, p.A232K) and *GHR* (rs385640152, p.F279Y) genes have large effects on the milk composition in many cattle breeds including FV. In 249 sequenced FV animals, the MAFs of rs109326954 and rs385640152 were 6.2 and 7.2%. Both variants were more frequent in the multi-breed reference population (17.3 and 12.1%, Table 3). Depending on the reference population and imputation algorithm used, the proportion of correctly imputed genotypes was between 99.3 and 99.9% and between 86.1 and 99.4% for rs109326954 and rs385640152, respectively (Table 3). The imputation error rates at both variants were lower with *Minimac* than *FImpute*. Using *Minimac* and a multi-breed reference population provided the most accurate genotypes at both variants (99.9 and 99.4%).

Association tests between the imputed genotypes and milk fat percentage revealed that the *DGAT1*:p.A232K-variant was among the most significantly associated variants across all scenarios tested, which reflected high accuracy of imputation regardless of the reference population and imputation algorithm used (Fig. 2; Table 3). However, several adjacent variants ($\pm$90 kb) were in complete or near complete LD ($r^2 > 0.99$) with rs109326954 and had *P* values that were slightly higher, identical, or slightly lower. When the genotypes were imputed using *Minimac* and considering a within-breed reference population, five variants in complete LD including rs109326954 were the most significantly associated (Fig. 2e, f).

The imputation algorithm and composition of the reference population had a large effect on the ability to detect an association between the *GHR*:p.F279Y-variant and milk fat percentage (Fig. 3). The *GHR*:p.F279Y-variant was not significantly associated ($P > 1 \times 10^{-8}$) when the genotypes were imputed using *FImpute* or *Minimac*$_{BG}$ and a within-breed reference population, which reflected high imputation error rates (Table 3). There were 350 and 395 variants detected, respectively, that had lower *P* values than rs385640152. Association tests with genotypes that were obtained using *Minimac*$_{DOS}$ (within-breed) or *FImpute* (multi-breed) revealed significant association of rs385640152 ($P = 1.0 \times 10^{-15}$, 3.0×10^{-10}). However, six and 1089 variants had lower *P* values than rs385640152. When the genotypes were inferred using *Minimac* and a multi-breed reference population, rs385640152 was the most significantly associated variant, which reflected higher accuracy of imputation (Fig. 3e, f; Table 3).

Fine-mapping of five fat percentage QTL with imputed sequence variants

The fine-mapping of two known QTL indicated that genotypes that are imputed using *Minimac* and considering a multi-breed reference population are an accurate source for association tests, particularly when predicted allele dosages rather than best-guess genotypes are used as explanatory variables. We imputed 23,256,743 sequence variants in 6958 FV animals using *Minimac* and a multi-breed reference population to perform association tests between imputed allele dosages and fat percentage. The mean and median accuracy of imputation for 23,256,743 sequence variants (r^2-values from *Minimac*) were 0.71 and 0.95, respectively (see Additional file 4: Figure S4); 78.3 and 73.2% of the imputed sequence variants had r^2-values greater than 0.3 and 0.5, respectively. Our association study revealed seven genomic regions with significantly associated variants (i.e., QTL), including two QTL on BTA14 and 20 that encompassed the *DGAT1* and *GHR* genes (Figs. 2f, 3f, 4a). The top variants were imputed sequence variants at all QTL.

A total of 239 variants located between 91,857,670 and 93,955,207 bp on BTA5 were significantly associated ($P < 2.1 \times 10^{-9}$) with fat percentage [Fig. 4b and Table S1 (see Additional file 5: Table S1)]. Thirteen variants in the first intron of the *MGST1* gene (*microsomal glutathione S-transferase 1*) were in high LD ($r^2 > 0.95$) with each other and had markedly lower *P* values ($P < 6.5 \times 10^{-21}$) than all other variants. These 13 variants were also associated with milk composition in New Zealand dairy cows [23]. The top variant (rs208248675) was located at 93,945,991 bp. The rs208248675 *A*-allele had a frequency of 19.4% in FV cattle and decreased fat percentage.

A QTL for fat percentage on BTA6 encompassed the casein gene complex [Fig. 4c and Table S1 (see Additional file 5: Table S1)]. One hundred and sixty-four variants with $P < 2.1 \times 10^{-9}$ were located between 87,084,144 and 87,296,017 bp. The most significantly associated variant (rs109193501 at 87,154,594 bp, $P = 7.6 \times 10^{-14}$) was located in an intron of the *CSN1S1* gene (*casein alpha s1*).

Two hundred and sixty-two variants located within a 50-kb interval (between 103,274,736 and 103,324,728 bp) on BTA11 were associated with fat percentage [Fig. 4d and Table S1 (see Additional file 5: Table S1)]. The top variant (rs381989107 at 103,296,192 bp, $P = 1.5 \times 10^{-15}$) was located 5472 bp upstream of the translation start site of the *LGB* gene [*beta-lactoglobulin*, also known as *progestagen associated endometrial protein* (*PAEP*)]. Two missense variants (p.G80D and p.V134A) in the *LGB* gene (rs110066229 at 103,303,475 bp and rs109625649 at 103,304,757 bp) that distinguish the LGB protein variants A and B [42] were in high LD with rs381989107 and had *P* values less than 4.5×10^{-12} (see Additional file 5: Table S1). A variant in the promoter of the *LGB* gene that causes an aberrant LGB expression in Brown Swiss cattle (OMIA 001437-9913) [43] was not polymorphic in the sequenced FV animals.

Table 3 Accuracy of imputation for two causal variants in Fleckvieh cattle

| Variant | Effect | Minor allele frequency | | | Proportion of correctly imputed genotypes/r_{IMPSEQ} | | | | | |
| | | | | | Fimpute | | Minimac$_{BG}$ | | Minimac$_{DOS}$ | |
		TaqMan	Within-breed	Multi-breed	Within-breed	Multi-breed	Within-breed	Multi-breed	Within-breed	Multi-breed
14:1,802,266[a]/rs109326954	p.A232K	0.074	0.062	0.183	0.993/0.976	0.998/0.992	0.999/0.996	0.999/0.996	–/0.979	–/0.995
20:31,909,478/rs385640152	p.F279Y	0.074	0.072	0.121	0.861/0.639	0.931/NA[b]	0.868/NA[b]	0.994/0.938	–/0.703	–/0.895

The minor allele frequencies were calculated using TaqMan-derived and sequence genotypes in either within- or multi-breed reference populations

[a] The p.A232K-variant in the *DGAT1* gene results from two adjacent SNPs in LD located at 1,802,265 and 1,802,266 bp (rs109234250 and rs109326954). In the current study, we considered only the variant at 1,802,266 bp

[b] The imputed variant was not polymorphic in the sample of 902 genotyped animals precluding the calculation of r_{IMPSEQ} values

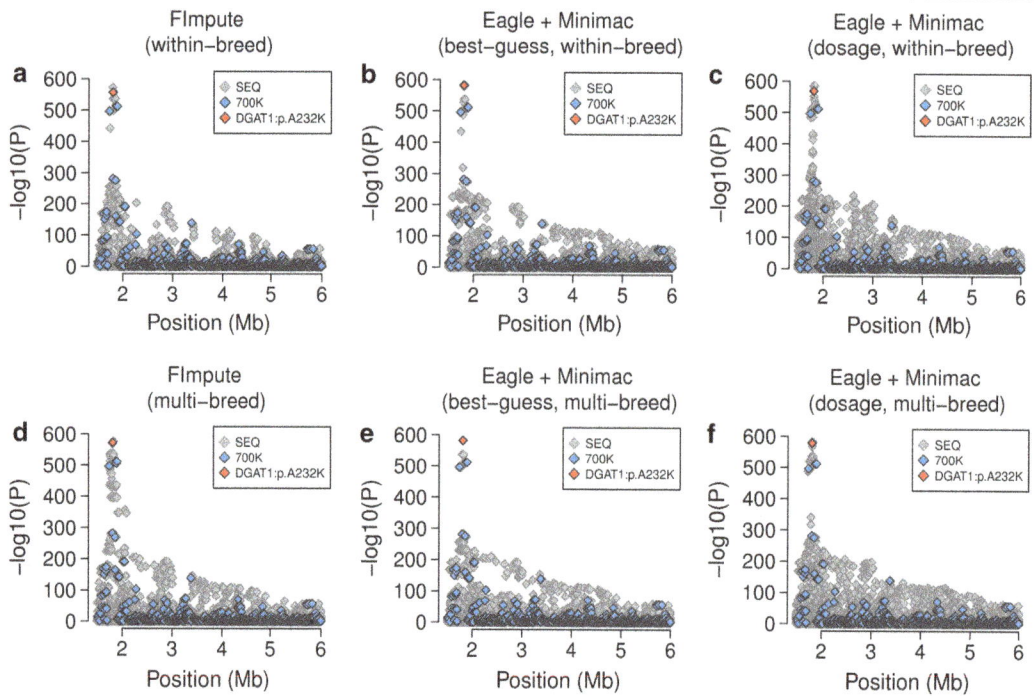

Fig. 2 Fine-mapping of a fat percentage QTL on bovine chromosome 14. **a–e** Association between 48,641 imputed sequence variants located at the proximal end of bovine chromosome 14 and milk fat percentage in 6958 Fleckvieh animals. Genotypes for the association studies were imputed using either *FImpute* (**a**, **d**) or *Minimac* (**b**, **c**, **e**, **f**) with either within- (**a–c**) or multi-breed reference populations (**d–f**). *Grey* and *blue colours* represent sequence and array-derived variants. The *red symbol* represents the p.A232K-mutation in the *DGAT1* gene

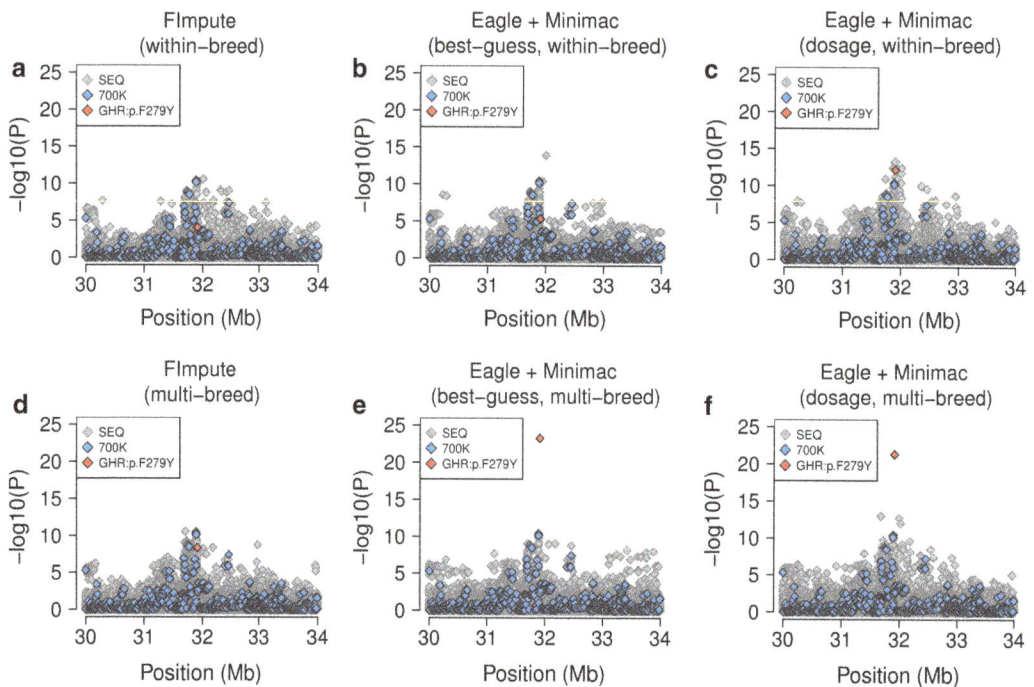

Fig. 3 Fine-mapping of a fat percentage QTL on bovine chromosome 20. **a–e** Association between 29,205 imputed sequence variants on bovine chromosome 20 and milk fat percentage in 6958 Fleckvieh animals. Genotypes for the association studies were imputed using either *FImpute* (**a**, **d**) or *Minimac* (**b**, **c**, **e**, **f**) with either (**a–c**) or multi breed reference populations (**d–f**). *Grey* and *blue colours* represent sequence and array-derived variants, respectively. The *red symbol* represents the p.F279Y-mutation in the *GHR* gene

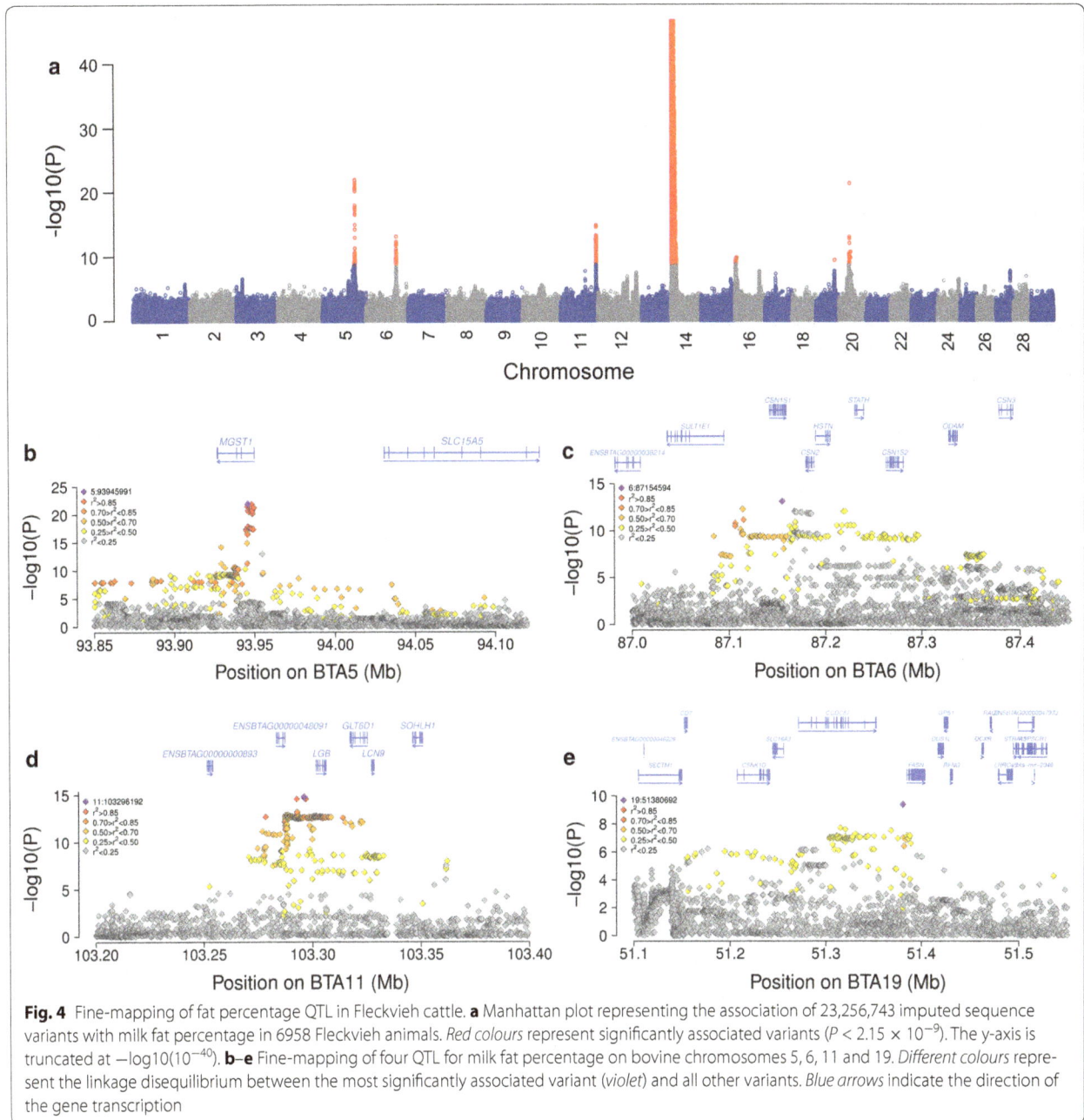

Fig. 4 Fine-mapping of fat percentage QTL in Fleckvieh cattle. **a** Manhattan plot representing the association of 23,256,743 imputed sequence variants with milk fat percentage in 6958 Fleckvieh animals. *Red colours* represent significantly associated variants ($P < 2.15 \times 10^{-9}$). The y-axis is truncated at $-\log10(10^{-40})$. **b–e** Fine-mapping of four QTL for milk fat percentage on bovine chromosomes 5, 6, 11 and 19. *Different colours* represent the linkage disequilibrium between the most significantly associated variant (*violet*) and all other variants. *Blue arrows* indicate the direction of the gene transcription

A QTL on BTA16 included 14 significantly associated variants located between 481,671 and 3,265,708 bp (see Additional file 5: Table S1). The top association signal (rs380194132, $P = 1.6 \times 10^{-10}$) resulted from an intergenic variant that was located at 3,242,688 bp. Since the genomic region that included associated sequence variants with similar P values extended over almost 3 Mb, the identification of putative candidate gene(s) underlying the QTL was not attempted (see Additional file 6: Figure S5).

On BTA19, a single variant located at 51,380,692 bp was significantly associated (rs208289132, $P = 4.1 \times 10^{-10}$) with fat percentage [Fig. 4e and Table S1 (see Additional file 5: Table S1)]. The associated variant was located 4230 bp upstream of the translation start site of the *FASN* gene (*fatty acid synthase*).

Computing resources required

We assessed the computing resources required to impute genotypes at 23,256,743 sequence variants for 6958 target animals using 1577 sequenced animals as a reference. *Eagle* and *Minimac* were run on 10 processors per chromosome, whereas *FImpute* was run on a single processor. *FImpute* ran out of memory and failed to impute genotypes for BTA12 and 23 likely because large structural variants are located on both chromosomes (see Additional file 7: Figure S6). The process (CPU) time to infer haplotypes and impute sequence variants for 27 chromosomes was 1037, 490 and 146 h, respectively, for *Eagle*, *Minimac* and *FImpute*.

The wall-clock time and RAM usage for *FImpute* was between 2.9 and 9.6 h and 6.7 and 29.7 GB per chromosome, respectively (see Additional file 8: Figure S7). The estimation of haplotype phases took between 1.6 and 5.9 h using *Eagle* and it required between 1.8 and 3.4 GB of RAM with between 11,178 and 38,009 SNPs per chromosome for 6958 animals. The imputation of sequence variants using *Minimac* took between 2.5 and 9.4 h and it required between 6.3 and 21 GB of RAM per chromosome.

The estimation of haplotype phases for 1577 sequenced reference animals using *Eagle* (multi-threaded on 10 CPUs) took between 4.2 and 15.1 h with between 413,371 and 1,444,299 sequence variants, respectively, and it required between 15.4 and 51.7 GB of RAM per chromosome.

Discussion

We evaluated the accuracy of imputation from dense genotypes to sequence variants using a population of 1577 sequenced animals which is a four- to eight-fold increase in sample size compared to previous studies in cattle [2, 17, 18]. Our findings are likely to be relevant for many bovine populations because we considered animals from two breeds with different effective population sizes as validation panel and inferred sequence variant genotypes on six chromosomes that reflect a broad spectrum of LD [44]. The accuracy of imputation was the correlation between imputed and sequenced genotypes ($r_{IMP,SEQ}$), which is a measure that depends less on the allele frequencies of the sequence variants than e.g., the proportion of correctly imputed genotypes [45]. However, the cross-validation procedure that was applied in our study yields reliable results only when the quality of the sequence data is high. Low-fold sequencing coverage and sequencing or assembly errors may result in flawed genotypes [46]. When genotypes at such positions are masked and subsequently imputed, genotype imputation might provide true genotypes that differ from the sequenced genotypes thereby underestimating the accuracy of imputation. Considering that the genotype error rates were low in the sequenced FV

and HOL animals [2, 47], it is unlikely that our findings are biased due to flawed genotypes at some sequence variants.

The $r_{IMP,SEQ}$-values varied considerably across chromosomes and dropped at several positions along the genome. Previous studies that used array-derived genotypes showed that a decline in imputation accuracy may indicate intra- or inter-chromosomal misplacement of SNPs [12, 48, 49]. We detected low $r_{IMP,SEQ}$-values at regions where the bovine genome contains large segmental duplications [40, 41]. The population-wide imputation of sequence variant genotypes for BTA12 and 23 was not possible using *Fimpute*. Assessing the accuracy of imputation along these chromosomes using cross-validation within the data of sequenced animals revealed large segments with low imputation quality on both chromosomes including the highly variable major histocompatibility complex [50]. We were able to eventually impute sequence variant genotypes for BTA12 and 23 using *FImpute* when those segments were excluded from the reference panel. Imputing genotypes within both segments was possible using *Minimac*. However, the inferred genotypes are flawed and must be treated with caution as our results show (see Additional file 7: Figure S6). While an improved assembly of the reference genome may resolve some of these problems, a better resolution of large structural elements is not possible using short paired-end DNA sequencing [51]. Since flawed genotypes compromise downstream analyses [24, 25], it is advisable to exclude sequence variants within segments of low imputation quality from the target population and retain the array-derived genotypes only.

We assessed the performance of *FImpute* and *Minimac* because both methods can impute millions of polymorphic sites in large populations within a reasonable time. Previous studies showed that *FImpute* is more accurate than other tools with similar computing costs [52, 53]. The population-based genotype imputation algorithm implemented in *Minimac* is highly accurate and fast because it takes previously phased genotypes as input [11, 54]. We phased the target population using *Eagle* because it enables timely and accurate haplotype inference [32]. Although the computing costs to impute 23 million sequence variants in 6958 animals were more than ten-fold higher with *Eagle* and *Minimac* than *FImpute*, the wall-clock time was only 1.67-fold higher because *Eagle* and *Minimac* enable multi-threading to reduce the process time [32, 54]. The difference in computing costs is greater when haplotypes are not available for the reference animals. Since many variant detection pipelines include phasing and imputation algorithms [2, 47, 55], the computing time required to phase the sequence data was not considered in our study. However, the phasing of the reference panel was necessary because the variant detection pipeline of the 1000 bull genomes project was applied to successive 5-Mb segments rather than whole chromosomes [2].

The accuracy of imputation was higher with *Minimac* than *FImpute*, although *Minimac* does not consider pedigree information. Most individuals of the 1000 bull genomes population were selected in such a way that they represent a large proportion of the genetic variation of current populations [2, 47, 56]. They are likely less related with each other than a random subset from the population, although some parent-offspring pairs were included. The accuracy of imputation might be higher when reference and target animals are closely related particularly when *FImpute* is used [53]. However, dense genotypes for reference and target animals, e.g., sequence and HD-derived genotypes, can be used to identify short shared haplotypes among apparently unrelated animals that originate from ancient ancestors possibly predating the separation of breeds [9, 57]. Including reference animals from various breeds increased the accuracy of imputation, particularly in FV cattle. The principal components analysis revealed that animals from many breeds clustered nearby the FV population indicating that they are distantly related whereas the HOL animals formed a cluster that was separated from all other breeds. The benefit of a multi-breed reference population on the accuracy of imputation was less pronounced in HOL cattle. Since the number of reference animals was nearly twice as large in HOL, our results may also corroborate the suggestion that multi-breed reference panels are particularly useful to impute genotypes when the reference population is small [18, 58, 59]. In agreement with previous studies in cattle [2, 18], a multi-breed reference population increased the accuracy of imputation at low MAF variants likely reflecting the shared ancestry of many bovine populations and the limited number of sequenced animals per breed. However, it seems advisable to periodically evaluate different imputation strategies because multi-breed reference populations can also compromise the imputation of sequence variants when the diversity of the reference panel increases [26].

The population-based imputation of sequence variant genotypes was particularly advantageous at variants with a low MAF corroborating previous findings [12, 53, 60]. The true genotypes were more correlated to predicted allele dosages than best-guess genotypes which is in line with the findings of Brøndum et al. [18]. Moreover, our results show that more sequence variants are polymorphic in the target population when allele dosages are considered instead of best-guess genotypes. An association study between allele dosages and milk fat percentage revealed that two causal variants with a MAF less than 10% in the *DGAT1* [34, 35] and *GHR* [36] genes were the most significantly associated variants at two QTL on

BTA14 and 20, respectively. This finding demonstrates that using predicted allele dosages in association studies with imputed sequence variants can pinpoint causal mutations which agrees with Zheng et al. [61] and Khatkar et al. [62]. So far, imputed sequence variants did not substantially improve the accuracies of genomic predictions in real data which might partly result from flawed genotypes at low-frequency variants [17, 63]. We show that allele dosages are more accurate than best-guess genotypes particularly at variants with a low MAF because they better reflect imputation uncertainty. However, the analysis of allele dosages for millions of sequence variants in tens of thousands of individuals for genomic predictions is computationally costly and has not been attempted so far [17, 64]. Since the proportion of variants with a low MAF is more than three-fold higher in sequence than HD variants, further research is warranted to investigate if allele dosages may enhance genomic predictions with imputed sequence variants.

Our findings show that the composition of the reference population and the choice of the imputation algorithm are critical to infer accurate genotypes thereby enabling us to pinpoint causal mutations in association studies with imputed sequence variants. When the accuracy of imputation is low, causal variants may be "buried" among other variants in LD (e.g., Fig. 3a, d). In such situations, association testing will not reveal true causal mutations because variants with lower P values are likely to be prioritized as candidate causal variants [23]. However, true causal variants are not necessarily the top variants in association studies even when their genotypes are almost perfectly imputed. Although the imputation accuracy for the causal p.A232K-polymorphism in the *DGAT1* gene was greater than 99.5%, several adjacent variants had identical or slightly lower P values, which likely indicates sampling errors [65] or synthetic associations [66, 67]. Nevertheless, the causal mutations in the *DGAT1* and *GHR* genes offer a convenient approach to evaluate the ability to detect causal mutations with imputed sequence variants in cattle, because both variants segregate in many breeds and phenotypes for fat percentage are readily available.

Our association study with more than 23 million sequence variants detected seven QTL that were significantly associated with milk fat percentage. Applying a less stringent significance threshold [68] would reveal additional QTL (see Additional file 9: Figure S8). Since we were able to pinpoint the causal mutations at two QTL, it is likely that causal mutations for other QTL are among the most significantly associated variants. However, the QTL identified in our study include several sequence variants in high LD and with similar P values, which renders the identification of causal sites difficult. Association studies in animals from multiple breeds may facilitate differentiation between causal and non-causal sites in LD [2]. Causal sites may be located on ancient haplotypes that still segregate across multiple breeds as is the case for the fat percentage QTL on BTA5. Our association study revealed thirteen candidate causal variants in high LD in an intron of the *MGST1* gene that had markedly lower *P* values than all other variants. It is very likely that one of those variants is the actual causal polymorphism because they were also highly significantly associated with milk composition in another dairy cattle breed [23]. An improved functional annotation of the bovine genome and a fine-mapping strategy that prioritizes sequence variants using functional information may facilitate pinpointing causal mutations at such QTL [22, 69].

Conclusions

Genotypes for millions of sequence variants can be accurately inferred in large bovine cohorts using population-based imputation algorithms. The accuracy of imputation for variants with a low MAF is higher with multi-breed reference populations. The accuracy of imputation is not constant across chromosomes and may drop at regions where the genome contains large segmental duplications. Considering predicted allele dosages rather than best-guess genotypes as explanatory variables is beneficial for association studies with imputed sequence variants. The composition of the reference population and the choice of the approach used to infer genotypes are critical to detect causal mutations in association studies with imputed sequence variant genotypes.

Additional files

Additional file 1: Figure S1. Principal component analysis in 1577 sequenced animals. **a–c** Plot of the top two principal components of the genomic relationship matrix. *Different colours* and *symbols* represent different breeds. The partners of the 1000 bull genomes consortium assigned the animals to breeds. **b** *Non-grey symbols* indicate 249 animals that were considered as Fleckvieh animals. **c** *Green symbols* indicate 450 Holstein animals.

Additional file 2: Figure S2. Sequence variants that were invariant in the validation populations. **a–d** Proportion of sequence variants that were invariant after imputation in the FV (**a, c**) and HOL (**b, d**) validation populations for six chromosomes analysed and different MAF classes. **e, f** The correlation between imputed and true genotypes was calculated using only sequence variants that were polymorphic in the FV (**e**) and HOL (**f**) validation populations.

Additional file 3: Figure S3. Imputation accuracy along six chromosomes in Fleckvieh cattle. **a–f** The correlation between true and imputed genotypes for sequence variants located within successive 1-Mb windows on six chromosomes. *Different colours* and *symbols* represent correlation coefficients obtained using different imputation scenarios. *Insets* HD SNP coverage and sequence variant density along the chromosome. *Black* and *red colours* represent the number of SNPs that were included in the BovineHD Bead Chip (HD) and sequence (Seq) variants (×1000, (K)) that were polymorphic in the multi-breed reference population, respectively, per million basepairs (Mb).

Additional file 4: Figure S4. Accuracy of imputation for 23,256,743 sequence variants in 6958 animals. The mean (*red*) and median (*blue*) accuracy of imputation (r^2-values were obtained using *Minimac*) for 23,256,743 sequence variants (*grey bars*) as a function of the minor allele frequency. The *dotted lines* represent the mean and median accuracy of imputation across all sequence variants.

Additional file 5: Table S1. Significantly associated variants at five fat percentage QTL. Variants located on chromosomes 5, 6, 11, 16 and 19 with *P* values less than 2.1×10^{-9}. The positions of the variants correspond to the UMD3.1 assembly of the bovine genome. The substitution effects (beta, standard error of beta) are given for the alternative allele. The R2 value is the estimated accuracy of imputation from *Minimac*.

Additional file 6: Figure S5. Detailed view of a milk fat percentage QTL on bovine chromosome 16. *Different colours* represent the linkage disequilibrium between the most significantly associated variant (*violet*) and all other variants. *Blue arrows* indicate the direction of the gene transcription.

Additional file 7: Figure S6. Imputation accuracy along chromosomes 12 and 23 in Fleckvieh cattle. **a, b** Correlation between true and imputed genotypes for sequence variants located within successive 1-Mb windows on chromosomes 12 and 23. Different colours and symbols represent correlation coefficients obtained using different imputation scenarios. **c, d** *Red colours* represent the number of SNPs that were included in the BovineHD Bead Chip (HD) and sequence (Seq) variants (×1000 (K)) that were polymorphic in the multi-breed reference population, respectively, per million basepairs (Mb). We were eventually able to impute sequence variants for BTA12 and 23 using *FImpute* when we discarded sequence variants that were located between 70 and 77 Mb and between 25 and 30 Mb, respectively, from the reference panel.

Additional file 8: Figure S7. Computing resources required to impute 23,256,742 sequence variants in 6958 animals. The wall-clock times (**a**) and random-access memory (RAM) usage (**b**) required to infer haplotypes and genotypes with *FImpute* (*green*), Eagle (*dark blue*) and *Minimac* (*light blue*) were assessed on 12-core Intel® Xeon® processors rated at 2.93 GHz with 96 GB of RAM. *FImpute* ran out of memory and did not finish when we attempted to infer genotypes for BTA12 and BTA23. *FImpute* was run on a single processor whereas Eagle and *Minimac* used 10 processors per chromosome.

Additional file 9: Figure S8. Detailed view of a milk fat percentage QTL on bovine chromosome 27. *Different colours* represent the linkage disequilibrium between the top variant (*violet*) and all other variants. *Blue arrows* indicate the direction of the gene transcription. The top variant (36,211,258 bp) was associated with fat percentage ($P = 1.9 \times 10^{-8}$) albeit not at the genome-scale. Twenty-two variants in high LD ($r^2 > 0.68$) with the top variant were located between 36,200,888 and 36,253,406 bp and had *P* values less than 7.7×10^{-7}. Among those were three candidate causal variants (36,211,252 bp with $P = 2.4 \times 10^{-8}$, 36,211,708 bp with $P = 2.8 \times 10^{-8}$, 36,209,319 bp with $P = 3.3 \times 10^{-8}$) for fat content in the early lactation that were reported in Daetwyler et al. [2].

Authors' contributions

HP, IMM, HDD and MEG designed the experiments, HP analyzed the data, PJB provided support in computing, RF and RE provided genotype and phenotype data, HP wrote the manuscript. All authors read and approved the final manuscript.

Author details

[1] Agriculture Victoria, AgriBio, Centre for AgriBiosciences, Bundoora, VIC 3083, Australia. [2] Chair of Animal Breeding, Technische Universitaet Muenchen, 85354 Freising, Germany. [3] Institute of Animal Breeding, Bavarian State Research Center for Agriculture, 85586 Grub, Germany. [4] School of Applied Systems Biology, La Trobe University, Bundoora, VIC 3083, Australia. [5] Faculty of Veterinary and Agricultural Sciences, University of Melbourne, Melbourne, VIC 3010, Australia.

Acknowledgements

We thank Arbeitsgemeinschaft Süddeutscher Rinderzüchter und Besamungsorganisationen e.V. (ASR), Arbeitsgemeinschaft österreichischer

Fleckviehzüchter (AGÖF) and Förderverein Bioökonomieforschung e.V. (FBF) for providing genotyping data. We acknowledge the 1000 bull genomes project for providing sequence variant genotypes for 1577 animals.

Competing interests
The authors declare that they have no competing interests.

Funding
HP was financially supported by a postdoctoral fellowship (Grant-ID: PA2789/1-1) from the Deutsche Forschungsgemeinschaft (DFG).

References
1. Wiggans GR, Cooper TA, VanRaden PM, Van Tassell CP, Bickhart DM, Sonstegard TS. Increasing the number of single nucleotide polymorphisms used in genomic evaluation of dairy cattle. J Dairy Sci. 2016;99:4504–11.
2. Daetwyler HD, Capitan A, Pausch H, Stothard P, van Binsbergen R, Brøndum RF, et al. Whole-genome sequencing of 234 bulls facilitates mapping of monogenic and complex traits in cattle. Nat Genet. 2014;46:858–65.
3. Scheet P, Stephens M. A fast and flexible statistical model for large-scale population genotype data: applications to inferring missing genotypes and haplotypic phase. Am J Hum Genet. 2006;78:629–44.
4. Burdick JT, Chen W-M, Abecasis GR, Cheung VG. In silico method for inferring genotypes in pedigrees. Nat Genet. 2006;38:1002–4.
5. Browning BL, Browning SR. A unified approach to genotype imputation and haplotype-phase inference for large data sets of trios and unrelated individuals. Am J Hum Genet. 2009;84:210–23.
6. Howie BN, Donnelly P, Marchini J. A flexible and accurate genotype imputation method for the next generation of genome-wide association studies. PLoS Genet. 2009;5:e1000529.
7. Howie B, Fuchsberger C, Stephens M, Marchini J, Abecasis GR. Fast and accurate genotype imputation in genome-wide association studies through pre-phasing. Nat Genet. 2012;44:955–9.
8. Hickey JM, Kinghorn BP, Tier B, van der Werf JH, Cleveland MA. A phasing and imputation method for pedigreed populations that results in a single-stage genomic evaluation. Genet Sel Evol. 2012;44:9.
9. Sargolzaei M, Chesnais JP, Schenkel FS. A new approach for efficient genotype imputation using information from relatives. BMC Genomics. 2014;15:478.
10. VanRaden PM, O'Connell JR, Wiggans GR, Weigel KA. Genomic evaluations with many more genotypes. Genet Sel Evol. 2011;43:10.
11. Marchini J, Howie B. Genotype imputation for genome-wide association studies. Nat Rev Genet. 2010;11:499–511.
12. Pausch H, Aigner B, Emmerling R, Edel C, Götz KU, Fries R. Imputation of high-density genotypes in the Fleckvieh cattle population. Genet Sel Evol. 2013;45:3.
13. Browning BL, Browning SR. Genotype imputation with millions of reference samples. Am J Hum Genet. 2016;98:116–26.
14. Kong A, Masson G, Frigge ML, Gylfason A, Zusmanovich P, Thorleifsson G, et al. Detection of sharing by descent, long-range phasing and haplotype imputation. Nat Genet. 2008;40:1068–75.
15. Druet T, Schrooten C, de Roos APW. Imputation of genotypes from different single nucleotide polymorphism panels in dairy cattle. J Dairy Sci. 2010;93:5443–54.
16. Zhang Z, Druet T. Marker imputation with low-density marker panels in Dutch Holstein cattle. J Dairy Sci. 2010;93:5487–94.
17. van Binsbergen R, Bink MC, Calus MP, van Eeuwijk FA, Hayes BJ, Hulsegge I, et al. Accuracy of imputation to whole-genome sequence data in Holstein Friesian cattle. Genet Sel Evol. 2014;46:41.
18. Brøndum RF, Guldbrandtsen B, Sahana G, Lund MS, Su G. Strategies for imputation to whole genome sequence using a single or multi-breed reference population in cattle. BMC Genomics. 2014;15:728.
19. Qanbari S, Pausch H, Jansen S, Somel M, Strom TM, Fries R, et al. Classic selective sweeps revealed by massive sequencing in cattle. PLoS Genet. 2014;10:e1004148.
20. Pausch H, Emmerling R, Schwarzenbacher H, Fries R. A multi-trait meta-analysis with imputed sequence variants reveals twelve QTL for mammary gland morphology in Fleckvieh cattle. Genet Sel Evol. 2016;48:14.
21. Meuwissen T, Goddard M. Accurate prediction of genetic values for complex traits by whole-genome resequencing. Genetics. 2010;185:623–31.
22. MacLeod IM, Bowman PJ, Vander Jagt CJ, Haile-Mariam M, Kemper KE, Chamberlain AJ, et al. Exploiting biological priors and sequence variants enhances QTL discovery and genomic prediction of complex traits. BMC Genomics. 2016;17:144.
23. Littlejohn MD, Tiplady K, Fink TA, Lehnert K, Lopdell T, Johnson T, et al. Sequence-based association analysis reveals an MGST1 eQTL with pleiotropic effects on bovine milk composition. Sci Rep. 2016;6:25376.
24. Khatkar MS, Moser G, Hayes BJ, Raadsma HW. Strategies and utility of imputed SNP genotypes for genomic analysis in dairy cattle. BMC Genomics. 2012;13:538.
25. Ertl J, Edel C, Emmerling R, Pausch H, Fries R, Götz KU. On the limited increase in validation reliability using high-density genotypes in genomic best linear unbiased prediction: observations from Fleckvieh cattle. J Dairy Sci. 2014;97:487–96.
26. Howie B, Marchini J, Stephens M. Genotype imputation with thousands of genomes. G3 (Bethesda). 2011;1:457–70.
27. Li H. Aligning sequence reads, clone sequences and assembly contigs with BWA-MEM. http://arxiv.org/abs/1303.3997. Accessed 4 July 2016.
28. Zimin AV, Delcher AL, Florea L, Kelley DR, Schatz MC, Puiu D, et al. A whole-genome assembly of the domestic cow, Bos taurus. Genome Biol. 2009;10:R42.
29. Li H, Handsaker B, Wysoker A, Fennell T, Ruan J, Homer N, et al. The sequence alignment/map format and SAMtools. Bioinformatics. 2009;25:2078–9.
30. Chang CC, Chow CC, Tellier LC, Vattikuti S, Purcell SM, Lee JJ. Second-generation PLINK: rising to the challenge of larger and richer datasets. Gigascience. 2015;4:7.
31. Yang J, Lee SH, Goddard ME, Visscher PM. GCTA: a tool for genome-wide complex trait analysis. Am J Hum Genet. 2011;88:76–82.
32. Loh PR, Danecek P, Palamara PF, Fuchsberger C, Reshef YA, Finucane HK, et al. Reference-based phasing using the haplotype reference consortium panel. Nat Genet. 2016;48:1443–8.
33. Pausch H, Kölle S, Wurmser C, Schwarzenbacher H, Emmerling R, Jansen S, et al. A nonsense mutation in TMEM95 encoding a nondescript transmembrane protein causes idiopathic male subfertility in cattle. PLoS Genet. 2014;10:e1004044.
34. Grisart B, Coppieters W, Farnir F, Karim L, Ford C, Berzi P, et al. Positional candidate cloning of a QTL in dairy cattle: identification of a missense mutation in the bovine DGAT1 gene with major effect on milk yield and composition. Genome Res. 2002;12:222–31.
35. Winter A, Krämer W, Werner FAO, Kollers S, Kata S, Durstewitz G, et al. Association of a lysine-232/alanine polymorphism in a bovine gene encoding acyl-CoA:diacylglycerol acyltransferase (DGAT1) with variation at a quantitative trait locus for milk fat content. Proc Natl Acad Sci USA. 2002;99:9300–5.
36. Blott S, Kim JJ, Moisio S, Schmidt-Küntzel A, Cornet A, Berzi P, et al. Molecular dissection of a quantitative trait locus: a phenylalanine-to-tyrosine substitution in the transmembrane domain of the bovine growth hormone receptor is associated with a major effect on milk yield and composition. Genetics. 2003;163:253–66.
37. Pausch H, Wurmser C, Reinhardt F, Emmerling R, Fries R. Short communication: validation of 4 candidate causative trait variants in 2 cattle breeds using targeted sequence imputation. J Dairy Sci. 2015;98:4162–7.
38. Kang HM, Sul JH, Service SK, Zaitlen NA, Kong S, Freimer NB, et al. Variance component model to account for sample structure in genome-wide association studies. Nat Genet. 2010;42:348–54.
39. VanRaden PM. Efficient methods to compute genomic predictions. J Dairy Sci. 2008;91:4414–23.
40. Liu G, Ventura M, Cellamare A, Chen L, Cheng Z, Zhu B, et al. Analysis of recent segmental duplications in the bovine genome. BMC Genomics. 2009;10:571.
41. Bickhart DM, Hou Y, Schroeder SG, Alkan C, Cardone MF, Matukumalli LK, et al. Copy number variation of individual cattle genomes using next-generation sequencing. Genome Res. 2012;22:778–90.

42. Ganai NA, Bovenhuis H, van Arendonk JA, Visker MH. Novel polymorphisms in the bovine beta-lactoglobulin gene and their effects on beta-lactoglobulin protein concentration in milk. Anim Genet. 2009;40:127–33.

43. Braunschweig MH, Leeb T. Aberrant low expression level of bovine beta-lactoglobulin is associated with a C to A transversion in the BLG promoter region. J Dairy Sci. 2006;89:4414–9.

44. Boitard S, Rodríguez W, Jay F, Mona S, Austerlitz F. Inferring population size history from large samples of genome-wide molecular data—an approximate bayesian computation approach. PLoS Genet. 2016;12:e1005877.

45. Hickey JM, Crossa J, Babu R, de los Campos G. Factors affecting the accuracy of genotype imputation in populations from several maize breeding orograms. Crop Sci. 2012;52:654–63.

46. Wall JD, Tang LF, Zerbe B, Kvale MN, Kwok PY, Schaefer C, et al. Estimating genotype error rates from high-coverage next-generation sequence data. Genome Res. 2014;24:1734–9.

47. Jansen S, Aigner B, Pausch H, Wysocki M, Eck S, Benet-Pagès A, et al. Assessment of the genomic variation in a cattle population by re-sequencing of key animals at low to medium coverage. BMC Genomics. 2013;14:446.

48. Erbe M, Hayes BJ, Matukumalli LK, Goswami S, Bowman PJ, Reich CM, et al. Improving accuracy of genomic predictions within and between dairy cattle breeds with imputed high-density single nucleotide polymorphism panels. J Dairy Sci. 2012;95:4114–29.

49. Utsunomiya ATH, Santos DJA, Boison SA, Utsunomiya YT, Milanesi M, Bickhart DM, et al. Revealing misassembled segments in the bovine reference genome by high resolution linkage disequilibrium scan. BMC Genomics. 2016;17:705.

50. Emam M, Tabatabaei S, Sargolzaei M, Cartwright SL, Schenkel FS, Miglior F, et al. Evaluating the accuracy of imputation in the highly polymorphic MHC region of genome. J Anim Sci. 2016;94:174–5.

51. English AC, Salerno WJ, Hampton OA, Gonzaga-Jauregui C, Ambreth S, Ritter DI, et al. Assessing structural variation in a personal genome—towards a human reference diploid genome. BMC Genomics. 2015;16:286.

52. Sun C, Wu X-L, Weigel KA, Rosa GJM, Bauck S, Woodward BW, et al. An ensemble-based approach to imputation of moderate-density genotypes for genomic selection with application to Angus cattle. Genet Res (Camb). 2012;94:133–50.

53. Ma P, Brøndum RF, Zhang Q, Lund MS, Su G. Comparison of different methods for imputing genome-wide marker genotypes in Swedish and Finnish Red cattle. J Dairy Sci. 2013;96:4666–77.

54. Das S, Forer L, Schönherr S, Sidore C, Locke AE, Kwong A, et al. Next-generation genotype imputation service and methods. Nat Genet. 2016;48:1284–7.

55. DePristo MA, Banks E, Poplin R, Garimella KV, Maguire JR, Hartl C, et al. A framework for variation discovery and genotyping using next-generation DNA sequencing data. Nat Genet. 2011;43:491–8.

56. Goddard ME, Hayes BJ. Genomic selection based on dense genotypes inferred from sparse genotypes. Proc Assoc Advmt Anim Breed Genet. 2009;18:26–9.

57. Schwarzenbacher H, Burgstaller J, Seefried FR, Wurmser C, Hilbe M, Jung S, et al. A missense mutation in TUBD1 is associated with high juvenile mortality in Braunvieh and Fleckvieh cattle. BMC Genomics. 2016;17:400.

58. Li H, Sargolzaei M, Schenkel F. Accuracy of whole- genome sequence genotype imputation in cattle breeds. In: Proceedings of the 10th world congress on genetics applied to livestock production: 18–22 August 2014; Vancouver; 2014.

59. Bouwman AC, Veerkamp RF. Consequences of splitting whole-genome sequencing effort over multiple breeds on imputation accuracy. BMC Genet. 2014;15:105.

60. Frischknecht M, Neuditschko M, Jagannathan V, Drögemüller C, Tetens J, Thaller G, et al. Imputation of sequence level genotypes in the Franches-Montagnes horse breed. Genet Sel Evol. 2014;46:63.

61. Zheng J, Li Y, Abecasis GR, Scheet P. A comparison of approaches to account for uncertainty in analysis of imputed genotypes. Genet Epidemiol. 2011;35:102–10.

62. Khatkar MS, Thomson PC, Raadsma HW. Utility of imputed SNP genotypes for genome-wide association studies in dairy cattle. Proc Assoc Advmt Anim Breed Genet. 2013;20:554–7.

63. Gonzalez-Recio O, Daetwyler HD, MacLeod IM, Pryce JE, Bowman PJ, Hayes BJ, et al. Rare variants in transcript and potential regulatory regions explain a small percentage of the missing heritability of complex traits in cattle. PLoS One. 2015;10:e0143945.

64. Wang T, Chen YPP, Bowman PJ, Goddard ME, Hayes BJ. A hybrid expectation maximisation and MCMC sampling algorithm to implement Bayesian mixture model based genomic prediction and QTL mapping. BMC Genomics. 2016;17:744.

65. Barendse W. The effect of measurement error of phenotypes on genome wide association studies. BMC Genomics. 2011;12:232.

66. Bickel RD, Kopp A, Nuzhdin SV. Composite effects of polymorphisms near multiple regulatory elements create a major-effect QTL. PLoS Genet. 2011;7:e1001275.

67. Kühn C, Thaller G, Winter A, Bininda-Emonds ORP, Kaupe B, Erhardt G, et al. Evidence for multiple alleles at the DGAT1 locus better explains a quantitative trait locus with major effect on milk fat content in cattle. Genetics. 2004;167:1873–81.

68. Gao X, Becker LC, Becker DM, Starmer JD, Province MA. Avoiding the high Bonferroni penalty in genome-wide association studies. Genet Epidemiol. 2010;34:100–5.

69. The FAANG Consortium, Andersson L, Archibald AL, Bottema CD, Brauning R, Burgess SC, et al. Coordinated international action to accelerate genome-to-phenome with FAANG, the Functional Annotation of Animal Genomes project. Genome Biol. 2015;16:57.

13

Predictive performance of genomic selection methods for carcass traits in Hanwoo beef cattle: impacts of the genetic architecture

Hossein Mehrban[1], Deuk Hwan Lee[2*], Mohammad Hossein Moradi[3], Chung IlCho[4], Masoumeh Naserkheil[5] and Noelia Ibáñez-Escriche[6]

Abstract

Background: Hanwoo beef is known for its marbled fat, tenderness, juiciness and characteristic flavor, as well as for its low cholesterol and high omega 3 fatty acid contents. As yet, there has been no comprehensive investigation to estimate genomic selection accuracy for carcass traits in Hanwoo cattle using dense markers. This study aimed at evaluating the accuracy of alternative statistical methods that differed in assumptions about the underlying genetic model for various carcass traits: backfat thickness (BT), carcass weight (CW), eye muscle area (EMA), and marbling score (MS).

Methods: Accuracies of direct genomic breeding values (DGV) for carcass traits were estimated by applying fivefold cross-validation to a dataset including 1183 animals and approximately 34,000 single nucleotide polymorphisms (SNPs).

Results: Accuracies of BayesC, Bayesian LASSO (BayesL) and genomic best linear unbiased prediction (GBLUP) methods were similar for BT, EMA and MS. However, for CW, DGV accuracy was 7% higher with BayesC than with BayesL and GBLUP. The increased accuracy of BayesC, compared to GBLUP and BayesL, was maintained for CW, regardless of the training sample size, but not for BT, EMA, and MS. Genome-wide association studies detected consistent large effects for SNPs on chromosomes 6 and 14 for CW.

Conclusions: The predictive performance of the models depended on the trait analyzed. For CW, the results showed a clear superiority of BayesC compared to GBLUP and BayesL. These findings indicate the importance of using a proper variable selection method for genomic selection of traits and also suggest that the genetic architecture that underlies CW differs from that of the other carcass traits analyzed. Thus, our study provides significant new insights into the carcass traits of Hanwoo cattle.

Background

Hanwoo (*Bos taurus coreanae*) is an indigenous cattle breed in Korea that has been intensively bred for meat during the last 30 years [1]. Until the 1980s, Hanwoo cattle were used extensively for farming, transportation and religious sacrifices [2] but they have now become popular for meat production owing to their rapid growth and

high-quality meat. It is now one of the most economically important species in Korea. The extensive marbling of the Hanwoo beef is an important factor that influences the perception of meat quality in commercial beef production [3]. Hanwoo beef is known for its marbled fat, tenderness, juiciness and characteristic flavor. In addition, it has a lower cholesterol content and higher omega 3 fatty acid content, which makes it healthier than the meat from other bovine breeds [4]. In spite of its high price, i.e. almost three times that of imported beef meat from other breeds [5], Hanwoo beef is very popular both

*Correspondence: dhlee@hknu.ac.kr
[2] Department of Animal Life and Environment Science, Hankyong National University, Jungang-ro 327, Anseong-si, Gyeonggi-do 456-749, Korea
Full list of author information is available at the end of the article

among Korean consumers and abroad because of these invaluable traits [6].

The main aim of the Hanwoo beef industry is to increase both the quality (marbling, tenderness and flavor) and the quantity (carcass weight) of the meat. Estimated breeding values for backfat thickness (BT), carcass weight (CW), eye muscle area (EMA), and marbling score (MS) are commonly used as selection criteria in attempts to increase meat yield and quality, and subsequently to improve the income generated from steer feedlots and calf sales [7]. The recently developed genomic selection approach is beginning to revolutionize animal breeding. It refers to a genetic evaluation method that uses phenotypic data and genotypes of dense single nucleotide polymorphisms (SNPs) to estimate effects of SNPs from a training population and subsequently to predict the genetic values of selection candidates based on their genotypes [8]. It has been widely applied to dairy cattle breeding [9–11] and is now beginning to be used in other livestock species [12, 13]. Genomic predictions for beef cattle are attractive because many traits that affect the profitability of beef production, such as carcass traits, are difficult to select for because they are expensive to measure or are measured only on the relatives of breeding bulls [14]. Accurate genomic estimated breeding values would lead to greater genetic gain for these traits [15].

Accuracy of genomic prediction is key to the success of genomic selection [13]. Several analytical approaches have been proposed to predict genetic values based on genomic data, among which genomic (ridge regression) best linear unbiased prediction (GBLUP or RRBLUP), Bayesian shrinkage (e.g. BayesA) and variable selection models [e.g. BayesB, BayesCπ, BayesC and BayesL (LASSO)] have been widely used [13, 16]. The main differences between these models are their assumptions concerning the distributions of the effects of genetic markers. GBLUP (or equivalent RRBLUP procedures) models assume that all effects of SNPs are drawn from the same normal distribution and thus, that all SNPs have small effects [8]. The Bayesian approaches allow the variances of the SNP effects to differ from one another. However, Gianola et al. [17]. argued that for BayesA and BayesB models there is a strong dependency on the prior distributions of the marker variance because, in this case, the posterior variance is estimated with only one marker, thus its posterior distribution has only one more degree of freedom than its prior distribution. BayesCπ, is less sensitive to the prior assumption of the marker variance compared with BayesA and BayesB models because all SNPs have a common variance and the proportion of SNPs with no effect (π) has a uniform prior distribution

that is estimated during the analysis [18]. In BayesC, π is considered to be a fixed value [19], which leads to more accurate detection of quantitative trait loci (QTL) than BayesCπ, especially for traits with a moderate to high heritability and when sufficient numbers of records are available [20]. However, one drawback of the Bayesian methods is the need for the definition of priors. The requirement of a prior for the parameter π is circumvented in the BayesL method, which requires less information [21, 22].

Several studies have compared the performance of statistical methods applied to genomic selection and reported that genomic evaluation is more accurate than conventional genetic evaluation, see for example in dairy cattle [23, 24], beef cattle [25–27], pigs [28], sheep [29] and chickens [13, 30]. However, to date the performance of genomic selection in Hanwoo cattle has not been investigated. In addition, genomic prediction methods may perform differently for different traits and, thus lead to results that may differ because the genetic architecture that underlies a trait varies with the trait considered [9, 18]. Several studies have shown that Bayesian approaches produce higher accuracies than linear models when traits are influenced by genes with large effects [16, 31–34].

The aim of our study was to evaluate methods for genomic prediction in Hanwoo cattle. Three different methods, GBLUP, BayesC and BayesL, which differed in assumptions about the genetic architecture of traits, were used to compare the accuracy of genomic predictions for the traits BT, CW, EMA and MS.

Methods
Phenotypic and pedigree data
Phenotypic data from 5218 purebred Hanwoo steers produced by 590 young bulls were collected by the Hanwoo Improvement Center of the National Agricultural Cooperative Federation (NACF) between 1996 and 2012 in South Korea during a progeny testing program. Pedigree data from 44,538 individuals were used in the animal model. The four carcass traits included in the analysis, BT, CW, EMA and MS, were recorded at about 24 months of age on samples collected 24 h postmortem between the 13th rib and the 1st lumbar vertebra, according to the Korean carcass grading procedure by the National Livestock Cooperatives Federation. MS was assessed using a categorical system of nine classes that range from 1 (no marbling) to 9 (abundant marbling). Because MS data were skewed, they were transformed by a natural logarithm to lnMS after adding 1 to all records. Table 1 summarizes the statistics used for each trait to estimate variance components.

Table 1 Summary statistics for the phenotypic data used to estimate variance components

Trait (unit)	Number of animals in the pedigree	Number of animals with records	Mean (SE)	Min.	Max.	SD
BT (mm)	44,538	5218	8.60 (0.05)	1	35	3.74
CW (kg)	44,538	5217	341.01 (0.63)	158	518	45.26
EMA (cm^2)	44,538	5213	78.73 (0.13)	40	123	9.18
lnMS (Score)	44,538	3382	1.38 (0.01)	0.69	2.30	0.37

BT backfat thickness, *CW* carcass weight, *EMA* eye muscle area, *MS* marbling score

Genotypes

A total of 1679 animals were genotyped using the Illumina BovineSNP50 K (n = 959) and HD 777 K (n = 720) Beadchips (Illumina Inc., San Diego, CA, USA). Common SNPs between the 50 K and 777 K SNP chips were selected which resulted in 43,852 SNPs. All animals with more than 10% missing data (N = 68) and those with an inconsistency between pedigree and genomic relationships (N = 5) were excluded from further analyses. Phenotypic records were available for 1183 of the remaining 1606 animals that were genotyped (Table 2). To ensure overall quality of the samples and a consistent set of genotypes, quality control procedures were applied to the initial data [35]. SNPs were excluded from further analyses if their minor allele frequency (MAF) was lower than 0.01 (6679 SNPs) or if the percentage of calls (the proportion of SNP genotypes over all animals, calculated by the Illumina GenCall analysis software) was less than 0.98 (2677 SNPs). For the remaining SNPs, any outliers [that departed from the Hardy–Weinberg equilibrium (p < 10^{-6}) across all animals from one breed] were used to identify genotyping errors (302 SNPs). Missing genotypes were imputed using BEAGLE [36]. Finally, 34,194 SNPs remained for analyses.

Statistical analysis

Estimation of heritability

Heritability for each carcass trait (Table 1) was estimated using the restricted maximum likelihood method (REML) for animal models, using BLUPF90 (AIREMLF90) software [37]. The mixed model used was:

$$y = Xb + Zu + e,$$

where **y** is the vector of observations; **b** is the vector of fixed effects including slaughter date and batch effects as a contemporary group (369, 369, 368 and 176 levels for BT, CW, EMA and MS, respectively), and slaughter age (days from birth to slaughter) as a covariate; **u** is the vector of random animal effects and is assumed to follow a normal distribution $N(\mathbf{0}, A\sigma_a^2)$, A and σ_a^2 are the numerator relationship matrix and polygenic variance, respectively; **e** is the vector of random residual effects and is assumed to follow a normal distribution $N(\mathbf{0}, I\sigma_e^2)$, where **I** is an identity matrix including all animals with records and σ_e^2 is the error variance; and **X** and **Z** are design matrices that relate records to fixed effects and random animal effects, respectively.

Genomic prediction

Genomic predictions were performed for animals that had both genotype and phenotype records using three different models, i.e. GBLUP, BayesL [38] and BayesC [19]. GBLUP was applied using AIREMLF90 software [37] as follows:

$$y_c = 1\mu + Zg + e,$$

where y_c is a vector of the trait of interest, which was adjusted for fixed effects (slaughter date and batch effects as a contemporary group, and slaughter age as a covariate) based on the full dataset (see, Table 1); 1 is a vector of 1 s; μ is the overall mean; **Z** is the incidence matrix of direct genomic breeding values (DGV) and **g** is the vector of DGV and is assumed to follow a normal distribution $N\left(\mathbf{0}, G\sigma_g^2\right)$, where **G** is the marker-based genomic relationship matrix as a genomic relationship matrix and σ_g^2 the genetic variance captured by the markers; **e** is a vector of random residual effects and is assumed to follow a normal distribution $N(\mathbf{0}, I\sigma_e^2)$, where **I** is an identity matrix; and σ_ε^2 is the residual variance.

The **G**-matrix was built using the information from genome-wide dense SNPs [39] with the default options (except for a MAF of 0.01) in the preGSf90 program [40]. In the Bayesian framework, genomic analyses were

Table 2 Summary statistics for the phenotypic data used in the genomic analysis

Trait (unit)	Number of animals	Mean (SE)	Min.	Max.	SD
BT (mm)	1183	8.24 (0.10)	2	24	3.53
CW (kg)	1183	360.18 (1.16)	183	476	39.85
EMA (cm^2)	1183	82.99 (0.26)	55	121	8.78
lnMS (Score)	1183	1.34 (0.01)	0.69	2.30	0.34

BT backfat thickness, *CW* carcass weight, *EMA* eye muscle area, *MS* marbling score

performed using GS3 software [38]. The allelic substitution effect of each SNP was estimated using BayesL and BayesC, which were fitted with values in the covariate codes as 0, 2 (for homozygotes) and 1 (for heterozygotes) using the following model:

$$\mathbf{y}_c = 1\mu + \sum_{i=1}^{m} \mathbf{z}_i \alpha_i \delta_i + \boldsymbol{\varepsilon},$$

where $\mathbf{y}_c$ is a vector of corrected phenotypes as defined before, 1 is a vector of 1s; μ is the overall mean, m is the number of SNPs; $\mathbf{z}_i$ is the vector of genotype covariates for SNP$_i$, α_i is the allelic substitution effect of SNP$_i$, δ_i is an indicator variable for the presence (1) or absence (0) of the ith SNP in the model (for the BayesL method, δ_i is equal to 1 for all (i); $\boldsymbol{\varepsilon}$ is the vector of random residual effects assumed to follow a normal distribution $N(\mathbf{0}, \mathbf{I}\sigma_\varepsilon^2)$, where $\mathbf{I}$ is an identity matrix; and σ_ε^2 is the residual variance.

In the BayesL method, the prior distribution for α_i (with $\delta_i = 1$) follows a normal distribution $N(\mathbf{0}, \mathbf{I}\sigma_\alpha^2)$ and the prior distribution was as follows [38]:

$$\Pr\left(\alpha_i | \tau^2\right) = N\left(0, \tau_i^2\right),$$

$$\Pr\left(\tau_i^2\right) = \frac{\lambda^2}{2} \exp\left(-\lambda^2 |\tau_i^2|\right).$$

The prior distribution for σ_α^2 for all methods, was an inverted χ^2 distribution with two degrees of freedom and expectation was equal to $\sigma_a^2/(1-\pi)\sum_{i=1}^{m} 2p_i q_i$ as proposed by Habier et al. [18] where σ_a^2 is the estimated additive genetic variance using the animal model and p and q are the allelic frequencies at the ith SNP. In the BayesC method, the value of π is fixed. To identify the most suitable proportion of SNPs with no effect, the parameter π was considered to be equal to 0.999 and π values ranging from 0.91 to 0.99 in 0.02 increments (six values of π) were used. The residual variance was also assigned an inverted χ^2 distribution with two degrees of freedom and the expected value was equal to the residual variance as estimated using the animal model. The Markov chain Monte Carlo (MCMC) process was run for 550,000 cycles with 50,000 iterations as burn-in with a thinning interval of 50, so the effect of SNPs was estimated as a posterior mean of 10,000 samples.

The DGV for each animal in the validation set was estimated as the sum of the cross-product of animal genotype and the estimated SNP effect over all SNPs.

To confirm results of Bayesian analyses, a single-marker regression was run by using the Wombat software [41] with the following model:

$$\mathbf{y}_c = 1\mu + \mathbf{w}_i \mathbf{s}_i + \mathbf{Z}\mathbf{u} + \mathbf{e},$$

where $\mathbf{y}_c$ is a vector with adjusted phenotypes as defined before, 1 is a vector of 1s; μ is the overall mean; $\mathbf{w}_i$ is the vector of genotype covariates for SNP$_i$; $\mathbf{s}_i$ is the allelic substitution effect of the ith SNP; $\mathbf{u}$ is the vector of random animal effects and is assumed to follow a normal distribution $N(\mathbf{0}, \mathbf{A}\sigma_a^2)$, where $\mathbf{A}$ and σ_a^2 are the numerator relationship matrix and polygenic variance, respectively; $\mathbf{e}$ is the vector of random residual effects and is assumed to follow a normal distribution $N(\mathbf{0}, \mathbf{I}\sigma_e^2)$, where $\mathbf{I}$ is an identity matrix including all animals with records and σ_e^2 is the error variance; and $\mathbf{Z}$ is a design matrix that relate records to random animal effects.

To adjust for multiple testing, a Bonferroni-corrected threshold of 0.05/N (=1.46 × 10^{-6}) was used, where N is the number of SNPs used for the analyses.

Validation of models

The dataset was randomly split into five approximately equal subsets (fivefold cross-validation). Four subsets were used as training populations (≈946) and the fifth subset as a validation sample (≈237). The animals for the various subsets were selected randomly, except that paternal half-sibs were always placed in the same subset [42]. Cross-validation was replicated 10 times. Pedigree relationships within folds were on average equal to 0.038 and between fivefolds ranged from 0.023 to 0.031, with an average relationship of 0.026 for 10 replications. The predictive ability of DGV was determined by calculating the correlation between the DGV and the adjusted phenotypes for each of the five subsets. To estimate the prediction accuracy for each trait, predictive ability was divided by the square root of the heritability for that trait [43]. The accuracy for each replicate was obtained as the mean of the accuracies for the fivefold cross-validations of the ten replicates. The slope of the regression of the adjusted phenotypes on DGV was calculated as a measurement of the bias of the DGV in each method and trait. In addition, the mean square error (MSE) was predicted as the mean of the square differences between corrected phenotypes and DGV. In order to investigate the impact of the size of reference population on accuracy of DGV, analyses were also performed with training population sizes of 473 (50%) and 710 (75%) animals that were randomly sampled from the original training set. The validation population size was kept constant for all training sample sizes as in [44]. The means of accuracies and biases for different traits and methods were computed using the 10 replicates of the same cross-validation structure previously described.

Estimation of genomic heritability

In GBLUP, the genomic variance (σ_g^2) is estimated by REML. However, for the BayesC and BayesL methods, σ_g^2 is estimated by $2\sigma_\alpha^2(1-\pi)\sum_{i=1}^{m} p_i q_i$ [38], where σ_α^2 is the common effect marker variance, π is the proportion of

SNPs with no effect, p_i and q_i are the allelic frequencies at SNP i. Genomic heritability (h_g^2) was estimated according to the following formula [45]:

$$h_g^2 = h^2 \frac{\sigma_g^2}{\sigma_a^2},$$

where h^2 and σ_a^2 are the pedigree-based heritability and additive genetic variance, respectively.

Estimation of effective population size and expected accuracy

The past effective population size (N_e) for the tth generation ($t = (2c_t)^{-1}$), was estimated using the following model [46]:

$$E\left(r^2 - \frac{1}{n}\right) = \frac{1}{4N_e c + \alpha}$$

where r^2 is the pair-wise linkage disequilibrium, n is the number of animals sampled (1606 animals), c is the recombination rate (Morgan) defined for a particular physical distance and α is a correction for the occurrence of mutations ($\alpha = 2$) [47]. Due to the sensitivity of the estimated effective population size to the threshold that is set for MAF [46], we considered two different MAF thresholds, i.e. 0.1 and 0.2.

The expected accuracy of the genomic prediction ($r_{g\hat{g}}$) in our population was calculated using the formula derived by Daetwyler et al. [32], i.e. $r_{g\hat{g}} = \sqrt{\frac{N_P h^2}{N_P h^2 + M_e}}$. This formula depends on h^2 (heritability of the trait), N_P (number of animals in the training population) and M_e (the number of independent chromosome segments). M_e was calculated by using two different approximations: (1) $M_{e1} = \frac{2N_e L}{\ln(4N_e l)}$ [48] and (2) $M_{e2} = 2N_e L$ [49], where N_e is the effective population size, L is the genome length and l is the average chromosome length. Therefore, these two approximations of M_e lead to two different estimates of $r_{g\hat{g}}$.

Results and discussion

Estimation of heritability

The pedigree-based estimates of variance components for the carcass traits are in Table 3. Medium to high heritabilities were estimated for carcass traits in Hanwoo cattle. Estimated heritabilities for CW and EMA agreed with those previously reported in Hanwoo cattle by Lee et al. [7]. However, estimated heritabilities for BT and MS were higher (+9 and +11.3, respectively) than those in the study of Lee et al. [7]. In Japanese Black cattle, Onogi et al. [50] reported similar heritabilities for EMA (0.43) and MS (0.66) but a higher heritability for CW (0.56) than our study. In a study on the Angus breed, Saatchi et al. [25] reported higher heritabilities for CW and EMA

Table 3 Variance components (standard error) estimated using pedigree and phenotypic data

Trait (unit)	σ_a^2	σ_e^2	σ_p^2	h^2
BT (mm)	5.57 (0.62)	5.75 (0.49)	11.32 (0.26)	0.49 (0.05)
CW (kg)	315.28 (46.76)	699.95 (40.51)	1015.23 (22.26)	0.31 (0.04)
EMA (cm²)	26.75 (3.27)	35.33 (2.67)	62.08 (1.42)	0.43 (0.05)
lnMS (Score)	0.08 (0.01)	0.05 (0.008)	0.13 (0.004)	0.61 (0.06)

BT backfat thickness, *CW* carcass weight, *EMA* eye muscle area, *MS* marbling score

$\sigma_a^2, \sigma_e^2, \sigma_p^2, h^2$: additive genetic variance, error variance, phenotypic variance and heritability, respectively

and lower heritabilities for BT and MS than those found here. Our estimated heritabilities for carcass traits were within the range of those obtained for multi-breed commercial beef cattle by Rolf et al. [16].

Estimation of effective population size

We used the average extent of linkage disequilibrium (LD) in the genome to estimate effective population sizes at various times in the past. Estimates of N_e were not influenced by the threshold set for MAF i.e. 0.10 or 0.20 [see Additional file 1: Figure S1]. Therefore, we used a threshold of 0.10 for MAF to estimate N_e. The results showed that N_e declined across generations to reach a value of 224 in the latest generation. The effective population size that was estimated here for Hanwoo cattle was not consistent with that reported by Lee et al. [51], who also found that it declined across generations but to 98, three generations ago. However, we used a sample size that was approximately 6 times larger than that used by Lee et al. [51] and also a much larger number of SNPs to estimate linkage disequilibrium (r^2). Moreover, Li and Kim [52] estimated an effective population size of 402, five generations ago, by using 547 Hanwoo bulls and a 50 K SNP chip, whereas our estimate for that generation was 298. With the exception of the N_e reported by Marquez et al. [53] ($N_e = 445$) for American Red Angus beef cattle and by Saatchi et al. [25] ($N_e = 654$) for American Angus beef cattle, most studies in beef and dairy cattle [54–58] have found smaller N_e than in the present study. According to Godard and Hayes [59], this implies that a larger reference population would be required for Hanwoo cattle than for the above-mentioned breeds [54–58] to obtain a similar accuracy in genomic prediction.

Comparison of models

The parameter π is a fixed value in the BayesC method [19]. We analyzed a range of π values from 0.91 to 0.999 to determine the most accurate π for the BayesC method for each trait. As shown in Fig. 1a, the realized accuracy for BT remained stable across a range of π values from

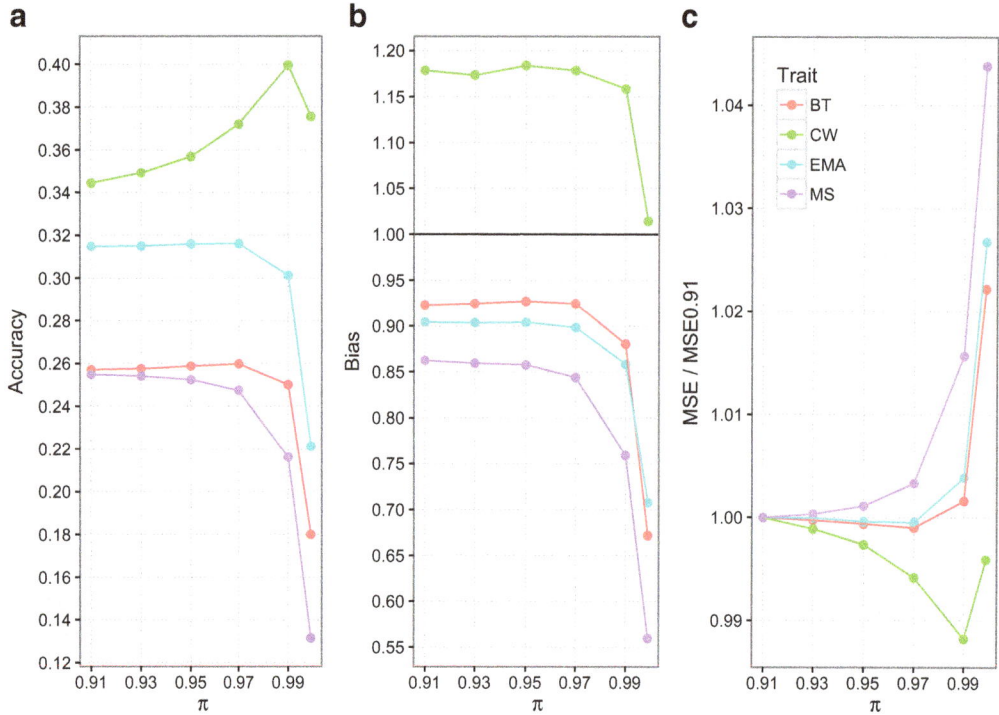

Fig. 1 Accuracy (**a**), bias (**b**), and mean square error (MSE) (**c**) of DGV obtained by different methods. Comparison of the accuracies, biases and MSE obtained with BayesC using different values of π for backfat thickness (BT), carcass weight (CW), eye muscle area (EMA), and marbling score (MS) traits. MSE are shown as the ratio of MSE to MSE of BayesC91

0.91 to 0.97, and then decreased for π values above 0.97. Similar patterns were observed for EMA and MS, with accuracies decreasing for π values above 0.97 and 0.91, respectively. In contrast, the accuracy of CW improved as π increased to reach a peak for a π value of 0.99 and then declined dramatically. Overall, the values of π for which the BayesC model provided the highest accuracy were 0.97 (BayesC97), 0.99 (BayesC99), 0.97 (BayesC97) and 0.91 (BayesC91) for BT, CW, EMA and MS traits, respectively (Fig. 1a). The lowest bias was obtained with π values of 0.95 for BT, 0.999 for CW, 0.95 for EMA, and 0.91 for MS (Fig. 1b). Thus, for CW there was a conflict between accuracy and bias to determine the most suitable π value. The highest accuracy and lowest bias for CW were obtained for π values of 0.99 and 0.999, respectively. Nevertheless, González-Recio et al. [60] showed that the MSE is a more flexible criterion than correlation and bias for comparing models because it takes both prediction bias and variability into account. Due to the fact that MSE depends on the trait, we used the MSE ratio (ratio between MSE and MSE of BayesC91) to compare across traits and models. The lowest MSE ratio was achieved when π was set to 0.97, 0.99, 0.97, and 0.91 for BT, CW, EMA and MS, respectively (Fig. 1c).

A comparison of the accuracy and bias obtained for CW with the BayesC99, BayesL and GBLUP methods, revealed the superiority of the BayesC99 model (Fig. 2a); the accuracy of this model was higher than those of GBLUP (+0.071) and BayesL (+0.070) and the bias was lower than those of GBLUP (−0.02) and BayesL (−0.11) (Fig. 2b). For the other carcass traits (BT, EMA and MS), the accuracy and bias of BayesC99, BayesL and GBLUP methods were similar.

In terms of MSE, BayesC99 exhibited the best performance (the lowest MSE) for CW, while for the other traits, the differences in MSE between the methods were trivial [see Additional file 2: Table S1].

The predictive performance of the models depended on the trait analyzed. The three methods performed similarly for BT, EMA and MS traits, whereas for CW BayesC clearly outperformed GBLUP and BayesL. This indicates that the infinitesimal model holds for BT, EMA and MS but not completely for CW. In other words, BT, EMA and MS traits would be controlled by several genes, each with a small effect, whereas one or more individual genes would have a large effect on CW. These findings were confirmed by the single-marker method used for the GWAS analysis, which detected genome-wide significant

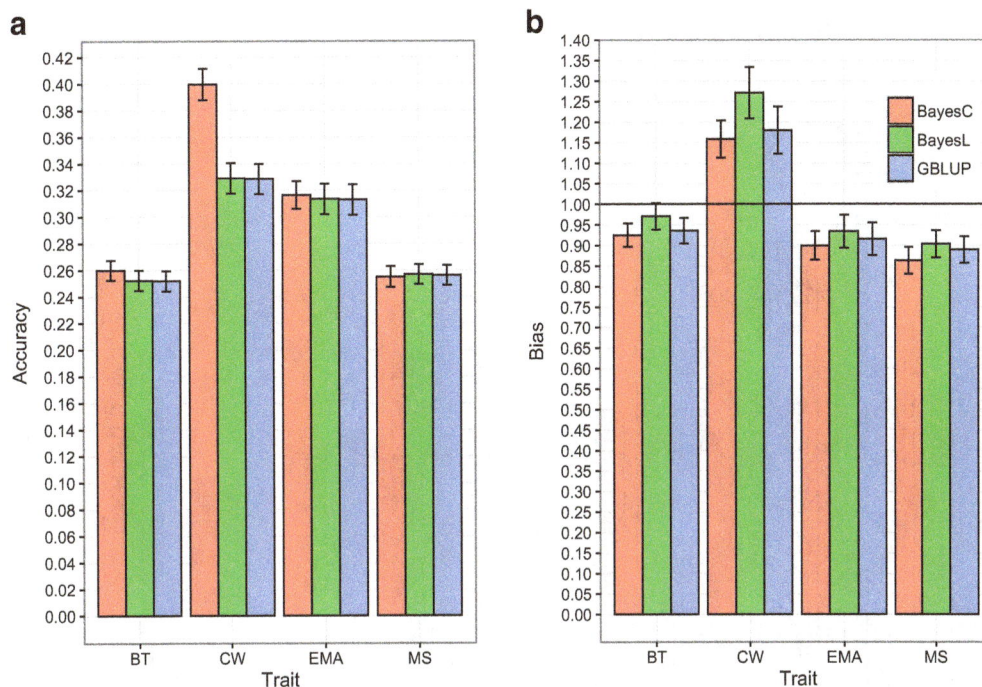

Fig. 2 Accuracy (±SE) (**a**) and bias (±SE) (**b**) of DGV obtained by different methods. In BayesC, π of 0.97, 0.99, 0.97 and 0.91 were considered for backfat thickness (BT), carcass weight (CW), eye muscle area (EMA), and marbling score (MS) traits, respectively

SNPs on chromosomes 6 and 14 for CW but not for MS, BT and EMA [see Additional file 3: Figure S2]. However, our results could be quite sensitive to the size of the reference population. Gao et al. [61] showed that by increasing the number of animals in the reference population, the difference in accuracy between Bayesian and GBLUP approaches decreased. Therefore, the impact of the size of the training population on accuracy was also investigated. As shown in Fig. 3, the accuracy of prediction for the traits and methods studied decreased as the size of the training population decreased, in agreement with the literature [32, 44, 59]. Nevertheless, the superiority of BayesC compared to GBLUP and BayesL was maintained in terms of accuracy regardless of the size of the training sample for CW but not for BT, EMA, and MS, regardless of the π value (Fig. 3).

Wolc et al. [62] pointed out that mixture models (i.e. BayesB and BayesC) were clearly better than GBLUP for genomic prediction in the presence of QTL with a large effect, especially for small datasets and resulted in more accurate and persistent predictions. In our study, the accuracy of genomic prediction clearly differed between a Bayesian model (BayesC99) and GBLUP for CW with varying sizes of the training population as was also reported by [32].

Our results support a previous study on Hanwoo cattle by Lee et al. [7] that aimed at identifying major loci

associated with several carcass traits (BT, CW, EMA and MS). They demonstrated that six highly significant SNPs on chromosome 14 were associated with CW, but no significant SNPs were identified for the other carcass traits. Another GWAS on Japanese black beef cattle also detected three QTL that had a relatively large effect on CW [63]. Ogawa et al. [64] reported that MS is controlled by QTL that have only relatively small effects compared with the CW trait in Japanese black beef cattle. Other studies have also reported conflicting results. For example, Chen et al. [27] showed that GBLUP and the Bayesian methods were very similar in terms of accuracy for BT, CW, EMA and MS traits in Angus cattle and for CW, EMA and MS traits in Charolais cattle. They found that the BayesB95 ($\pi = 0.95$) model performed more accurately (3%) than GBLUP for BT in Charolais, whereas in contrast, Rolf et al. [16] found that the accuracy of BayesB95 ($\pi = 0.95$) was 3.4% lower than that of RRBLUP for the same trait in multi-breed commercial beef cattle. They showed that RRBLUP was more accurate than BayesB for BT, CW and MS, whereas, for EMA, the accuracy of DGV was the same using either method. Júnior et al. [65] obtained similar results for BT, CW, and EMA in terms of accuracy and MSE using RRBLUP, BayesC and BayesL in Nellore cattle. These observations may also support the argument that the genetic architecture of these traits may differ among breeds because of different population histories. Saatchi

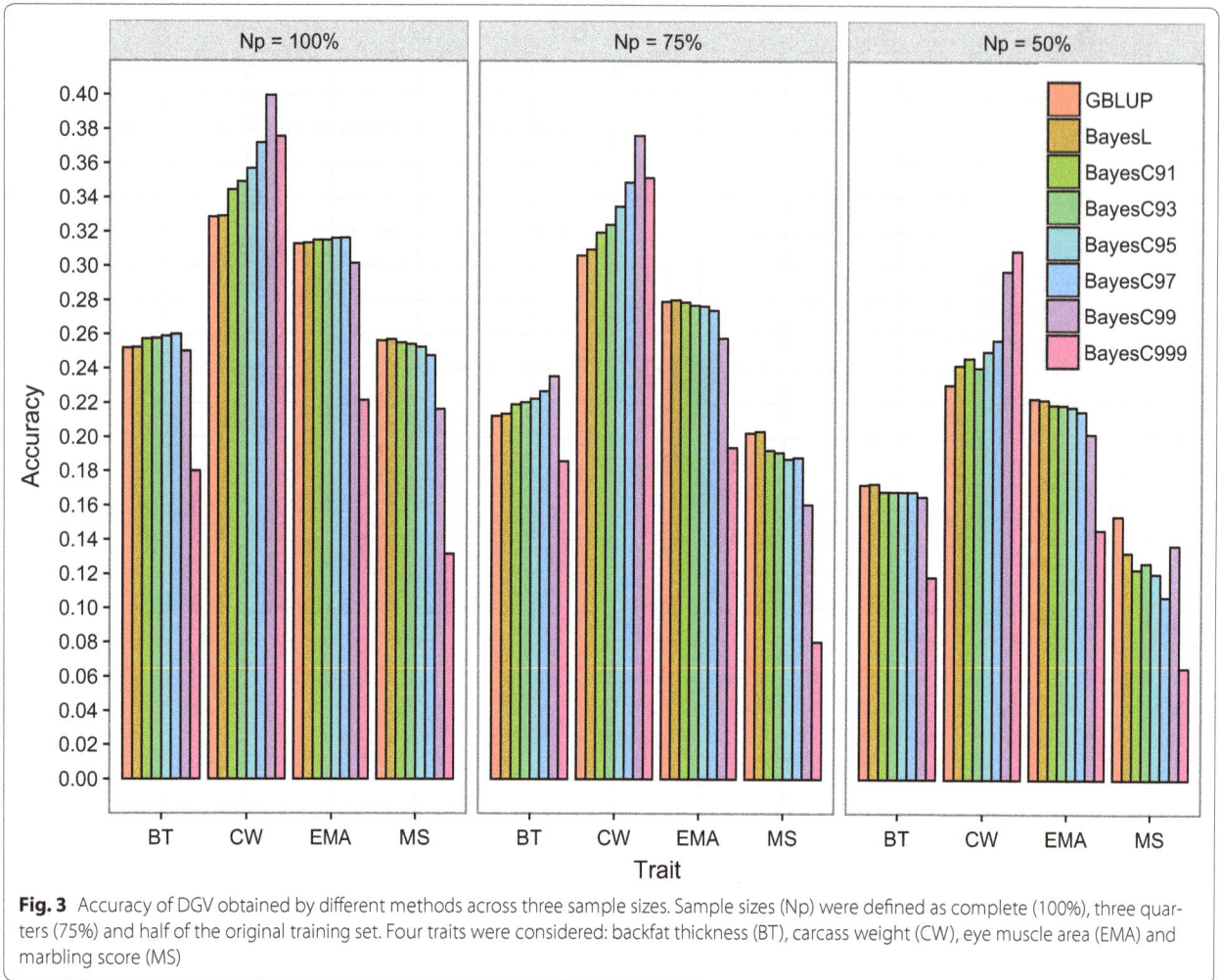

Fig. 3 Accuracy of DGV obtained by different methods across three sample sizes. Sample sizes (Np) were defined as complete (100%), three quarters (75%) and half of the original training set. Four traits were considered: backfat thickness (BT), carcass weight (CW), eye muscle area (EMA) and marbling score (MS)

et al. [66] showed that one reason that explains the differences in the QTL identified among different populations could be that the genetic architecture that underlies trait variation varies among breeds.

Comparison between the traits analyzed

In spite of their high heritabilities, prediction accuracies for BT and MS were lower than those for CW and EMA (Table 3; Figs. 1a, 2a), which is consistent with the results of Onogi et al. [50]. To investigate further the low prediction accuracy for BT and MS, genomic heritability (h_g^2) was estimated for each trait and with each method (Table 4). The proportion of genomic heritability to pedigree-based heritability (h_g^2/h^2) represents the proportion of genetic variance that was explained by the markers (σ_g^2/σ_a^2) [45]. Our results indicated that the estimated genomic variance (σ_g^2) was lower than the additive genetic variance σ_a^2 (Tables 3, 4) for all traits and with all methods except for CW using BayesC, which was slightly

larger. However, given the large standard error obtained for σ_g^2 (72.12) and σ_a^2 (46.76), the differences between σ_a^2 and σ_g^2 were not significant. Compared to CW and EMA, genomic heritabilities for BT and MS differed largely from pedigree-based heritabilities, regardless of the method (Table 4). With the GBLUP model, the proportion of genetic variance captured by SNPs for BT and MS was equal to 65 and 66%, respectively. In other words, for BT and MS, 35 and 34% of the genetic variance was not explained by SNPs, while for EMA and CW, only 15% and just 5% of the additive genetic variance was unexplained.

This finding may explain the lower prediction accuracies obtained for BT and MS compared with EMA and CW, in spite of their higher heritability. In addition, it was expected that the DGV for MS would be more accurate than those for BT because MS had a higher heritability (Table 3), possibly because MS is a categorical trait. Kizilkaya et al. [67] showed that the accuracy of DGV for an ordinal categorical trait was substantially lower than for a continuous trait

Table 4 Genomic variance (σ_g^2), marker variance explained (σ_g^2/σ_a^2) and genomic heritability (h_g^2) by fully corrected phenotype and medium-density SNP

Trait (unit)	Method[a]	σ_g^2 (SE)[b]	σ_g^2/σ_a^2	$h_g^2 = h^2\frac{\sigma_g^2}{\sigma_a^2}$
BT (mm)	BayesC[2]	3.71 (0.75)	0.67	0.33
	BayesL	3.63 (0.75)	0.65	0.32
	GBLUP	3.62 (0.73)	0.65	0.32
CW (kg)	BayesC	330.73 (72.12)	1.05	0.33
	BayesL	299.73 (72.96)	0.95	0.30
	GBLUP	300.70 (69.013)	0.95	0.30
EMA (cm²)	BayesC	23.19 (4.04)	0.87	0.37
	BayesL	23.00 (4.16)	0.86	0.37
	GBLUP	22.84 (4.14)	0.85	0.37
lnMS (Score)	BayesC	0.055 (0.009)	0.69	0.42
	BayesL	0.054 (0.009)	0.68	0.41
	GBLUP	0.053 (0.009)	0.66	0.40

BT backfat thickness, *CW* carcass weight, *EMA* eye muscle area, *MS* marbling score

[a] For BayesC, π values of 0.97, 0.99, 0.97 and 0.91 (the highest accuracy) were considered for BT, CW, EMA and MS, respectively

[b] SE in Bayesian methods were estimated as the standard deviation of the posterior distribution

under the same conditions of heritability, effective and training population sizes, and number of categories.

The low genomic heritabilities achieved for BT and MS indicate that more animals (with genotypes and phenotypes) are necessary to accurately estimate the effects of SNPs compared with CW and EMA. We also observed that the SNPs on the 50 K SNP chip could not capture all the genetic variability for those traits (BT and MS). Therefore, a high-density SNP chip could be used to adequately assess LD and potentially capture a larger proportion of the additive genetic variance than the medium-density chip (i.e. 50,000 SNPs). In order to investigate the performance of SNP density, 570,969 SNPs were imputed from the 50 K chip. Our findings indicate that the genomic variance σ_g^2 and (σ_g^2/σ_a^2) increased as the SNP density increased [see Additional file 4: Table S2]. The accuracy of DGV increased by 4% for BT and 12% for MS; however, for CW and EMA, the accuracy did not improve. Many studies using simulation and real data confirmed that the accuracy of genomic selection improves only slightly when a high-density SNP chip or whole-sequence data are used [34, 68–71].

In general, the realized accuracies of DGV for the four carcass traits, regardless of the method used, were low compared with results from other studies [16, 25, 50]. One of the main reasons for the lower accuracies observed in our study could be due to the small training

population size ($N \approx 946$) and the large effective population size ($N_e = 224$) for the Hanwoo breed. Theoretical studies have shown that, to obtain the same accuracy, the number of animals needed in the reference population increases with increasing effective population size [32, 59]. Using the K-means method, Saatchi et al. [25] estimated DGV accuracies of 0.60, 0.47, 0.60 and 0.69 for BT, CW, EMA and MS, respectively, using a training population of approximately 2200 Angus beef cattle. Using a training population of about 2000 animals in multi-breed commercial beef cattle, Rolf et al. [16] observed that the highest accuracies of DGV for BT, CW, EMA and MS were equal to 0.51, 0.78, 0.60 and 0.76, respectively. Onogi et al. [50] reported a predicted ability (correlation between the DGV and the adjusted phenotypes) of 0.44, 0.42 and 0.39 for CW, EMA and MS, respectively. In our study, the genetic relationship between the validation and reference populations was close to zero. This is the most challenging scenario for genomic prediction because a large part of the accuracy of DGV results from genetic relationships captured by SNPs [72]. This could explain that our prediction accuracies were lower than those reported by Onogi et al. [50] for which the number of genotyped animals was larger and the effective population size was smaller [64] than in our study.

An alternative for improving prediction accuracy for Hanwoo cattle, with a deep pedigree, is to apply single-step GBLUP (ssGBLUP) [73, 74]. In this method, accuracy is increased by using information from the pedigree and SNPs simultaneously [73]. However, as we have shown, GBLUP cannot be the best method for genomic prediction in the presence of QTL with a large effect such as the CW trait in our study. Thus, an alternative to increase the prediction accuracy for CW in single-step evaluation could be to use genomic relationship matrices weighted by marker realized variance as suggested by [75, 76].

Comparison of realized and expected accuracy

As shown in Fig. 4, the observed accuracies were lower than the expected accuracies according to the formula derived by Daetwyler et al. [32] when the approximation for M_e (i.e. number of independent chromosome segments) was $M_{e1} = 2N_eL/\ln(4N_eL)$ [48] but greater than the expected accuracy when M_e was $M_{e2} = 2N_eL$ [49]. Our results agree with those of Neves et al. [77] who reported that expected accuracies based on M_{e1} were higher than realized accuracies across traits; however, expected accuracies using M_{e2} were lower than realized accuracies in the case of within-family predictions.

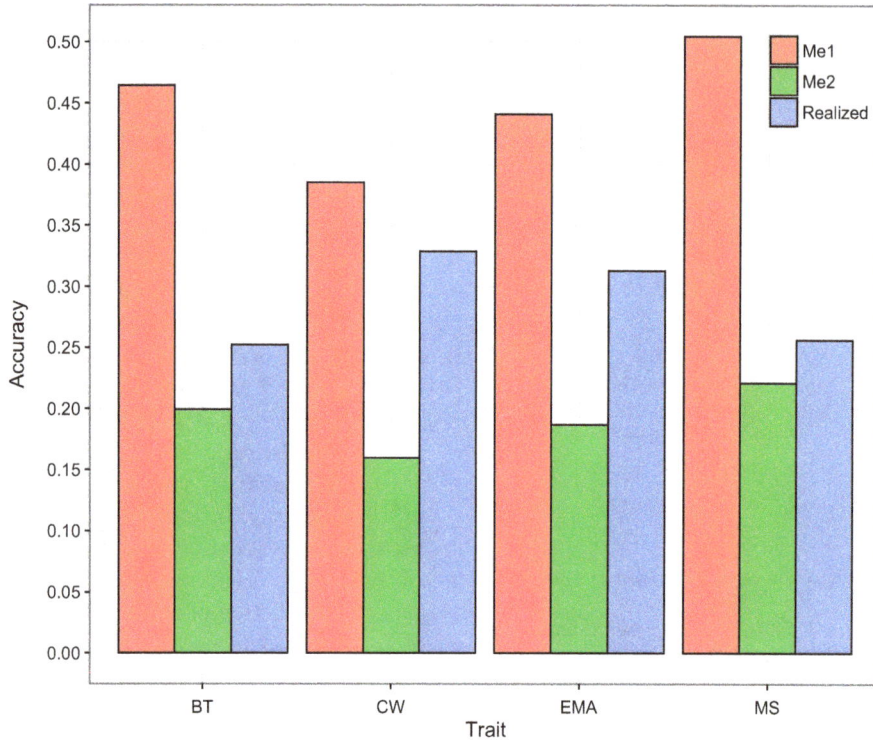

Fig. 4 Realized and expected accuracy of genomic predictions with GBLUP. Expected accuracies were calculated according to Daetwyler et al. [32] using two different approximations for the number of independent chromosome segment M_e ($M_{e1} = 2N_eL/\ln(4N_el)$ and $M_{e2} = 2N_eL$). The realized accuracies were averaged over 10 replicates for each trait [backfat thickness (BT), carcass weight (CW), eye muscle area (EMA) and marbling score (MS)]

Hayes et al. [49] pointed out that M_{e1} does not take into account that the small segments may still contain as many mutations in the QTL as the larger segments. Thus, Hayes et al. [49] recommended the use of $M_{e2} = 2N_eL$, which is a compromise between the number of segments (4 N_eL) and the number of segments weighted by length ($2N_eL/\log(4N_eL)$ per chromosome). However, M_{e2} is not an optimal approximation and based on our results as well as those of Neves et al. [76], it seems to underestimate the genomic prediction accuracy. However, the formula of Daetwyler et al. [32] assumes that all the genetic variance of the trait is explained by SNPs. Therefore, the formula is expected to overestimate prediction accuracy when SNPs cannot capture all the genetic variability. In our study, the genomic variance was smaller than the additive genetic variance (see Table 4), especially for BT and MS. Consequently, this could explain the differences between expected (M_{e1}) and realized accuracy for BT (0.21) and MS (0.25) and for EMA (0.13) and CW (0.06). This would indicate that when nearly all the total genetic variance is explained by the SNP array, the realized accuracies of GBLUP are closer to the expected values based on M_{e1} than on M_{e2}.

Conclusions

The performance of the statistical methods used depended on the trait analyzed. The results showed a clear superiority of BayesC compared with GBLUP and BayesL for CW, whereas for the other traits all methods performed similarly. The prediction accuracy of DGV for CW using BayesC was around 7% higher than that obtained with the GBLUP and BayesL methods. This indicates the importance of using a proper variable selection method for genomic selection of traits. In addition, the results also suggest that the genetic architecture underlying CW may differ from that underlying the other carcass traits. This could be due to the fact that BT, EMA and MS seem to be controlled by several genes, each with a small effect, whereas for CW, there are probably several individual genes that each have a large effect. Overall, our results provide the first information for implementing genomic prediction in Hanwoo beef cattle.

Predictive performance of genomic selection methods for carcass traits in Hanwoo beef cattle: impacts...

171

Additional files

Additional file 1: Figure S1. Estimates of effective population size (N_e) in the past generations. Thresholds of 0.1 and 0.2 were considered for minor allelic frequency (MAF). The figure describes the changes of N_e over generations for two different minor allelic frequencies (0.1 and 0.2).

Additional file 2: Table S1. Mean square error (SE) of genomic prediction for four carcass traits in Hanwoo beef cattle. This table provides the mean square errors of GBLUP, BayesL and BayesC genomic prediction for backfat thickness (BT), carcass weight (CW), eye muscle area (EMA), and marbling score (MS) traits.

Additional file 3: Figure S2. Manhattan plots of genome-wide association analyses for four carcass traits. This figure provides the log10 p-values of the SNPs analyzed in the genome-wide association analyses for backfat thickness (BT), carcass weight (CW), eye muscle area (EMA), and marbling score (MS) traits. The horizontal lines represent the 5% significance level with a p value threshold of 1.46×10^{-6} for backfat thickness (BT), carcass weight (CW), eye muscle area (EMA), and marbling score (MS) traits.

Additional file 4: Table S2. Genomic variance (σ_g^2), marker variance explained (σ_g^2/σ_a^2) and genomic heritability (h_g^2) obtained when using fully corrected phenotypes and high-density SNPs with the GBLUP method. Description: This table provides the results for the genomic variance (σ_g^2), marker variance explained (σ_g^2/σ_a^2) and genomic heritability (h_g^2) obtained when a high-density chip (777 k) was used to analyze backfat thickness (BT), carcass weight (CW), eye muscle area (EMA), and marbling score (MS) traits under the GBLUP method.

Authors' contributions

DHL conceived and designed the study and contributed to the discussion of the results. HM conceived the study, analyzed the data and drafted the manuscript. MHM contributed to the discussion of the results and drafted the manuscript. CIC and MN participated in analyzing the data. NIE conceived the study, evaluated the experiments, contributed to the discussion of the results and edited the manuscript. All authors read and approved the final manuscript.

Author details

[1] Department of Animal Science, Shahrekord University, P.O. Box 115, Shahrekord 88186-34141, Iran. [2] Department of Animal Life and Environment Science, Hankyong National University, Jungang-ro 327, Anseong-si, Gyeonggi-do 456-749, Korea. [3] Department of Animal Science, Faculty of Agriculture and Natural Resources, Arak University, Arâk 38156-8-8349, Iran. [4] Hanwoo Improvement Center, National Agricultural Cooperative Federation, Haeun-ro 691, Unsan-myeon, Seosan-si, Chungnam-do 356-831, Korea. [5] Department of Animal Science, University College of Agriculture and Natural Resources, University of Tehran, P.O. Box 4111, Karaj 31587-11167, Iran. [6] The Roslin Institute, Royal (Dick) School of Veterinary Studies, University of Edinburgh, Roslin, UK.

Acknowledgements

This work was supported by a Grant from the IPET Program (No. 20093068), Ministry of Agriculture, Food and Rural Affairs, Republic of Korea. We are also grateful to all the staff of the Korean Hanwoo Improvement Center of the National Agricultural Cooperative Federation (NACF) for supplying data as well as semen and blood samples of Hanwoo cattle.

Competing interests

The authors declare that they have no competing interests.

References

1. Choi JW, Choi BH, Lee SH, Lee SS, Kim HC, Yu D, et al. Whole-genome resequencing analysis of Hanwoo and Yanbian cattle to identify genome-wide SNPs and signatures of selection. Mol Cells. 2015;38:466–73.
2. Lee SH, Choi BH, Cho SH, Lim D, Choi TJ, Park BH, et al. Genome-wide association study identifies three loci for intramuscular fat in Hanwoo (Korean cattle). Livest Sci. 2014;165:27–32.
3. Choi Y, Davis ME, Chung H. Effects of genetic variants in the promoter region of the bovine adiponectin (ADIPOQ) gene on marbling of Hanwoo beef cattle. Meat Sci. 2015;105:57–62.
4. Jo C, Cho SH, Chang J, Nam KC. Keys to production and processing of Hanwoo beef: a perspective of tradition and science. Anim Front. 2012;2:32–8.
5. Korea Rural Economic Institute (KREI). Outlook and Agricultural Statistics Information System(OASIS). http://oasis.krei.re.kr (2015). Accessed 20 Oct 2015.
6. Lee SH, Park BH, Sharma A, Dang CG, Lee SS, Choi TJ, et al. Hanwoo cattle: origin, domestication, breeding strategies and genomic selection. J Anim Sci Technol. 2014;56:2.
7. Lee SH, Choi BH, Lim D, Gondro C, Cho YM, Dang CG, et al. Genome-wide association study identifies major loci for carcass weight on BTA14 in Hanwoo (Korean cattle). PLoS One. 2013;8:e74677.
8. Meuwissen THE, Hayes BJ, Goddard ME. Prediction of total genetic value using genome-wide dense marker maps. Genetics. 2001;157:1819–29.
9. Hayes BJ, Bowman PJ, Chamberlain AJ, Goddard ME. Invited review: genomic selection in dairy cattle: progress and challenges. J Dairy Sci. 2009;92:433–43.
10. VanRaden PM, Sullivan PG. International genomic evaluation methods for dairy cattle. Genet Sel Evol. 2010;42:7.
11. Colombani C, Legarra A, Fritz S, Guillaume F, Croiseau P, Ducrocq V, et al. Application of Bayesian least absolute shrinkage and selection operator (LASSO) and BayesCpi methods for genomic selection in French Holstein and Montbeliarde breeds. J Dairy Sci. 2013;96:575–91.
12. Duchemin SI, Colombani C, Legarra A, Baloche G, Larroque H, Astruc JM, et al. Genomic selection in the French Lacaune dairy sheep breed. J Dairy Sci. 2012;95:2723–33.
13. Liu T, Qu H, Luo C, Shu D, Wang J, Lund MS, et al. Accuracy of genomic prediction for growth and carcass traits in Chinese triple-yellow chickens. BMC Genet. 2014;15:110.
14. Boerner V, Johnston DJ, Tier B. Accuracies of genomically estimated breeding values from pure-breed and across-breed predictions in Australian beef cattle. Genet Sel Evol. 2014;46:61.
15. Bolormaa S, Pryce JE, Kemper K, Savin K, Hayes BJ, Barendse W, et al. Accuracy of prediction of genomic breeding values for residual feed intake and carcass and meat quality traits in Bos taurus, Bos indicus, and composite beef cattle. J Anim Sci. 2013;91:3088–104.
16. Rolf MM, Garrick DJ, Fountain T, Ramey HR, Weaber RL, Decker JE, et al. Comparison of Bayesian models to estimate direct genomic values in multi-breed commercial beef cattle. Genet Sel Evol. 2015;47:23.
17. Gianola D, de los Campos G, Hill WG, Manfredi E, Fernando R. Additive genetic variability and the Bayesian alphabet. Genetics. 2009;183:347–63.
18. Habier D, Fernando RL, Kizilkaya K, Garrick DJ. Extension of the bayesian alphabet for genomic selection. BMC Bioinformatics. 2011;12:186.
19. Fernando RL, Garrick D. Bayesian methods applied to GWAS. Methods Mol Biol. 2013;1019:237–74.
20. van den Berg I, Fritz S, Boichard D. QTL fine mapping with Bayes C(pi): a simulation study. Genet Sel Evol. 2013;45:19.
21. de los Campos G, Naya H, Gianola D, Crossa J, Legarra A, Manfredi E, et al. Predicting quantitative traits with regression models for dense molecular markers and pedigree. Genetics. 2009;182:375–85.
22. Legarra A, Robert-Granie C, Croiseau P, Guillaume F, Fritz S. Improved Lasso for genomic selection. Genet Res. 2011;93:77–87.
23. Su G, Guldbrandtsen B, Gregersen VR, Lund MS. Preliminary investigation on reliability of genomic estimated breeding values in the Danish Holstein population. J Dairy Sci. 2010;93:1175–83.
24. Lund MS, Roos AP, Vries AG, Druet T, Ducrocq V, Fritz S, et al. A common reference population from four European Holstein populations increases reliability of genomic predictions. Genet Sel Evol. 2011;43:43.
25. Saatchi M, McClure MC, McKay SD, Rolf MM, Kim J, Decker JE, et al. Accuracies of genomic breeding values in American Angus beef cattle using K-means clustering for cross-validation. Genet Sel Evol. 2011;43:40.
26. Neves HH, Carvalheiro R, O'Brien AM, Utsunomiya YT, do Carmo AS, Schenkel FS, et al. Accuracy of genomic predictions in Bos indicus (Nellore) cattle. Genet Sel Evol. 2014;46:17.

27. Chen L, Vinsky M, Li C. Accuracy of predicting genomic breeding values for carcass merit traits in Angus and Charolais beef cattle. Anim Genet. 2015;46:55–9.

28. Tribout T, Larzul C, Phocas F. Efficiency of genomic selection in a purebred pig male line. J Anim Sci. 2012;90:4164–76.

29. Daetwyler HD, Swan AA, van der Werf JH, Hayes BJ. Accuracy of pedigree and genomic predictions of carcass and novel meat quality traits in multi-breed sheep data assessed by cross-validation. Genet Sel Evol. 2012;44:33.

30. Wolc A, Stricker C, Arango J, Settar P, Fulton JE, O'Sullivan NP, et al. Breeding value prediction for production traits in layer chickens using pedigree or genomic relationships in a reduced animal model. Genet Sel Evol. 2011;43:5.

31. Hayes BJ, Pryce J, Chamberlain AJ, Bowman PJ, Goddard ME. Genetic architecture of complex traits and accuracy of genomic prediction: coat colour, milk-fat percentage, and type in Holstein cattle as contrasting model traits. PLoS Genet. 2010;6:e1001139.

32. Daetwyler HD, Pong-Wong R, Villanueva B, Woolliams JA. The impact of genetic architecture on genome-wide evaluation methods. Genetics. 2010;185:1021–31.

33. Zhang Z, Liu J, Ding X, Bijma P, de Koning DJ, Zhang Q. Best linear unbiased prediction of genomic breeding values using a trait specific marker derived relationship matrix. PLoS One. 2010;5:e12648.

34. Clark SA, Hickey JM, van der Werf JH. Different models of genetic variation and their effect on genomic evaluation. Genet Sel Evol. 2011;43:18.

35. Moradi MH, Nejati-Javaremi A, Moradi-Shahrbabak M, Dodds KG, McEwan JC. Genomic scan of selective sweeps in thin and fat tail sheep breeds for identifying of candidate regions associated with fat deposition. BMC Genet. 2012;13:10.

36. Browning SR, Browning BL. Rapid and accurate haplotype phasing and missing data inference for whole genome association studies by use of localized haplotype clustering. Am J Hum Genet. 2007;81:1084–97.

37. Misztal I, Tsuruta S, Strabel T, Auvray B, Druet T, Lee DH. BLUPF90 and related programs (BGF90). In: Proceedings of the 7th world congress on genetics applied to livestock production: 19–23 August 2002; Montpellier. CD-ROM Communication No 28-27. 2002.

38. Legarra A RA, Filangi O. GS3 genomic Selection–Gibbs Sampling–Gauss Seidel (and BayesCπ) (2011). https://qgsp.jouy.inra.fr/index.php?option=com_content&view=article&id=60&Itemid=67.

39. VanRaden PM. Efficient methods to compute genomic predictions. J Dairy Sci. 2008;91:4414–23.

40. Aguilar I, Misztal I, Tsuruta S, Legarra A, Wang H. PREGSF90–POSTGSF90: computational tools for the implementation of single-step genomic selection and genome-wide association with ungenotyped individuals in BLUPF90 programs. In: Proceedings of the 10th world congress on genetics applied to livestock production: 18–22 August 2014; Vancouver. 2014.

41. Meyer K, Tier B. SNP Snappy: a strategy for fast genome-wide association studies fitting a full mixed model. Genetics. 2012;190:275–7.

42. Khansefid M, Pryce JE, Bolormaa S, Miller SP, Wang Z, Li C, et al. Estimation of genomic breeding values for residual feed intake in a multibreed cattle population. J Anim Sci. 2014;92:3270–83.

43. Legarra A, Robert Granié C, Manfredi E, Elsen JM. Performance of genomic selection in mice. Genetics. 2008;180:611–8.

44. Abdollahi-Arpanahi R, Morota G, Valente BD, Kranis A, Rosa GJ, Gianola D. Assessment of bagging GBLUP for whole-genome prediction of broiler chicken traits. J Anim Breed Genet. 2015;132:218–28.

45. de los Campos G, Sorensen D, Gianola D. Genomic heritability: what is it? PLoS Genet. 2015;11:e1005048.

46. Corbin LJ, Liu AY, Bishop SC, Woolliams JA. Estimation of historical effective population size using linkage disequilibria with marker data. J Anim Breed Genet. 2012;129:257–70.

47. Tenesa A, Navarro P, Hayes BJ, Duffy DL, Clarke GM, Goddard ME, et al. Recent human effective population size estimated from linkage disequilibrium. Genome Res. 2007;17:520–6.

48. Rabier CE, Barre P, Asp T, Charmet G, Mangin B. On the accuracy of genomic selection. PLoS One. 2016;11:e0156086.

49. Hayes BJ, Visscher PM, Goddard ME. Increased accuracy of artificial selection by using the realized relationship matrix. Genet Res. 2009;91:47–60.

50. Onogi A, Ogino A, Komatsu T, Shoji N, Simizu K, Kurogi K, et al. Genomic prediction in Japanese Black cattle: application of a single-step approach to beef cattle. J Anim Sci. 2014;92:1931–8.

51. Lee SH, Cho YM, Lim D, Kim HC, Choi BH, Park HS, et al. Linkage disequilibrium and effective population size in Hanwoo Korean cattle. Asian Aust J Anim Sci. 2011;24:1660–5.

52. Li Y, Kim JJ. Effective population size and signatures of selection using bovine 50 K SNP chips in Korean native cattle (Hanwoo). Evol Bioinform Online. 2015;11:143–53.

53. Marquez GC, Speidel SE, Enns RM, Garrick DJ. Genetic diversity and population structure of American Red Angus cattle. J Anim Sci. 2010;88:59–68.

54. Cleveland MA, Blackburn HD, Enns RM, Garrick DJ. Changes in inbreeding of US Herefords during the twentieth century. J Anim Sci. 2005;83:992–1001.

55. Sorensen AC, Sorensen MK, Berg P. Inbreeding in Danish dairy cattle breeds. J Dairy Sci. 2005;88:1865–72.

56. Kim ES, Kirkpatrick BW. Linkage disequilibrium in the North American Holstein population. Anim Genet. 2009;40:279–88.

57. de Roos AP, Hayes BJ, Spelman RJ, Goddard ME. Linkage disequilibrium and persistence of phase in Holstein–Friesian, Jersey and Angus cattle. Genetics. 2008;179:1503–12.

58. Ni GY, Zhang Z, Jiang L, Ma PP, Zhang Q, Ding XD. Chinese Holstein cattle effective population size estimated from whole genome linkage disequilibrium. Yi Chuan. 2012;34:50–8.

59. Goddard ME, Hayes BJ. Mapping genes for complex traits in domestic animals and their use in breeding programmes. Nat Rev Genet. 2009;10:381–91.

60. González-Recio O, Rosa GJ, Gianola D. Machine learning methods and predictive ability metrics for genome-wide prediction of complex traits. Livest Sci. 2014;166:217–31.

61. Gao N, Li J, He J, Xiao G, Luo Y, Zhang H, et al. Improving accuracy of genomic prediction by genetic architecture based priors in a Bayesian model. BMC Genet. 2015;16:20.

62. Wolc A, Arango J, Settar P, Fulton JE, O'Sullivan NP, Dekkers JC, et al. Mixture models detect large effect QTL better than GBLUP and result in more accurate and persistent predictions. J Anim Sci Biotechnol. 2016;7:7.

63. Nishimura S, Watanabe T, Mizoshita K, Tatsuda K, Fujita T, Watanabe N, et al. Genome-wide association study identified three major QTL for carcass weight including the PLAG1-CHCHD7 QTN for stature in Japanese Black cattle. BMC Genet. 2012;13:40.

64. Ogawa S, Matsuda H, Taniguchi Y, Watanabe T, Nishimura S, Sugimoto Y, et al. Effects of single nucleotide polymorphism marker density on degree of genetic variance explained and genomic evaluation for carcass traits in Japanese Black beef cattle. BMC Genet. 2014;15:15.

65. Fernandez Júnior GA, Rosa GJ, Valente BD, Carvalheiro R, Baldi F, Garcia DA, et al. Genomic prediction of breeding values for carcass traits in Nellore cattle. Genet Sel Evol. 2016;48:7.

66. Saatchi M, Schnabel RD, Taylor JF, Garrick DJ. Large-effect pleiotropic or closely linked QTL segregate within and across ten US cattle breeds. BMC Genomics. 2014;15:442.

67. Kizilkaya K, Fernando RL, Garrick DJ. Reduction in accuracy of genomic prediction for ordered categorical data compared to continuous observations. Genet Sel Evol. 2014;46:37.

68. Jensen J, Su G, Madsen P. Partitioning additive genetic variance into genomic and remaining polygenic components for complex traits in dairy cattle. BMC Genet. 2012;13:44.

69. Khatkar MS, Moser G, Hayes BJ, Raadsma HW. Strategies and utility of imputed SNP genotypes for genomic analysis in dairy cattle. BMC Genomics. 2012;13:538.

70. Druet T, Macleod IM, Hayes BJ. Toward genomic prediction from whole-genome sequence data: impact of sequencing design on genotype imputation and accuracy of predictions. Heredity. 2014;112:39–47.

71. Heidaritabar M, Calus MP, Megens HJ, Vereijken A, Groenen MA, Bastiaansen JW. Accuracy of genomic prediction using imputed whole genome sequence data in white layers. J Anim Breed Genet. 2016;133:167–79.

72. Habier D, Fernando RL, Dekkers JC. The impact of genetic relationship information on genome-assisted breeding values. Genetics. 2007;177:2389–97.

73. Legarra A, Aguilar I, Misztal I. A relationship matrix including full pedigree and genomic information. J Dairy Sci. 2009;92:4656–63.

74. Misztal I, Aggrey SE, Muir WM. Experiences with a single-step genome evaluation. Poult Sci. 2013;92:2530–4.

75. Wang H, Misztal I, Aguilar I, Legarra A, Muir WM. Genome-wide association mapping including phenotypes from relatives without genotypes. Genet Res. 2012;94:73–83.

76. Tiezzi F, Maltecca C. Accounting for trait architecture in genomic predictions of US Holstein cattle using a weighted realized relationship matrix. Genet Sel Evol. 2015;47:24.

77. Neves HH, Carvalheiro R, Queiroz SA. A comparison of statistical methods for genomic selection in a mice population. BMC Genet. 2012;13:100.

14

Comparison of alternative approaches to single-trait genomic prediction using genotyped and non-genotyped Hanwoo beef cattle

Joonho Lee[1†] , Hao Cheng[1,2†], Dorian Garrick[1,3,4], Bruce Golden[4], Jack Dekkers[1], Kyungdo Park[5], Deukhwan Lee[6] and Rohan Fernando[1*]

Abstract

Background: Genomic predictions from BayesA and BayesB use training data that include animals with both phenotypes and genotypes. Single-step methodologies allow additional information from non-genotyped relatives to be included in the analysis. The single-step genomic best linear unbiased prediction (SSGBLUP) method uses a relationship matrix computed from marker and pedigree information, in which missing genotypes are imputed implicitly. Single-step Bayesian regression (SSBR) extends SSGBLUP to BayesB-like models using explicitly imputed genotypes for non-genotyped individuals.

Methods: Carcass records included 988 genotyped Hanwoo steers with 35,882 SNPs and 1438 non-genotyped steers that were measured for back-fat thickness (BFT), carcass weight (CWT), eye-muscle area, and marbling score (MAR). Single-trait pedigree-based BLUP, Bayesian methods using only genotyped individuals, SSGBLUP and SSBR methods were compared using cross-validation.

Results: Methods using genomic information always outperformed pedigree-based BLUP when the same phenotypic data were modeled from either genotyped individuals only or both genotyped and non-genotyped individuals. For BFT and MAR, accuracies were higher with single-step methods than with BayesB, BayesC and BayesCπ. Gains in accuracy with the single-step methods ranged from +0.06 to +0.09 for BFT and from +0.05 to +0.07 for MAR. For CWT, SSBR always outperformed the corresponding Bayesian methods that used only genotyped individuals. However, although SSGBLUP incorporated information from non-genotyped individuals, prediction accuracies were lower with SSGBLUP than with BayesC ($\pi = 0.9999$) and BayesB ($\pi = 0.98$) for CWT because, for this particular trait, there was a benefit from the mixture priors of the effects of the single nucleotide polymorphisms.

Conclusions: Single-step methods are the preferred approaches for prediction combining genotyped and non-genotyped animals. Alternative priors allow SSBR to outperform SSGBLUP in some cases.

Background

Since breeding technologies using genome-wide single nucleotide polymorphism (SNP) panels became available, genomic selection was rapidly adopted for improvement of livestock and has replaced the traditionally used pedigree-based best linear unbiased prediction (PBLUP). The BayesA and BayesB hierarchical Bayesian models with locus-specific variances were proposed by Meuwissen et al. [1]. BayesB can accommodate mixture models in which SNPs have zero effects with probability π [2, 3]. When $\pi = 0$, BayesB is known as BayesA. BayesC is another widely-used Bayesian mixture model, in which a

*Correspondence: rohan@iastate.edu
†Joonho Lee and Hao Cheng contributed equally to this work
[1] Department of Animal Science, Iowa State University, Ames, IA 50011, USA
Full list of author information is available at the end of the article

common variance is used for all SNPs instead of locus-specific variances [4], and a modification of that method known as BayesCπ treats π as an unknown parameter with a uniform prior distribution [5].

In general, the number of individuals with genomic information is a small subset of the individuals represented in the population with pedigree and phenotypic information. "Single-step" methodologies were developed to take advantage of all pedigree, phenotypic and genomic information simultaneously [6, 7]. The single-step genomic BLUP (SSGBLUP) method uses a relationship matrix that is computed from marker and pedigree information. SSGBLUP was shown to yield a similar or higher accuracy compared to methods using only genotyped individuals [8–10]. Fernando et al. [7] proposed a class of single-step Bayesian regression methods (SSBR) to extend SSGBLUP to incorporate BayesB-like models for SNP effects (SSBR-B). Similar extensions of SSGBLUP with BayesC-like models result in SSBR-C and SSBR-Cπ. SSBR methods may promise higher prediction accuracies and provide computational benefits when many animals are genotyped. In SSGBLUP, the distribution of marker effects conditional on the variance of marker effects is assumed univariate normal, whereas in SSBR, the prior for marker effects can follow a t-distribution, a double exponential distribution or mixture distributions, which may be advantageous in some situations.

In this paper, prediction accuracies from PBLUP, BayesB, BayesC, BayesCπ, SSGBLUP and SSBR-B, SSBR-C, SSBR-Cπ were compared in terms of cross-validation accuracies.

Methods
Data
Young Hanwoo bulls are routinely progeny-tested in batches at the Hanwoo Improvement Center (Seo-San, Chungnam, South Korea). DNA samples were collected from steers that included the progeny-tested offspring from the 46th to 51st selection batches. SNP genotypes were determined using Illumina Bovine SNP50 v1 (50 k) or Bovine HD (778 k) beadchips (Illumina, CA).

Carcass records were recorded at harvest at about 24 months of age. The carcass traits used in the analyses were back-fat thickness (BFT), carcass weight (CWT), eye-muscle area (EMA), and marbling score (MAR). Park et al. [11] reported heritabilities of 0.50, 0.30, 0.42 and 0.63 for BFT, CWT, EMA and MAR, respectively. Approval from the ethics committee was not required for these data since they were obtained from an existing industry database.

Of the 44 k SNPs that are included on both the 50 and 778 k beadchips, only autosomal SNPs with known map location were used. For quality control, SNPs that departed from the Hardy–Weinberg equilibrium ($p < 10^{-6}$) based on a Chi square test, or had a minor allele frequency (MAF) lower than 0.01, or a missing rate higher than 0.1 were excluded from further analysis. For the genotyped animals, SNPs with missing genotypes were imputed using Beagle 3.3 [12]. After these quality controls, 35,882 SNPs remained for analyses.

The numerator relationship matrix (NRM) based on pedigree information and the genomic relationship matrix (GRM) based on SNP genotypes were compared. Nineteen individuals, which showed unreasonable deviations between the NRM and GRM coefficients that were probably due to errors in the DNA sampling, were eliminated. Among these 19 individuals, five appeared to have been genotyped twice with different ID since their GRM relationship coefficients were near 1.0 while their NRM relationship coefficients were close to 0. For the other 14 individuals, either the GRM relationship coefficients were near 0 while those of the NRM were near 0.25 as would be the case for mistakenly recorded half-sib individuals, or the GRM relationship coefficients were near 0.25 while those of the NRM were near 0 as would be the case for half-sibs mistakenly recorded as unrelated. After elimination of these suspect individuals, the correlation coefficient between NRM and GRM increased from 0.856 to 0.866. Finally, 988 genotyped individuals remained for genomic prediction with a mean MAF of 0.243 and mean observed heterozygosity of 0.326.

Additional carcass records for 1438 non-genotyped progeny-tested steers were collected from the 39th to the 51st selection batches for the single-step and PBLUP analyses. Ancestors of the 2426 individuals with carcass records contributed to an 11-generation pedigree file that included 9637 animals.

Genotyped individuals were assigned to five mutually exclusive groups for cross-validation. K-means clustering based on pedigree relationship coefficients was used to minimize the relatedness between training and validation sets [13]. The five groups included 172, 280, 199, 139 and 198 individuals, respectively. Each group was used as the validation set while the remaining genotyped individuals were included in the training set. In SSGBLUP, SSBR and PBLUP with phenotypes on all animals, non-genotyped individuals were included in the training set. Phenotypes were pre-adjusted for contemporary group and age effects using multiple-trait PBLUP because animals from some progeny-test batches were assigned to different groups and because some analyses included additional non-genotyped animals from the same batches as genotyped animals.

Single-trait statistical models

Pedigree-based BLUP

In these analyses, the adjusted phenotypes were modeled as:

$$\mathbf{y} = \mathbf{1}\mu + \mathbf{Z}\mathbf{u} + \mathbf{e},$$

where $\mathbf{y}$ is a vector of adjusted phenotypic records from n_y animals, $\mathbf{1}$ is a vector of 1s, μ is the overall mean, $\mathbf{Z}$ is the design matrix allocating records to breeding values, $\mathbf{u}$ is the vector of breeding values, $\mathbf{e}$ is a random vector of residuals. It was assumed that $\mathbf{u} \sim N(\mathbf{0}, \mathbf{A}\sigma_g^2)$, where $\mathbf{A}$ is the numerator relationship matrix and σ_g^2 is the additive genetic variance. Residuals were assumed to be independently and identically distributed (iid) with null means and variance σ_e^2. Pedigree-based BLUP with phenotypes either on all animals or only on genotyped animals were referred to as PBLUP (n_y = 2426 minus validation animals) and PBLUP-G (n_y = 988 minus validation animals), respectively. Adjusted phenotypes were used to account for fixed effects in the validation set.

Bayesian methods using only genotyped animals

In these analyses, the adjusted phenotypes were modeled as:

$$\mathbf{y} = \mathbf{1}\mu + \mathbf{M}_g\boldsymbol{\alpha} + \mathbf{e},$$

where $\mathbf{y}$, $\mathbf{1}$ and $\mathbf{e}$ are $n_y \times 1$ vectors for $n_y = 988$ minus genotyped validation animals, μ is as defined before, $\mathbf{M}_g$ is the $n_y \times p$ matrix of SNP covariates at p loci, and $\boldsymbol{\alpha}$ is a $p \times 1$ random vector of allele substitution effects. A flat prior was used for μ. The prior for $\mathbf{e}$ was $\mathbf{e}|\sigma_e^2 \sim N(0, \mathbf{I}\sigma_e^2)$ with $(\sigma_e^2|v_e, S_e^2) \sim v_e S_e^2 \chi_{v_e}^2$. Priors for SNP effects were a mixture of a point mass at zero and a t-distribution in BayesB or a mixture of a point mass at zero and a normal distribution conditional on a common variance of SNP effects in BayesC and BayesCπ methods [2]. These methods were referred to as BayesB, BayesC or BayesCπ, and ignored adjusted phenotypes on non-genotyped animals, as for PBLUP-G.

Single-step GBLUP

In the single-step GBLUP analyses, the adjusted phenotypes were modeled as:

$$\mathbf{y} = \mathbf{1}\mu + \mathbf{Z}\mathbf{u} + \mathbf{e},$$

where $\mathbf{y}$ is the vector of adjusted phenotypes as before except that it includes both genotyped and non-genotyped individuals i.e. n_y = 2426 minus validation animals, μ and $\mathbf{e}$ are as defined before, with residuals that are iid with null means and variance σ_e^2, $\mathbf{Z}$ is the design matrix allocating records to breeding values, $\mathbf{u}$ is the vector of breeding values for both genotyped and non-genotyped individuals but now $\mathbf{u} \sim N(\mathbf{0}, \mathbf{H}\sigma_g^2)$, where:

$$\mathbf{H} = \begin{bmatrix} \mathbf{A}_{ng}\mathbf{A}_{gg}^{-1}\mathbf{G}\mathbf{A}_{gg}^{-1}\mathbf{A}_{gn} + \left(\mathbf{A}_{nn} - \mathbf{A}_{ng}\mathbf{A}_{gg}^{-1}\mathbf{A}_{gn}\right) & \mathbf{A}_{ng}\mathbf{A}_{gg}^{-1}\mathbf{G} \\ \mathbf{G}\mathbf{A}_{gg}^{-1}\mathbf{A}_{gn} & \mathbf{G} \end{bmatrix},$$

and $\mathbf{A}_{gg}$ is the 988 order partition of the numerator relationship matrix $\mathbf{A}$ that corresponds to genotyped animals, $\mathbf{A}_{nn}$ is the 11,075 order partition of $\mathbf{A}$ that corresponds to non-genotyped animals, $\mathbf{A}_{ng}$ or $\mathbf{A}_{gn}$ are partitions of $\mathbf{A}$ corresponding to relationships between non-genotyped and genotyped animals or vice versa, and $\mathbf{G}$ is a GRM of order 988. We applied three methods to construct the GRM. The standard $\mathbf{G}$ was constructed as $\mathbf{G} = \frac{\mathbf{T}\mathbf{T}'}{\sum 2q_i(1-q_i)}$ (SSGBLUP-I) with $\mathbf{T}$ being the centered matrix of SNP covariates ($\mathbf{T} = \mathbf{M}_g - \frac{1}{n}\mathbf{1}\mathbf{1}'\mathbf{M}_g$), q_i representing the allele frequency of the ith SNP. This is the same $\mathbf{G}$ as previously used to compare relationship coefficients between NRM and GRM and eliminate the 19 individuals with genotype-pedigree conflicts, except that 19 rows and corresponding columns were deleted. In the standard $\mathbf{G}$, the additive genetic variance attributed to each SNP genotype is equally important and GRM are identical for all traits. Recently, methodologies for constructing $\mathbf{G}$ with weighting factors to account for locus-specific variances were proposed [14–16]. The method reported by Wang et al. [14] calculates SNP effects from the solution of SSGBLUP-I and then reconstructs a new GRM using weights that are obtained from the previously calculated SNP effects. This can be repeated iteratively to obtain a sequence of GRM. In this approach, GRM will differ for each trait.

The prediction model based on the GRM constructed from one iteration was referred to as SSGBLUP-II and the GRM constructed from five iterations was referred to as SSGBLUP-III. To remove singularity, GRM can be blended with NRM [17] but this was not done in our study, nor were residual polygenic effects separately modeled in either SSGBLUP or SSBR. Instead, diagonal and off-diagonal elements of $\mathbf{G}$ were separately scaled so that their means equal the corresponding means of $\mathbf{A}_{gg}$, which is expected to remove the singularity of GRM in SSGBLUP that is introduced by centering the SNPs.

Single-step Bayesian regression methods

In the single-step Bayesian regression analyses, the adjusted phenotypes were modeled as:

$$\mathbf{y} = \mathbf{X}\boldsymbol{\beta} + \mathbf{Z}\mathbf{M}\boldsymbol{\alpha} + \mathbf{Z}_n\boldsymbol{\epsilon} + \mathbf{e},$$

where $\mathbf{y}$ is the adjusted phenotypic vector for both genotyped and non-genotyped individuals, $\mathbf{X} = \begin{bmatrix} \mathbf{1} & -\mathbf{Z}_n\mathbf{A}_{ng}\mathbf{A}_{gg}^{-1}\mathbf{1} \\ \mathbf{1} & -\mathbf{Z}_g\mathbf{1} \end{bmatrix}$, $\boldsymbol{\beta} = \begin{bmatrix} \mu \\ \mu_g \end{bmatrix}$, μ is the overall mean, and μ_g represents the difference in breeding values between genotyped and non-genotyped animals, $\mathbf{Z}$ is the design matrix, $\mathbf{M} = \begin{bmatrix} \widehat{\mathbf{M}_n} \\ \mathbf{M}_g \end{bmatrix}$, where $\mathbf{M}_g$ is the matrix of SNP covariates for genotyped animals and $\widehat{\mathbf{M}_n} = \mathbf{A}_{ng}\mathbf{A}_{gg}^{-1}\mathbf{M}_g$, representing imputed SNP covariates for non-genotyped animals that are derived from genotyped relatives, $\boldsymbol{\epsilon}$ is the imputation residual, $\mathbf{Z}_n$ and $\mathbf{Z}_g$ are the design matrices allocating records to breeding values of non-genotyped animals and genotyped animals. Flat priors were used for μ and μ_g. The prior for e_i is $e_i|\sigma_e^2 \sim_{iid} N(0, \sigma_e^2)$ with $(\sigma_e^2|v_e, S_e^2) \sim v_e S_e^2 \chi_{v_e}^2$. The prior for $\boldsymbol{\epsilon}$ is $\boldsymbol{\epsilon}|\sigma_g^2 \sim N(0, (\mathbf{A}_{nn} - \mathbf{A}_{ng}\mathbf{A}_{gg}^{-1}\mathbf{A}_{gn})\sigma_g^2)$ with $(\sigma_g^2|v_g, S_g^2) \sim v_g S_g^2 \chi_{v_g}^2$. The same priors for SNP effects as in BayesB, BayesC and BayesCπ were used in single-step Bayesian regression methods and were referred to as SSBR-B, SSBR-C, or SSBR-Cπ.

The π values in the subsequent analyses for BayesB, BayesC, SSBR-B and SSBR-C were chosen such that they provided the highest accuracies from fivefold cross-validation. Accuracies in BayesB and BayesC were compared using various π values i.e. 0.9999, 0.999, 0.995, 0.99, 0.98 and, then, in steps from 0.95 to 0.6 decreasing by 0.05.

Analyses were performed with GenSel [5] for BayesB, BayesC and BayesCπ methods using only genotyped animals. Estimated breeding values of PBLUP and SSGBLUP were obtained using the software BLUPF90 [18] modified for genomic analyses [17]. For SSBR methods, JWAS the Julia package for whole-genome analyses [19] was used.

Validation

For each validation set, prediction accuracy was calculated as the correlation between the vector of adjusted phenotypes and the vector of estimated breeding values, divided by the square root of trait heritability. Prediction accuracies from these fivefold cross-validation sets were pooled to obtain a single prediction accuracy that was relevant to the method and trait by weighting each of the five validation correlations by the number of individuals in that set. Regressions of adjusted phenotype on estimated breeding value were calculated for all prediction methods.

Genome-wide association studies

Genome-wide association studies (GWAS) were performed using the BayesB method with the π value that had given the highest prediction accuracy, in order to describe the genetic architecture for different traits in terms of window variance [20].

Results

Predictive accuracies for the four traits obtained with BayesB and BayesC for different π values are in Fig. 1. For BFT, EMA and MAR, predictive accuracies of BayesB and BayesC were similar, but decreased as π increased, and fewer SNPs were assumed to have non-zero effects. For CWT, we observed a different pattern with accuracies increasing as π increased and accuracies of BayesB being always higher than those of BayesC. These two results suggest that CWT is influenced by a few quantitative trait loci (QTL) that explain a large proportion of the genetic variance. The proportions of genetic variance explained by 1-Mb non-overlapping genomic windows are in Fig. 2, and demonstrate that the QTL for CWT were larger than those for the other traits.

The π values that maximized the cross-validation accuracies in BayesB were 0.95, 0.98, 0.95, and 0.6 for BFT, CWT, EMA and MAR, respectively, and were used in SSBR-B. The corresponding π values in BayesC were 0.98, 0.9999, 0.98, and 0.6 for BFT, CWT, EMA and MAR, respectively, and were used in SSBR-C.

Several windows showed distinctly larger effects than the rest of the genome for BFT and EMA, but the window with the largest effect explained only about 1% of the genetic variance. For MAR, the windows showed smaller effects than those for BFT and EMA with the most significant window explaining less than 0.3% of the genetic variance. These results show that, for BFT, EMA and MAR, many QTL each with a small effect are widely distributed across the whole genome, which is consistent with the infinitesimal model. In contrast, for CWT, one window on chromosome 4 and two windows on chromosome 14 explained together more than 15% of the genetic variance while the other windows showed small effects. Using single-SNP association tests, Lee et al. [21] found similar results that indicated that SNPs on chromosome 14 were strongly associated with CWT in Hanwoo beef cattle. These differences in genomic architecture between the four traits probably explain the difference in the pattern of prediction accuracy between CWT and the three other traits as shown in Fig. 1. BayesB, which shrinks QTL with small effects to a greater extent than BayesC, may capture QTL with large effects better and therefore yield higher prediction accuracies [22]. BayesB and BayesC methods with a high π value tend to capture the same few QTL with large effects, thus their similar prediction accuracies.

Prediction accuracies of models SSGBLUP-I and SSBR-C ($\pi = 0$) without estimated variances were identical and equal to 0.351 for BFT, 0.415 for CWT, 0.413 for EMA

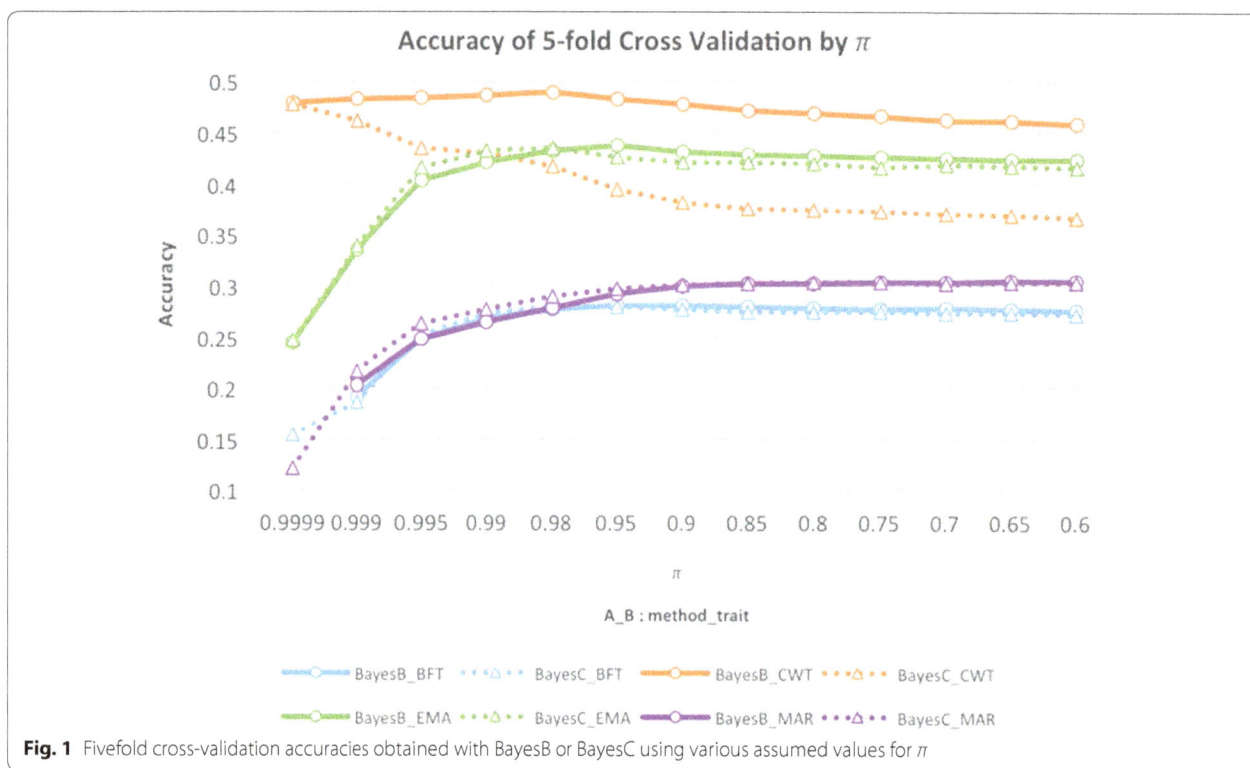

Fig. 1 Fivefold cross-validation accuracies obtained with BayesB or BayesC using various assumed values for π

and 0.377 for MAR as expected since these models with assumed variance parameters are equivalent in terms of prediction of breeding values [7]. In practice, variance components are often treated as unknown and are estimated in a separate analysis, e.g. restricted maximum likelihood (REML) followed by GBLUP, or jointly with an informative prior, e.g. BayesB, SSBR-B, etc. The variances of additive genetic effects, SNP effects and residual effects were estimated in the subsequent analyses described below.

To compare methods that use all individuals with those that use only genotyped individuals, prediction accuracies (Fig. 3) were calculated using PBLUP (all animals) and PBLUP-G (PBLUP using only phenotypes on genotyped animals), BayesB, BayesC, BayesC ($\pi = 0$), and BayesCπ, SSGBLUP-I and SSGBLUP-II and SSBR-B, SSBR-C, SSBR-C ($\pi = 0$), and SSBR-Cπ.

Genomic methods versus pedigree-based BLUP
Methods using genomic information always outperformed PBLUP with the same phenotypic data. Using data from only genotyped animals, accuracies were higher with BayesB, BayesC and BayesCπ than with PBLUP-G for all traits. When data from both genotyped and non-genotyped individuals were used, prediction accuracies of the single-step methods were higher than those of PBLUP for all traits.

Single-step methods versus BayesB, BayesC and BayesCπ
For BFT and MAR, prediction accuracies of the single-step methods were higher than those of BayesB, BayesC and BayesCπ. Gains in accuracy with the single-step methods ranged from +0.06 to +0.09 for BFT and from +0.05 to +0.07 for MAR, whereas for EMA, there was no advantage and only a slight gain in accuracy was observed in PBLUP versus PBLUP-G. For CWT, SSBR always outperformed the corresponding Bayesian methods using only genotyped individuals and the gains in accuracy were +0.05 (SSBR-C ($\pi = 0$) vs. BayesC ($\pi = 0$)), +0.01 (SSBR-C ($\pi = 0.9999$) vs. BayesC ($\pi = 0.9999$)), +0.10 (SSBR-Cπ vs. BayesCπ) and +0.04 (SSBR-B ($\pi = 0.98$) vs. BayesB ($\pi = 0.98$)). However, although information from non-genotyped individuals was incorporated, for CWT prediction accuracy of SSGBLUP was lower than that of BayesC ($\pi = 0.9999$) and BayesB ($\pi = 0.98$) due to the benefits of mixture priors of the SNP effects for this particular trait.

Comparisons between single-step methods
The differences in accuracies between single-step methods (yellow and blue bars in Fig. 3) were small for BFT, EMA and MAR, and a similar pattern was found between Bayesian methods (red bars in Fig. 3) using only genotyped individuals. For the CWT trait for which the GWAS detected a small number of regions with large

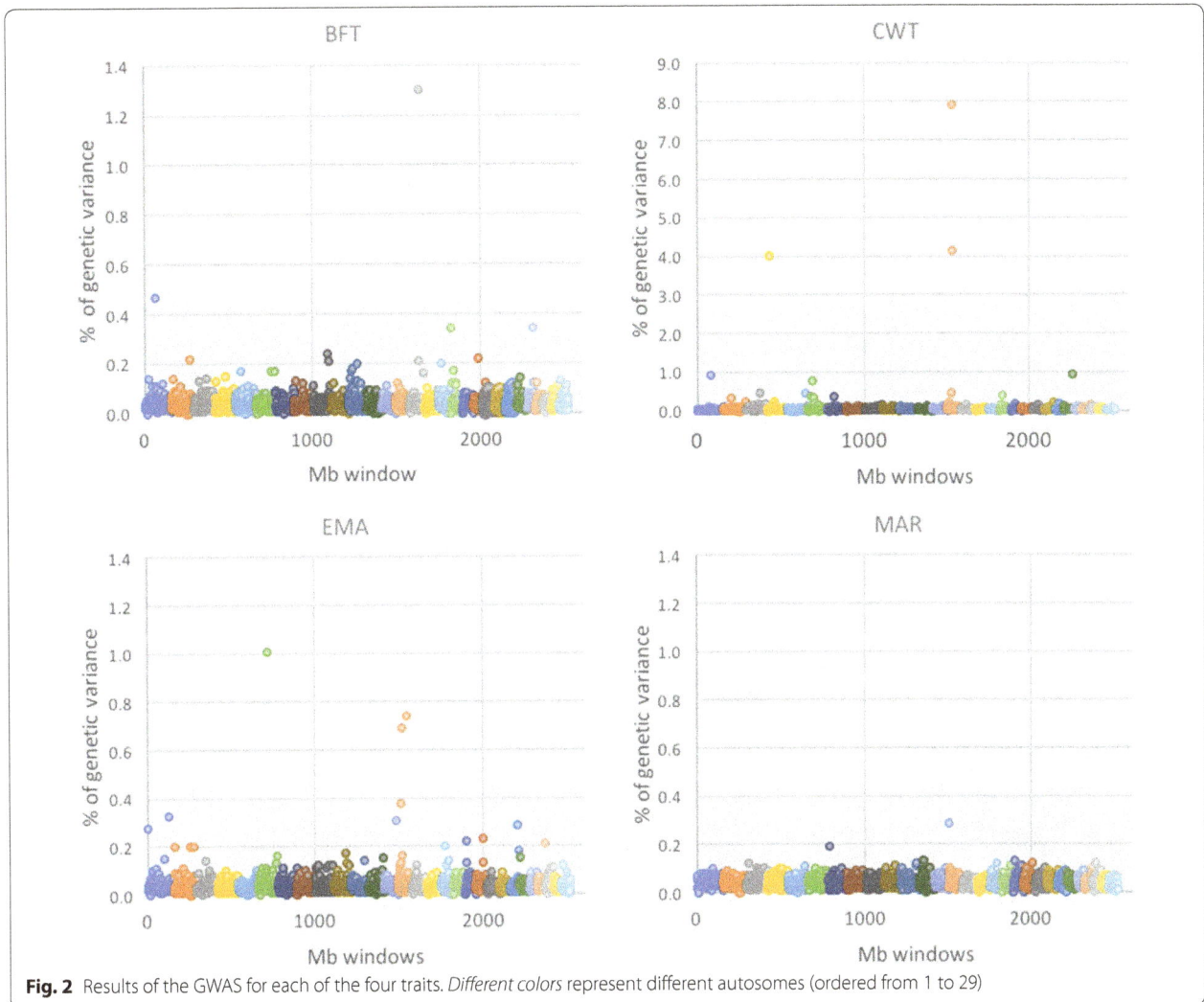

Fig. 2 Results of the GWAS for each of the four traits. *Different colors* represent different autosomes (ordered from 1 to 29)

effects, prediction accuracies differed with the method used. With the benefits of mixture priors and information from non-genotyped individuals, prediction accuracies of the SSBR methods, especially SSBR-B, were higher (+0.09) than those of the SSGBLUP methods. As for the SSBR methods with mixture priors, the SSGBLUP methods, which use weighted GRM (SSGBLUP-II and SSGB-LUP-III), showed higher accuracies than SSGBLUP-I for CWT. Prediction accuracy of SSGBLUP-II was similar to that of SSGBLUP-I for EMA and MAR but lower for BFT. Prediction accuracy of SSGBLUP-III was lower than that of SSGBLUP-I for EMA, MAR and BFT. Regressions of adjusted phenotype on estimated breeding value did not show large differences among methods, but SSGBLUP-II and SSGBLUP-III had the lowest coefficients for all traits, much lower than 1, which indicates that their genomic predictions are biased upwards (Table 1).

Discussion

Prediction accuracies of all methods using genomic information were higher than those of pedigree-based BLUP. However, the degree of superiority of genomic selection differed between methods and traits.

We hypothesize that the advantage of including phenotypic observations from non-genotyped animals into an analysis using phenotypic observations from genotyped animals would be similar for pedigree methods (PBLUP compared to PBLUP-G) and for genomic methods (SSBR-C compared to BayesC). Simultaneous use of all pedigree, phenotypic and genomic information in single-step methods improved prediction accuracy relative to methods that only use data from genotyped animals for all traits, except EMA. For EMA, there was similarly little benefit from including the extra data in the PBLUP analyses (compared to PBLUP-G).

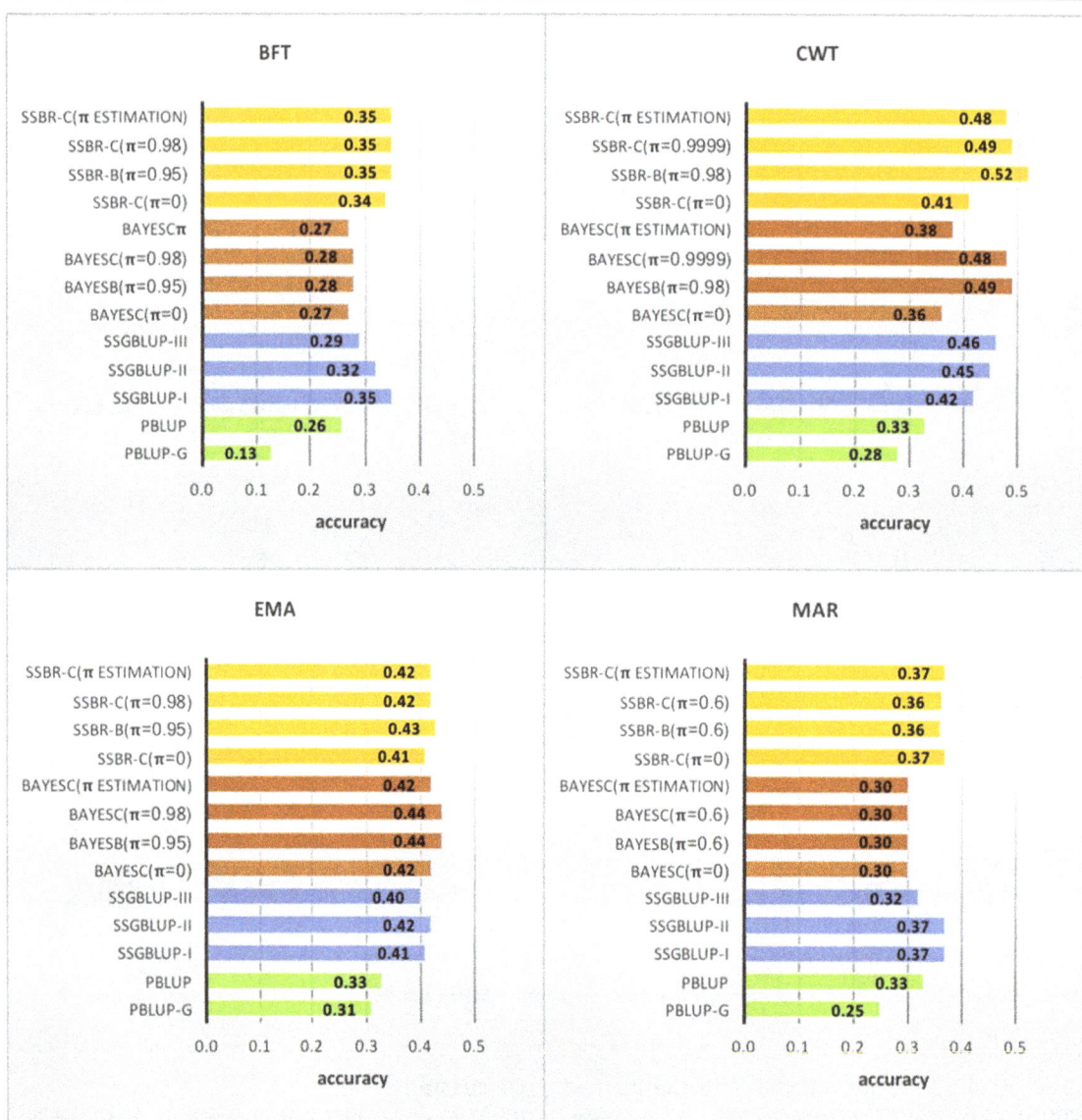

Fig. 3 Prediction accuracies by cross-validation for a variety of methods applied to backfat (BFT), carcass weight (CWT), eye-muscle area (EMA) and marbling (MAR). Conventional PBLUP based on only genotyped individuals (PBLUP-G) or using all animals (PBLUP), BayesB with chosen π (BAYESC(π = chosen value)), BayesC with chosen π (BAYESC (π = chosen value)) BayesC with π = 0 (BAYESC (π = 0)) or BayesC estimating π (BAYESC (π ESTIMATION)), single-step genomic BLUP constructing two different genomic relationship matrix (SSGBLUP-I and SSGBLUP-II) and single-step Bayesian regression corresponding to Bayesian methods (SSBR-B (π = chosen value), SSBR-C (π = chosen value), SSBR-C (π = 0), and SSBR-C (π ESTIMATION))

Both SSBR and SSGBLUP methods showed similar prediction accuracies when the genetic architecture appeared to approach the infinitesimal model as was the case for BFT, EMA, and MAR. However, for CWT, prediction accuracies of the SSBR methods were higher than those of SSGBLUP when there were only a few QTL with large effects. For that trait, the SSBR methods benefited from the use of the mixture priors.

The largest benefit of the SSBR methods was reached when an appropriate π was applied. However, it is computationally intensive to find this value of π through cross-validation. Methods for estimating π are beneficial, but they require large datasets. An appropriate π was more critical for the Bayesian methods that only used genotyped individuals than for the SSBR methods. For example, differences in prediction accuracies between

Table 1 Regression coefficient of adjusted phenotype on estimated breeding values for backfat (BFT), carcass weight (CWT), eye-muscle area (EMA) and marbling (MAR) traits

Prediction methods	Trait			
	BFT	CWT	EMA	MAR
SSBR-C (π estimation)	0.85	0.97	0.99	0.88
SSBR-B (π = chosen)[a]	0.88	1.08	1.07	0.74
SSBR-C (π = chosen)[b]	0.88	1.02	1.04	0.89
SSBR-C (π = 0)	0.86	1.21	1.00	0.87
BayesC (π estimation)	0.82	1.05	1.05	0.86
BayesB (π = chosen)[a]	0.82	1.03	1.26	0.70
BayesC (π = chosen)[b]	0.88	1.06	1.12	0.87
BayesC (π = 0)	0.86	1.20	1.09	0.88
SSGBLUP-I	0.73	1.15	0.97	0.79
SSGBLUP-II	0.54	0.84	0.75	0.64
SSGBLUP-III	0.52	0.90	0.79	0.61
PBLUP	0.76	1.12	1.02	0.93
PBLUP-G	0.61	1.33	1.30	0.92

[a] Chosen π of BayesB and SSBR-B for BFT, CWT, EMA and MAR were 0.95, 0.98, 0.95 and 0.6, respectively

[b] Chosen π of BayesC and SSBR-C for BFT, CWT, EMA and MAR were 0.98, 0.9999, 0.98 and 0.6, respectively

BayesC (π = 0.9999) and BayesCπ reached values of 0.10 but only of 0.01 between SSBR-C (π = 0.9999) and SSBR-Cπ. Presumably, priors become less important in the single-step analyses where more data are used.

Three factors can result in increased accuracy. First, the inclusion of genomic information, which was revealed when genomic methods were compared to pedigree-based BLUP. Second, the use of additional phenotypic information from including non-genotyped individuals, which was shown by comparing Bayesian methods using only genotyped animals with their corresponding single-step methods. Third, the use of methods that exploit genomic regions with large effects, as was found for one of the four traits using either mixture priors or iterative weighted methods for computing GRM.

SSGBLUP with iterative calculation of weighted genomic matrices had the disadvantage that it reduced prediction accuracy and increased bias for traits that were not associated with genomic regions with large effects, whereas the Bayesian models with mixture priors performed comparably regardless of the genomic architecture. SSGBLUP with iterative calculation of weighted genomic matrices shrinks small effects to zero, and more so with each additional iteration. There is no statistical basis to determine the optimal number of iterations except by trial and error, and neither one nor five iterations resulted in improvements in this dataset.

In this study, which is based on a small population of Hanwoo cattle, prediction accuracy was higher for all genomic evaluations compared to pedigree-based BLUP. In such a situation, where the genomic reference population is relatively small, single-step methods, which can routinely account for genomic regions with large effects when they are present, are recommended for additional gains in accuracy.

Conclusions

The "single-step" methodologies, which take advantage of all pedigree, phenotypic and genomic information simultaneously, give similar or higher prediction accuracies compared to methods using only genotyped individuals. Compared to SSGBLUP, the SSBR methods showed additional benefit for the CWT trait, which is associated with QTL with large effects. There is no disadvantage in using SSBR methods for all traits.

Authors' contributions
JL and HC conceived the study, undertook the analysis and wrote the draft. DG, RF, JD, BG contributed to the analysis. DG, RF contributed to the final version of manuscript. All authors read and approved the final manuscript.

Author details
[1] Department of Animal Science, Iowa State University, Ames, IA 50011, USA. [2] Department of Statistics, Iowa State University, Ames, IA 50011, USA. [3] Institute of Veterinary, Animal and Biomedical Sciences, Massey University, Palmerston North, New Zealand. [4] ThetaSolutions, LLC, Atascadero, CA, USA. [5] Department of Animal Biotechnology, Chonbuk National University, Chonju, Jeollabuk-do, South Korea. [6] Department of Animal Science, Hankyong National University, Anseong, Gyeonggi-do, South Korea.

Acknowledgements
This work was supported by a grant from the Next Generation BioGreen 21 Program (PJ01111502), Rural Development Administration, Republic of Korea. This work was supported by the US Department of Agriculture, Agriculture and Food Research Initiative National 2 Institute of Food and Agriculture Competitive Grant No. 2015-67015-22947.

Competing interests
The authors declare that they have no competing interests.

References
1. Meuwissen THE, Hayes BJ, Goddard ME. Prediction of total genetic value using genome-wide dense marker maps. Genetics. 2001;157:1819–29.
2. Garrick DJ, Dekkers JCM, Fernando RL. The evolution of methodologies for genomic prediction. Livest Sci. 2014;166:10–8.
3. Cheng H, Qu L, Garrick DJ, Fernando RL. A fast and efficient Gibbs sampler for BayesB in whole-genome analyses. Genet Sel Evol. 2015;47:80.
4. Kizilkaya K, Fernando RL, Garrick DJ. Genomic prediction of simulated multibreed and purebred performance using observed fifty thousand single nucleotide polymorphism genotypes. J Anim Sci. 2010;88:544–51.
5. Habier D, Fernando RL, Kizilkaya K, Garrick DJ. Extension of the Bayesian alphabet for genomic selection. BMC Bioinformatics. 2011;12:186.
6. Legarra A, Aguilar I, Misztal I. A relationship matrix including full pedigree and genomic information. J Dairy Sci. 2009;92:4656–63.

7. Fernando RL, Dekkers JCM, Garrick DJ. A class of Bayesian methods to combine large numbers of genotyped and non-genotyped animals for whole-genome analyses. Genet Sel Evol. 2014;46:50.

8. Misztal I, Aggrey SE, Muir WM. Experiences with a single-step genome evaluation. Poult Sci. 2013;92:2530–4.

9. Lourenco DAL, Misztal I, Tsuruta S, Aguilar I, Ezra E, Ron M, et al. Methods for genomic evaluation of a relatively small genotyped dairy population and effect of genotyped cow information in multiparity analyses. J Dairy Sci. 2014;97:1742–52.

10. Lourenco DA, Tsuruta S, Fragomeni BO, Masuda Y, Aguilar I, Legarra A, et al. Genetic evaluation using single-step genomic best linear unbiased predictor in American Angus. J Anim Sci. 2015;93:2653–62.

11. Park B, Choi T, Kim S, Oh SH. National genetic evaluation (system) of Hanwoo (Korean native cattle). Asian Australas J Anim Sci. 2013;26:151–6.

12. Browning SR, Browning BL. Rapid and accurate haplotype phasing and missing-data inference for whole-genome association studies by use of localized haplotype clustering. Am J Hum Genet. 2007;81:1084–97.

13. Saatchi M, McClure MC, McKay SD, Rolf MM, Kim J, Decker JE, et al. Accuracies of genomic breeding values in American Angus beef cattle using K-means clustering for cross-validation. Genet Sel Evol. 2011;43:40.

14. Wang H, Misztal I, Aguilar I, Legarra A, Muir WM. Genome-wide association mapping including phenotypes from relatives without genotypes. Genet Res (Camb). 2012;94:73–83.

15. Su G, Christensen OF, Janss L, Lund MS. Comparison of genomic predictions using genomic relationship matrices built with different weighting factors to account for locus-specific variances. J Dairy Sci. 2014;97:6547–59.

16. Calus MPL, Schrooten C, Veerkamp RF. Genomic prediction of breeding values using previously estimated SNP variances. Genet Sel Evol. 2014;46:52.

17. Aguilar I, Misztal I, Johnson DL, Legarra A, Tsuruta S, Lawlor TJ. Hot topic: a unified approach to utilize phenotypic, full pedigree, and genomic information for genetic evaluation of Holstein final score. J Dairy Sci. 2010;93:743–52.

18. Misztal I, Tsuruta S, Strabel T, Auvray B, Druer T, Lee DH. BLUPF90 and related programs (BGF90). In: Proceedings of the 7th world congress on genetics applied to livestock production: 19–23 August 2002; Montpellier. 2002.

19. Cheng H, Garrick DJ, Fernando RL. JWAS: Julia implementation of whole-genome analyses software using univariate and multivariate Bayesian mixed effects model. http://QTL.rocks (2016).

20. Wolc A, Arango J, Jankowski T, Dunn I, Settar P, Fulton JE, et al. Genome-wide association study for egg production and quality in layer chickens. J Anim Breed Genet. 2014;131:173–82.

21. Lee SH, Choi BH, Lim D, Gondro C, Cho YM, Dang CG, et al. Genome-wide association study identifies major loci for carcass weight on BTA14 in Hanwoo (Korean cattle). PLoS One. 2013;8:e74677.

22. Wolc A, Arango J, Settar P, Fulton JE, O'Sullivan NP, Dekkers JC, et al. Mixture models detect large effect QTL better than GBLUP and result in more accurate and persistent predictions. J Anim Sci Biotechnol. 2016;7:7.

15

The use of genomic information increases the accuracy of breeding value predictions for sea louse (*Caligus rogercresseyi*) resistance in Atlantic salmon (*Salmo salar*)

Katharina Correa[1,2], Rama Bangera[2], René Figueroa[2], Jean P. Lhorente[2] and José M. Yáñez[1,2]*

Abstract

Sea lice infestations caused by *Caligus rogercresseyi* are a main concern to the salmon farming industry due to associated economic losses. Resistance to this parasite was shown to have low to moderate genetic variation and its genetic architecture was suggested to be polygenic. The aim of this study was to compare accuracies of breeding value predictions obtained with pedigree-based best linear unbiased prediction (P-BLUP) methodology against different genomic prediction approaches: genomic BLUP (G-BLUP), Bayesian Lasso, and Bayes C. To achieve this, 2404 individuals from 118 families were measured for *C. rogercresseyi* count after a challenge and genotyped using 37 K single nucleotide polymorphisms. Accuracies were assessed using fivefold cross-validation and SNP densities of 0.5, 1, 5, 10, 25 and 37 K. Accuracy of genomic predictions increased with increasing SNP density and was higher than pedigree-based BLUP predictions by up to 22%. Both Bayesian and G-BLUP methods can predict breeding values with higher accuracies than pedigree-based BLUP, however, G-BLUP may be the preferred method because of reduced computation time and ease of implementation. A relatively low marker density (i.e. 10 K) is sufficient for maximal increase in accuracy when using G-BLUP or Bayesian methods for genomic prediction of *C. rogercresseyi* resistance in Atlantic salmon.

Background

Sea lice are common marine external parasites that belong to the order of Copepoda. *Caligus rogercresseyi* is the major sea lice species of interest in the southern hemisphere, while *Lepeophtheirus salmonis* is the major species of interest in the northern hemisphere. Both species affect Atlantic salmon (*Salmo salar*) and rainbow trout (*Oncorhynchus mykiss*) [1]. *C. rogercresseyi* parasitize farmed salmonids and wild fish in Chile and constitute one of the main threats for salmon aquaculture in one of the major salmon producing countries [2, 3].

Sea lice can cause skin lesions, osmotic imbalance, and increased susceptibility to bacterial and viral infections through suppression of host immune responses and damage to the host skin [4, 5]. Sea lice may also play a role in the transmission of different fish pathogens [1, 6]. Large economic losses occur as a result of reduced feed conversion and growth, indirect mortality, loss of product value, and treatment costs [1]. The worldwide cost of the control of this parasite in the salmon farming industry was estimated in 2009 to be US$480M [7].

Different chemicals are used to control sea lice, but increasing resistance to antiparasitic drugs has been reported [8]. An alternative method to control sea lice is selective breeding, which has been proposed as a feasible option to improve disease resistance in several livestock and aquaculture species [9–13]. To be selected efficiently, a trait must exhibit significant genetic variation. Genetic parameters have been estimated for *C. rogercresseyi* resistance in Atlantic salmon. Lhorente et al. [14] estimated a heritability (h^2) of 0.32 for sessile *C. rogercresseyi*

*Correspondence: jmayanez@uchile.cl
[1] Facultad de Ciencias Veterinarias y Pecuarias, Universidad de Chile, 'Av Santa Rosa 11735, La Pintana, Chile
Full list of author information is available at the end of the article

count on fins in Atlantic salmon, while Yáñez et al. [15] and Correa et al. [16] estimated a h^2 of 0.12 for the same trait using pedigree and molecular information, respectively. These low to moderate heritabilities indicate that it is feasible to include these traits in the breeding program for Atlantic salmon.

Using a 50 K genotyping array, a recent study conducted by Correa et al. [16] provided evidence for a polygenic architecture of resistance to *C. rogercresseyi*. Similarly, Houston et al. [17, 18] used a 200 K single nucleotide polymorphism (SNP) genotyping array to assess the genetic architecture of resistance to *L. salmonis* and observed polygenic inheritance for this trait also. The polygenic nature of these traits suggests that major quantitative trait loci (QTL) contributing to the genetic variance are unlikely and thus, that marker-assisted selection (MAS) may not be the most appropriate way to include molecular information for selection [19]. In such cases, it is possible to use a dense SNP genotyping panel to obtain genomic estimated breeding values (GEBV) of animals that lack own phenotypic records, an approach known as genomic selection (GS) [20]. The difference between MAS and GS is that MAS uses only single nucleotide polymorphisms (SNPs) that are significant in an association analysis, whereas GS uses all SNPs without having to set a significance threshold [19]. Genetic gain can be increased in salmon breeding programs through the use of molecular markers to calculate GEBV, which are more accurate than EBV calculated using pedigree-based methods [21–23].

Several GS methodologies to predict GEBV have been proposed, which differ in prior assumptions on the distribution of the effects of the SNPs [19]. For example, in genomic best linear unbiased prediction (G-BLUP), marker effects are assumed to be normally distributed, which implies that there is a large number of QTL underlying the trait, each with a small effect [20]. In Bayesian Lasso, marker effects are assumed to follow a double exponential distribution [24], which implies that a large proportion of the markers have an effect close to zero and a small proportion have moderate to large effects [19]. In Bayes C, only a fraction of the markers is assumed to have an effect and, all these are assumed to have a common variance, instead of locus-specific variances [25].

Genomic selection has been tested with real data in salmonid species in a few studies [26–29] that have evaluated growth, fillet color and disease resistance traits. In this study, we compared accuracies of pedigree-based BLUP EBV with those obtained with Bayesian and G-BLUP methods for *C. rogercresseyi* resistance, using data from 2404 individuals and a 50 K genotyping array.

Methods
Phenotypic records
Phenotype data were available for 2404 Atlantic salmon smolts from 118 families (i.e. progeny of 118 dams and 40 sires) from the breeding population of Salmones Chaicas, Xth Region, Chile. These individuals were experimentally challenged with *C. rogercresseyi*. The average number of fish per family was 22 and ranged from 9 to 24, and the average weight of fish was 274.9 g (SD = 90.6 g). The fish were tagged with passive integrated transponders, acclimated, and then distributed to three replicate tanks. The challenge test was carried out as described previously [14–16]. Briefly, infestation with the parasite was carried out by using 13 to 24 copepodids per fish and stopping the water flow for 6 h after infestation. The challenge lasted 6 days; then, fish were euthanized and fins from each fish were collected and fixed for processing and lice counting. The resistance trait was defined as the count of sessile lice per fish on all fins after the infestation period, since that is highly representative of the total lice count on the fish [14]. Experimental tank and final body weight were recorded for each fish.

Genotypes
Genotype data were available from a previous study [16]. Briefly, genomic DNA was extracted from fin clips using a commercial kit (DNeasy Blood and Tissue, Qiagen), quality controlled and quantified. All phenotyped fish were genotyped using a 50 K SNP Affymetrix® Axiom® myDesign™ Genotyping Array designed by Aquainnovo and the University of Chile. More details about the SNPs included in the 50 K array are in Correa et al. [30] and Yáñez et al. [31, 32].

The genotypes were quality controlled using the Affymetrix Genotyping Console and the SNPolisher R package following the Axiom® Genotyping Solution Data Analysis Guide [33]. Additional quality control steps were conducted by filtering out SNPs with a Hardy–Weinberg equilibrium test p value less than 0.00001, an SNP call rate lower than 0.95 and a minor allele frequency lower than 0.01.

Estimation of breeding values
The conventional pedigree-based approach, P-BLUP, was used as the control for genomic evaluations, and EBV for each individual were obtained using a linear mixed model implemented in BLUPF90 [34]. The model used was as follows:

$$\mathbf{y} = \mathbf{X\beta} + \mathbf{Tg} + \mathbf{e},$$

where $\mathbf{y}$ is a vector of phenotypes (lice count on fins), $\mathbf{\beta}$ is a vector of fixed effects (mean, tank and body weight effects), $\mathbf{g}$ is a vector of random additive polygenic genetic effects that follows a normal distribution $\sim N(\mathbf{0}, \mathbf{A}\sigma_\mathbf{g}^2)$,

X and **T** are incidence matrices, **A** is the pedigree additive relationship matrix, and **e** is the residual [35].

The genomic EBV (GEBV) for each individual were estimated using three different GS models: G-BLUP, Bayesian Lasso and Bayes C. G-BLUP was as implemented in the BLUPF90 software [34]. G-BLUP is a modification of the BLUP method, where the numerator relationship matrix **A** is replaced by a genomic relationship matrix **G**, as described by VanRaden [36].

For Bayesian methods, marker and additive polygenic effects were estimated jointly using the following model implemented in the GS3 software [37]:

$$\mathbf{y} = \mathbf{X}\boldsymbol{\beta} + \mathbf{Z}\boldsymbol{\alpha} + \mathbf{Tg} + \mathbf{e},$$

where $\boldsymbol{\alpha}$ is a vector of random marker allele substitution effects and **Z** is the corresponding incidence matrix. The Gibbs sampler was run using 100,000 iterations with a burn-in of 20,000 iterations. Priors were drawn from double—exponential and scaled—inverse χ^2 distributions, for Bayesian Lasso and Bayes C, respectively. Bayesian GEBV were estimated as the sum of the polygenic and marker effects.

Pedigree (P-BLUP) and genomic (G-BLUP) heritabilities were calculated using the AIREMLF90 software [34] as follows: $h^2 = \sigma_{\mathbf{g}}^2/(\sigma_{\mathbf{g}}^2 + \sigma_{\mathbf{e}}^2)$, where $\sigma_{\mathbf{g}}^2$ is the estimate of the additive genetic variance and $\sigma_{\mathbf{e}}^2$ is the estimate of the residual variance.

Comparison of models

The different models were compared using a fivefold cross validation scheme. Briefly, all genotyped and phenotyped animals were randomly separated into five validations sets, which were predicted one at a time by masking their phenotypes and using the remaining animals as training set to estimate the marker effects. Thus, for each validation run, the dataset was split into a training set (80%) and a validation set (20%). To reduce the stochastic effects, this cross-validation analysis was replicated 10 times. Accuracy was used to assess the performance of each model and was estimated as:

$$R_{EBV,BV} = \frac{R_{EBV,y}}{h},$$

where $R_{EBV,y}$ is the correlation between the EBV of a given model (predicted for the validation set using information from the training set) and the actual phenotype, while h is the square root of the pedigree-based estimate of heritability [26, 38]. To test prediction accuracies obtained by using various SNP densities, 0.5, 1, 5, 10 and 25 K SNPs were selected from the full set as follows. First, 500 SNPs that had a level of linkage disequilibrium (LD, measured as r^2) less than 0.2 and were homogeneously distributed across the genome were selected among the SNPs that passed quality control. Then, SNPs with a homogeneous distribution across the genome were added to the first 500 SNPs to create the 1 K panel. The same procedure was reiterated to create the 5, 10 and 25 K panels. Accuracies were then calculated for each model and SNP density, compared to those obtained with the P-BLUP model, and the relative increase in accuracy was assessed.

Results and discussion

A total of 36,616 SNPs passed the genotyping quality control. An average lice count on fins of 5.1 (SD = 4.4) and an average final body weight of 281 g (SD = 92.8) were obtained. The average lice number per family ranged from 2.3 to 11.3. The pedigree-based estimate of heritability of lice count on fins was equal to 0.10, which is consistent with a previous study on the same population [15], while the marker-based estimate of heritability using the 37 K full set was equal to 0.11.

In all cases, prediction accuracies obtained with the GS models were higher than those obtained with the pedigree-based model, which had an accuracy of 0.41 (Table 1). It is interesting to note that as few as 500 SNPs were sufficient to increase the accuracy of breeding value predictions. Further tests using even lower marker densities may be interesting to assess at which marker density accuracy of genomic prediction drops below that based on pedigree information. The relative increases in accuracy of the different methods for each SNP density are shown on Fig. 1. Comparing the results for all SNP densities shows that the different GS models behaved similarly, with accuracies ranging from 0.45 to 0.5. In general, accuracies increased moderately with increasing SNP density and achieved asymptotic values when 10 K or more SNPs were used. Bayesian Lasso performed slightly better than the other methods at lower SNP densities, but at higher densities, prediction accuracy was similar for all three GS methods. We observed a slight decrease in accuracy with Bayes C when using the full SNP dataset, which was not observed with G-BLUP or Bayesian Lasso. G-BLUP mostly captures genetic relationships between animals, whereas Bayes C uses the LD between SNPs and QTL to calculate predictions. In Bayes C, only a fraction of the SNPs is assumed to have an effect on the trait, and thus when adding more SNPs for calculating predictions, the use of redundant information (SNPs in high LD with each other) [39] might lead to incorrect selection of the subset of SNPs with an effect from the whole set. The accuracies obtained in our study on Atlantic salmon were not as high as those reported for other livestock species, which can be even higher than 0.85 in some cases. Nevertheless, accuracies were similar to those determined in previous studies on Atlantic salmon for *L. salmonis* resistance, for

Table 1 Accuracies for sea lice (*Caligus rogercresseyi*) resistance obtained using different models and different marker densities

Marker density	P-BLUP	G-BLUP	Bayesian Lasso	Bayes C
0	0.41	–	–	–
500	–	0.45	0.47	0.45
1000	–	0.48	0.49	0.48
10,000	–	0.50	0.50	0.50
25,000	–	0.50	0.50	0.50
50,000	–	0.50	0.50	0.49

P-BLUP accuracy only applies to 0 markers as it does not use marker information

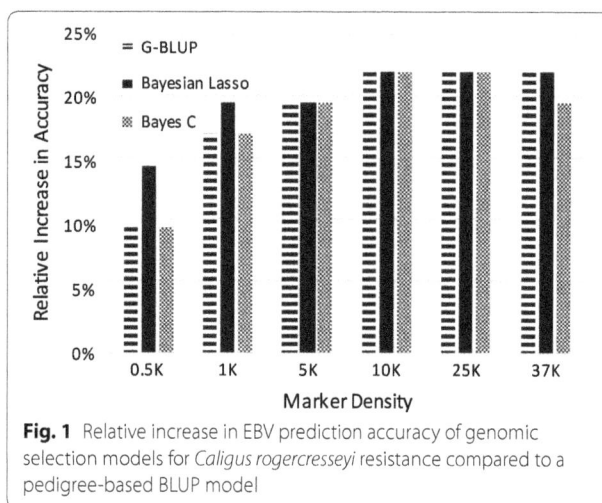

Fig. 1 Relative increase in EBV prediction accuracy of genomic selection models for *Caligus rogercresseyi* resistance compared to a pedigree-based BLUP model

example, Tsai et al. [29] reported accuracies ranging from 0.4 to 0.6. To further increase the accuracy of GEBV predictions in Atlantic salmon breeding programs, it may be necessary to increase the number of individuals and generations in the training population [40].

We found that the three GS methods had similar GEBV prediction accuracies, in spite of the different priors used in each model. This is the first study that aims at evaluating the performance of GS methods for *C. rogercresseyi* resistance in Atlantic salmon. Moreover, it is the first study that evaluates Bayesian method performance for sea lice resistance using a high-density SNP array and different SNP densities in this species. We demonstrate that it is possible to increase the accuracy of breeding value predictions for *C. rogercresseyi* resistance using SNP information.

Prediction accuracies obtained by using genome-wide marker information were higher than those obtained by using only pedigree information to account for the relationship between individuals. These results are in line with other studies that evaluated the performance of GS for different traits in Atlantic salmon

[26, 27]. For example, using only pedigree information, Ødegård et al. [26] estimated an EBV reliability (defined as $R^2_{EBV,y}/h^2$) of 0.34 and 0.36, which increased by up to 50 and 20% for *L. salmonis* resistance and fillet color when using 22 K SNPs in an admixed population. For growth traits, Tsai et al. [27] achieved an increase in accuracy by up to 20% when using 5 K SNPs. Similarly, Tsai et al. [29] achieved an increase in accuracy of 27% when using 5 K SNPs for *L. salmonis* resistance. A recent study in rainbow trout [28] reported similar prediction accuracies between P-BLUP and GS methods using a 57 K genotyping array for bacterial cold water disease resistance.

Our results show that SNP density had a moderate impact on prediction accuracy for all GS methods tested in this study, and that as few as 500 SNPs were sufficient to increase EBV accuracy over P-BLUP. Moreover, 10 K may suffice to obtain maximum increases in accuracy for this trait and this population.

This study used real data from a breeding population and the results suggest that GS may be effective in improving resistance to this sea lice species. Further studies aimed at evaluating the use of low-density panels and imputation strategies are necessary to determine the minimum required SNP density for GS and thus reduce genotyping costs.

Conclusions

It is possible to improve prediction accuracy for *C. rogercresseyi* resistance in Atlantic salmon by using different densities of SNPs selected from a 37 K panel. By applying different genomic prediction approaches, we showed that as few as 500 SNPs were sufficient to increase the accuracy of EBV to a higher value than that obtained from pedigree-based methods. We found that the maximum increase in accuracy was obtained with 10 K SNPs when using G-BLUP and Bayesian methods.

Authors' contributions

KC performed the data analysis, contributed with discussion and wrote the initial manuscript. RB contributed with data analysis and discussion. RF contributed with data management and the creation of informatics tools. JPL contributed with the design of the study and discussion. JMY conceived and designed the study, supervised KC's work, contributed with discussion and with the writing of the manuscript. All authors read and approved the final manuscript.

Author details

[1] Facultad de Ciencias Veterinarias y Pecuarias, Universidad de Chile, Av Santa Rosa 11735, La Pintana, Chile. [2] Aquainnovo, Cardonal s/n Lote B, Puerto Montt, Chile.

Acknowledgements

The authors thank Francisca Erranz for her valuable collaboration in the experimental challenges.

Competing interests

The authors declare that they have no competing interests.

Funding
This project was funded by CORFO-Innova Chile 11IEI-12843. KC acknowledges CONICYT for the funding through the National PhD Grant (21130669).

References

1. Johnson SC, Treasurer JW, Bravo S, Nagasawa K, Kabata Z. A review of the impact of parasitic copepods on marine aquaculture. Zool Stud. 2004;43:229–43.
2. Guo FC, Woo PTK. Selected parasitosis in cultured and wild fish. Vet Parasitol. 2009;163:207–16.
3. Boxshall GA, Bravo S. On the identity of the common Caligus (Copepoda: Siphonostomatoidea: Caligidae) from salmonid netpen systems in Southern Chile. Contrib Zool. 2000;69:137–46.
4. Boxaspen K. A review of the biology and genetics of sea lice. ICES J Mar Sci. 2006;63:1304–16.
5. Frazer LN, Morton A, Krkosek M. Critical thresholds in sea lice epidemics: evidence, sensitivity and subcritical estimation. Proc Biol Sci. 2012;279:1950–8.
6. Oelckers K, Vike S, Duesund H, Gonzalez J, Wadsworth S, Nylund A. *Caligus rogercresseyi* as a potential vector for transmission of Infectious Salmon Anaemia (ISA) virus in Chile. Aquaculture. 2014;420–421:126–32.
7. Costello MJ. The global economic cost of sea lice to the salmonid farming industry. J Fish Dis. 2009;32:115–8.
8. Aaen SM, Helgesen KO, Bakke MJ, Kaur K, Horsberg TE. Drug resistance in sea lice: a threat to salmonid aquaculture. Trends Parasitol. 2015;31:72–81.
9. Ødegård J, Baranski M, Gjerde B, Gjedrem T. Methodology for genetic evaluation of disease resistance in aquaculture species: challenges and future prospects. Aquac Res. 2011;42:103–14.
10. Bishop SC, Woolliams JA. Genomics and disease resistance studies in livestock. Livest Sci. 2014;166:190–8.
11. Yáñez JM, Newman S, Houston RD. Genomics in aquaculture to better understand species biology and accelerate genetic progress. Front Genet. 2015;6:128.
12. Yáñez JM, Martínez V. Factores genéticos que inciden en la resistencia a enfermedades infecciosas. Arch Med Vet. 2010;42:1–13.
13. Yáñez JM, Houston RD, Newman S. Genetics and genomics of disease resistance in salmonid species. Front Genet. 2014;5:415.
14. Lhorente JP, Gallardo JA, Villanueva B, Araya AM, Torrealba DA, Toledo XE, et al. Quantitative genetic basis for resistance to *Caligus rogercresseyi* sea lice in a breeding population of Atlantic salmon (*Salmo salar*). Aquaculture. 2012;324–325:55–9.
15. Yáñez JM, Lhorente JP, Bassini LN, Oyarzún M, Neira R, Newman S. Genetic co-variation between resistance against both *Caligus rogercresseyi* and *Piscirickettsia salmonis*, and body weight in Atlantic salmon (*Salmo salar*). Aquaculture. 2014;433:295–8.
16. Correa K, Lhorente JP, Bassini L, López ME, Di Génova A, Maass A, et al. Genome wide association study for resistance to *Caligus rogercresseyi* in Atlantic salmon (*Salmo salar* L.) using a 50 K SNP genotyping array. Aquaculture. 2016. doi:10.1016/j.aquaculture.2016.04.008.
17. Houston RD, Taggart JB, Cézard T, Bekaert M, Lowe NR, Downing A, et al. Development and validation of a high density SNP genotyping array for Atlantic salmon (*Salmo salar*). BMC Genomics. 2014;15:90.
18. Houston RD, Bishop SC, Guy DR, Tinch AE, Taggart JB, Bron JE, et al. Genome wide association analysis for resistance to sea lice in Atlantic salmon: application of a dense SNP array. In: Proceedings of the 10th world congress on genetics applied to livestock production: 18–22 August 2014; Vancouver. 2014. https://asas.org/docs/default-source/wcgalp-proceedings-oral/265_paper_9597_manuscript_751_0.pdf?sfvrsn=2.
19. Hayes B, Goddard M. Genome-wide association and genomic selection in animal breeding. Genome. 2010;53:876–83.
20. Meuwissen THE, Hayes BJ, Goddard ME. Prediction of total genetic value using genome-wide dense marker maps. Genetics. 2001;157:1819–29.
21. Nirea KG, Sonesson AK, Woolliams JA, Meuwissen TH. Strategies for implementing genomic selection in family-based aquaculture breeding schemes: double haploid sib test populations. Genet Sel Evol. 2012;44:30.
22. Lillehammer M, Meuwissen THE, Sonesson AK. A low-marker density implementation of genomic selection in aquaculture using within-family genomic breeding values. Genet Sel Evol. 2013;45:39.
23. Taylor JF. Implementation and accuracy of genomic selection. Aquaculture. 2014;420–421:S8–14.
24. Yi N, Xu S. Bayesian LASSO for quantitative trait loci mapping. Genetics. 2008;179:1045–55.
25. Habier D, Fernando RL, Kizilkaya K, Garrick DJ. Extension of the bayesian alphabet for genomic selection. BMC Bioinformatics. 2011;12:186.
26. Ødegård J, Moen T, Santi N, Korsvoll SA, Kjøglum S, Meuwissen THE. Genomic prediction in an admixed population of Atlantic salmon (*Salmo salar*). Front Genet. 2014;5:402.
27. Tsai HY, Hamilton A, Tinch AE, Guy DR, Gharbi K, Stear M, et al. Genome wide association and genomic prediction for growth traits in juvenile farmed Atlantic salmon using a high density SNP array. BMC Genomics. 2015;16:969.
28. Vallejo RL, Leeds TD, Fragomeni BO, Gao G, Hernandez AG, Misztal I, et al. Evaluation of genome-enabled selection for bacterial cold water disease resistance using progeny performance data in rainbow trout: insights on genotyping methods and genomic prediction models. Front Genet. 2016;7:96.
29. Tsai HY, Hamilton A, Tinch AE, Guy DR, Bron JE, Taggart JB, et al. Genomic prediction of host resistance to sea lice in farmed Atlantic salmon populations. Genet Sel Evol. 2016;48:47.
30. Correa K, Lhorente JP, López ME, Bassini L, Naswa S, Deeb N, et al. Genome-wide association analysis reveals loci associated with resistance against *Piscirickettsia salmonis* in two Atlantic salmon (*Salmo salar* L.) chromosomes. BMC Genomics. 2015;16:854.
31. Yáñez JM, Naswa S, López ME, Bassini L, Correa K, Gilbey J, et al. Genome-wide single nucleotide polymorphism (SNP) discovery in Atlantic salmon (*Salmo salar*): validation in wild and farmed American and European populations. Mol Ecol Resour. 2016;16:1002–11.
32. Yáñez JM, Naswa S, López ME, Bassini L, Cabrejos ME, Gilbey J, et al. Development of a 200 K SNP array for Atlantic salmon: exploiting across continents genetic variation. In: Proceedings of the 10th world congress on genetics applied to livestock production: 18–22 August 2014; Vancouver. 2014. https://asas.org/docs/default-source/wcgalp-proceedings-oral/263_paper_9678_manuscript_833_0.pdf?sfvrsn=2.
33. Affymetrix I. SNPolisher user guide (version 1.5.0). 2014.
34. Misztal I, Tsuruta S, Lourenco D, Masuda Y, Aguilar I, Legarra A, et al. Manual for BLUPF90 family of programs. 2016. http://nce.ads.uga.edu/wiki/lib/exe/fetch.php?media=blupf90_all4.pdf.
35. Lynch M, Walsh B. Genetics and analysis of quantitative traits. Sunderland: Sinauer Associates; 1998.
36. VanRaden PM. Efficient methods to compute genomic predictions. J Dairy Sci. 2008;91:4414–23.
37. Legarra A, Ricardi A, Filangi O. GS3: genomic selection, gibbs sampling, gauss seidel (and BayesCπ). 2014. http://snp.toulouse.inra.fr/~alegarra/manualgs3_last.pdf.
38. Legarra A, Robert-Granie C, Manfredi E, Elsen JM. Performance of genomic selection in mice. Genetics. 2008;180:611–8.
39. Thavamanikumar S, Dolferus R, Thumma BR. Comparison of genomic selection models to predict flowering time and spike grain number in two hexaploid wheat doubled haploid populations. G3 (Bethesda). 2015;5:1991–8.
40. Sonesson AK, Meuwissen TH. Testing strategies for genomic selection in aquaculture breeding programs. Genet Sel Evol. 2009;41:37.
41. Di Génova A, Aravena A, Zapata L, González M, Maass A, Iturra P. SalmonDB: a bioinformatics resource for *Salmo salar* and *Oncorhynchus mykiss*. Database (Oxford). 2011;2011:bar050.

16

Potential of gene drives with genome editing to increase genetic gain in livestock breeding programs

Serap Gonen[1], Janez Jenko[1], Gregor Gorjanc[1], Alan J. Mileham[2], C. Bruce A. Whitelaw[1] and John M. Hickey[1*]

Abstract

Background: This paper uses simulation to explore how gene drives can increase genetic gain in livestock breeding programs. Gene drives are naturally occurring phenomena that cause a mutation on one chromosome to copy itself onto its homologous chromosome.

Methods: We simulated nine different breeding and editing scenarios with a common overall structure. Each scenario began with 21 generations of selection, followed by 20 generations of selection based on true breeding values where the breeder used selection alone, selection in combination with genome editing, or selection with genome editing and gene drives. In the scenarios that used gene drives, we varied the probability of successfully incorporating the gene drive. For each scenario, we evaluated genetic gain, genetic variance (σ_A^2), rate of change in inbreeding (ΔF), number of distinct quantitative trait nucleotides (QTN) edited, rate of increase in favourable allele frequencies of edited QTN and the time to fix favourable alleles.

Results: Gene drives enhanced the benefits of genome editing in seven ways: (1) they amplified the increase in genetic gain brought about by genome editing; (2) they amplified the rate of increase in the frequency of favourable alleles and reduced the time it took to fix them; (3) they enabled more rapid targeting of QTN with lesser effect for genome editing; (4) they distributed fixed editing resources across a larger number of distinct QTN across generations; (5) they focussed editing on a smaller number of QTN within a given generation; (6) they reduced the level of inbreeding when editing a subset of the sires; and (7) they increased the efficiency of converting genetic variation into genetic gain.

Conclusions: Genome editing in livestock breeding results in short-, medium- and long-term increases in genetic gain. The increase in genetic gain occurs because editing increases the frequency of favourable alleles in the population. Gene drives accelerate the increase in allele frequency caused by editing, which results in even higher genetic gain over a shorter period of time with no impact on inbreeding.

Background

This paper uses simulation to explore how gene drives increase genetic gain in livestock breeding programs. Genetic gain is brought about by increasing the frequency of favourable alleles. In most breeding programs, the increase in frequency is achieved slowly by selecting high merit individuals as the parents of the next generation based on phenotype and/or genotype information. The efficacy and efficiency of this type of breeding program depends on four factors: the ability to accurately identify high merit individuals, the intensity of selection, the time taken to replace one generation with another and the way in which the existing genetic diversity is maintained and converted into short- and long-term genetic gain.

Recent advances in genome editing have increased interest in using this technology to accelerate genetic gain in breeding programs [1]. Genome editing allows the precise deletion, addition or change of alleles at specific locations in the genome of a cell. These changes are

*Correspondence: john.hickey@roslin.ed.ac.uk
[1] The Roslin Institute and Royal (Dick) School of Veterinary Studies, The University of Edinburgh, Easter Bush, Midlothian, Scotland, UK
Full list of author information is available at the end of the article

permanent and heritable if they are made in zygotes or germline cells.

There are over 300 examples of the use of genome editing in plants and livestock [2], including edits for herbicide resistance in oilseed rape [3], in the *myostatin* gene for "double muscling" in pigs, cattle and sheep [4], the introduction of the polled gene into dairy cattle [5], and edits to confer resistance to porcine reproductive and respiratory syndrome virus (PRRS) and African swine fever virus (ASFV) in pigs [4, 6–8].

To date, all applications of genome editing in livestock used single edits to address simple traits that are controlled by a small number of causal variants with large effects. However, in livestock breeding programs, the majority of traits of interest are quantitative and are likely affected by thousands of causal variants, each with small effect. However, although there are many causal variants for each trait, a recent simulation study using an editing strategy called PAGE (promotion of alleles by genome editing) showed that discovering and editing relatively small numbers of causal variants can double the rate of both short- and long-term genetic gain compared to selection alone [1].

Although the increase in genetic gain from PAGE was impressive, many generations of editing were needed to fix favourable alleles [1]. This is because unfavourable alleles continue to segregate within the non-edited parents (i.e., dams) of each generation. Methods that can fix favourable alleles more quickly would be valuable within breeding programs. One such method is genome editing with gene drives.

Gene drives are naturally occurring phenomena that cause a mutation on one chromosome to copy itself onto its homologous chromosome. The copying process occurs because the gene drive initiates a double-stranded DNA break on the homologous chromosome. The DNA break is repaired by cellular pathways such as homology-directed repair, which uses the sequence of the chromosome that contains the gene drive elements as a repair template [9, 10]. An example of a naturally occurring gene drive is the so-called P-element, which invaded the fruit fly *Drosophila melanogaster* in the 1950s and has since spread worldwide [11].

With advances in genome editing technology, a gene drive can be incorporated with a genome edit made either on the germline cell of a parent or on the parent itself at the zygote stage to ensure that all offspring are homozygous for the edited allele. The possibility of using gene drives to promote the spread of alleles through a population was first proposed by Burt in 2003 [12]. This concept, now recently termed the 'mutagenic chain reaction', was empirically demonstrated in *Drosophila* through modification of the CRISPR/Cas9 system originally identified in bacteria [13–15]. In this case, the CRISPR/Cas9 gene drive system was used to induce a change in the *Drosophila* body colour from wild type to yellow by copying the gene drive-linked *yellow* gene onto the homologous chromosome of offspring inheriting one copy of the *yellow* gene [15].

Since this demonstration, artificially constructed gene drives have gained renewed interest as a way of quickly spreading alleles in natural populations [12]. Targeted gene drive mechanisms based on CRISPR/Cas9 editors have been reported to have conversion efficacies of more than 98% [16], demonstrating the potential of this technology in spreading alleles in populations. One recent proposal is to use gene drives to spread a deleterious allele in populations of mosquito hosts of the malaria parasite. The deleterious allele reduces the fitness of the mosquito, thus eliminating the mosquito population as well as the parasite [16].

Gene drives could be combined with genome editing for quantitative traits to fix edited alleles more quickly in livestock populations. Each edited allele could have a gene drive based on a CRISPR/Cas9 editor. As shown in Fig. 1, the gene drive would be co-inherited with the edited allele across generations. This would ensure complete homozygosity for the favourable allele amongst all descendants of an edited individual, regardless of the genotype of the other parent.

The objective of this study was to quantify the potential of using gene drives with genome editing to increase the genetic gain for quantitative traits in livestock breeding.

Methods

Simulation was used to evaluate the use of gene drives with genome editing in increasing the genetic gain for quantitative traits in livestock breeding. A variety of scenarios were tested, each using different editing strategies within the breeding program. All scenarios followed a common overall structure, where the simulation scheme was divided into historical and future components. The historical component was split into two parts: (1) evolution under the assumption that livestock populations have been evolving for tens of thousands of years prior to domestication; and (2) 21 recent generations of modern animal breeding with selection based on breeding values. The future component consisted of a further 20 generations of modern animal breeding. In each generation, parents of the next generation were selected based on true breeding values (TBV). Within a given scenario, the breeder was given the choice of using only selection, selection and genome editing, or selection and genome editing with gene drives. Recent historical animal breeding generations were denoted −20 to 0 and future animal breeding generations were denoted 1 to 20.

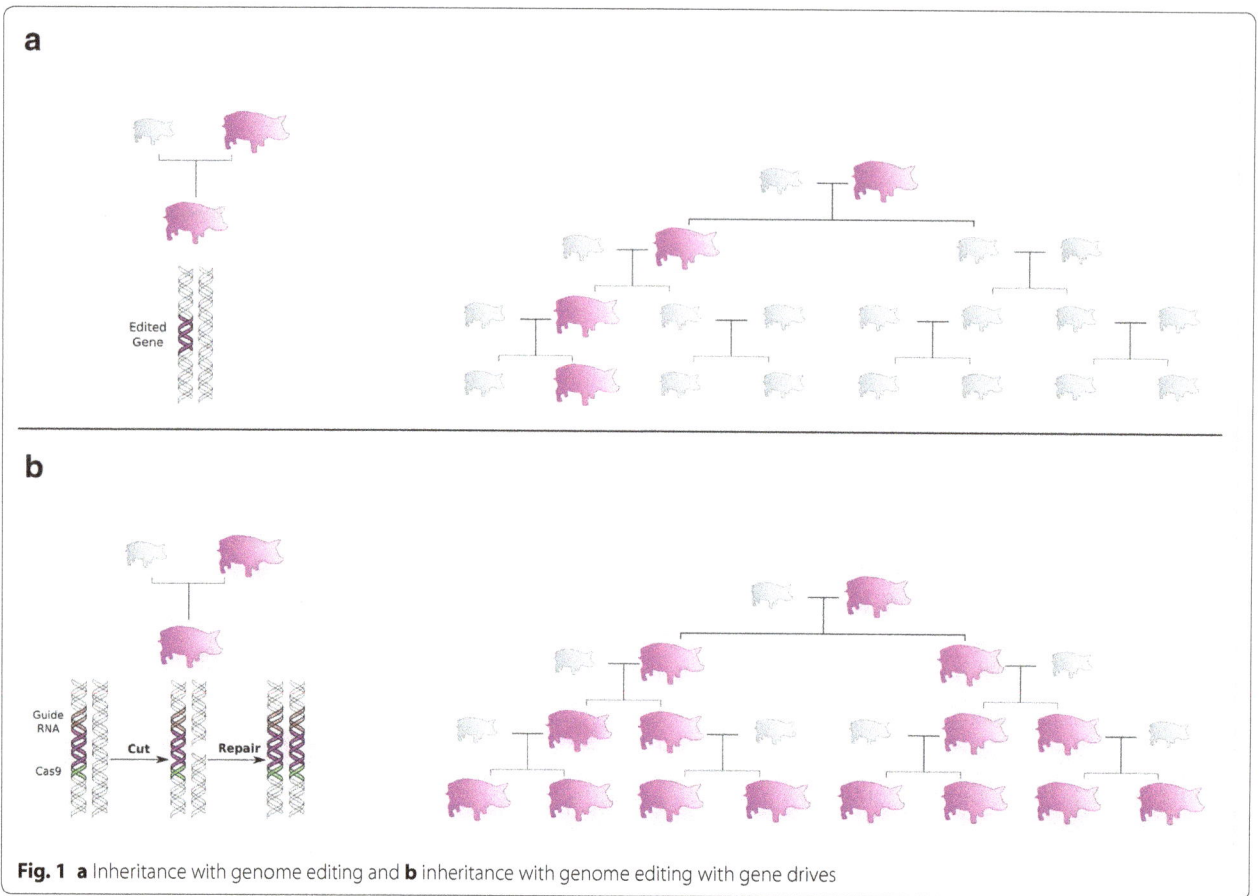

Fig. 1 **a** Inheritance with genome editing and **b** inheritance with genome editing with gene drives

The simulations were designed to: (1) generate whole-genome sequence data; (2) generate quantitative trait nucleotides (QTN) affecting phenotypes; (3) generate pedigree structures for a typical livestock population; (4) perform selection; and (5) perform genome editing with and without gene drives. For each scenario, the genetic gain, genetic variance (σ_A^2), rate of change in inbreeding (ΔF), number of distinct QTN edited, rate of increase in favourable allele frequencies of edited QTN and the time to fix favourable alleles were evaluated. Results are presented as the mean of ten replicates for each scenario on a per generation and/or cumulative basis.

Whole-genome sequence data, historical evolution
Sequence data was generated using the Markovian Coalescent Simulator (MaCS) [17] and AlphaSim [18, 19] for 1000 base haplotypes for each of ten chromosomes. Chromosomes were each 1 Morgan long comprising 10^8 base pairs and were simulated using a per site mutation rate of 2.5×10^{-8}, a per site recombination rate of 1.0×10^{-8} and an effective population size (N_e) that varied over time in accordance with estimates for the Holstein cattle population [20]. N_e was set to 100 in the final generation of the coalescent simulation, to $N_e = 1256$,

1000 years ago, to $N_e = 4350$, 10,000 years ago, and to $N_e = 43,500$, 100,000 years ago, with linear changes in between these time-points. The resulting sequence had approximately 650,000 segregating sites in total.

Quantitative trait variants
A quantitative trait was simulated by randomly sampling 10,000 QTN from the segregating sequence sites in the base population, with the restriction that 1000 QTN were sampled from each of the ten chromosomes. QTN had their allele substitution effect randomly sampled from a normal distribution with a mean of 0 and standard deviation of 0.01 (1.0 divided by the square root of the number of QTN). QTN effects were used to compute TBV for an individual.

Pedigree structure, gamete inheritance and selection strategies
A pedigree of 41 generations of 1000 individuals in equal sex ratio was simulated. In the first generation of the recent historical animal breeding population (denoted generation −20), individuals had their chromosomes sampled from the 1000 base haplotypes. In each subsequent generation (i.e., generations −19 to 20), the

chromosomes of each individual were sampled from parental chromosomes by recombination. A recombination rate of 1 Morgan per chromosome was simulated, resulting in a 10-Morgan genome. Recombination locations were simulated ignoring interference. In each generation, 25 males were selected to become the sires of the next generation using truncation selection on TBV. No selection was performed on females, and all 500 were used as dams.

Genetic gain

Genetic gain was calculated in units of the standard deviation of TBV in the base generation (generation −20 or generation 1) as $(\overline{TBV_{curr}} - \overline{TBV_{base}})/\sigma_{TBV_{base}}$, where $\overline{TBV_{curr}}$ is the mean TBV of the current generation and $\overline{TBV_{base}}$ and $\sigma_{TBV_{base}}$ are the mean and standard deviation of TBV in the base generation, respectively. Generation −20 was used as the base generation in order to observe the genetic improvement since the start of the recent historical breeding. Generation 1 was used for ease of presentation of some of the results. The genetic variance in each generation was calculated as: $\sigma_A^2 = a'a/(n-1)$, where a is a zero mean vector of TBV of the n individuals in that generation.

Efficiency of turning genetic variation into genetic gain

The efficiency of turning genetic variation into genetic gain at set generations was calculated by relating average genetic gain per generation to the rate of change in inbreeding of the future breeding component. The rate of change in inbreeding, ΔF, was estimated by fitting a linear regression model $\log(1-F_t) = \mu - \Delta F g_t$, which is a linearization of formula $\Delta F = (F_t - F_{t-1})/(1 - F_{t-1})$ [21] and where g_t is the mean breeding value at generation t. The efficiency of turning genetic variation into genetic gain was then calculated as the ratio of the average genetic gain per generation to ΔF as $(100 \times [(G_g - G_0)/g])/\Delta F$, where G_0 is generation 0 and G_g is generation g of the future breeding component.

Scenarios

Three main scenarios were simulated: (1) selection alone; (2) selection and genome editing; or (3) selection and genome editing with gene drives. When editing with gene drives, the probability of successfully incorporating a gene drive with an edited allele, i.e., the conversion efficacy of the gene drive mechanism, was modelled. Three conversion efficacies of 0.5, 0.75 and 1.0 were compared.

When applying genome editing, a maximum of 500 edits per generation were allowed. In each generation, 25 sires were selected based on TBV and then either all 25 were edited at 20 QTN each or the top 5 were edited

for 100 QTN each. For each sire, the QTN with the largest effect (i.e., α) on phenotype for which the sire was not already homozygous for the favourable allele was edited, assuming that QTN effects were a priori known.

Unless explicitly mentioned, all results showing the effect of gene drives were run with the gene drive conversion efficacy set to 1.00 (i.e., 100% efficacy).

Results

This paper uses simulation to examine how gene drives enhance the benefit of genome editing in breeding programmes with selection. The results highlight seven ways in which gene drives enhance the benefits of genome editing. Gene drives: (1) amplify the increase in genetic gain brought about by genome editing; (2) amplify the rate of increase in the frequency of favourable alleles and reduce the time it takes to fix them; (3) enable more rapid targeting of QTN with lesser effect for genome editing; (4) distribute fixed editing resources across a larger number of distinct QTN across generations; (5) focus editing on a smaller number of QTN within a given generation; (6) reduce the level of inbreeding when editing a subset of the sires; and (7) increase the efficiency of converting genetic variation into genetic gain.

Genetic gain

Gene drives amplify the increase in genetic gain brought about by genome editing. This is shown in Fig. 2, which plots the overall genetic gain against time for generations −20 to 20 when all 25 sires were edited. Generations −20 to 0 were identical for all scenarios and represent the recent historical breeding, in which selection was used without editing. Generations 0 to 20 represented the future breeding where the breeder had the choice of using selection alone, selection and genome editing, or selection and genome editing with gene drives. Since generations −20 to 0 were identical for all scenarios and no editing was performed, all results presented are standardised to generation 0. Standardised genetic gain is given on the y-axis on the right in Fig. 2.

Figure 2 shows that by generation 20, gene drives achieved 1.43 times more genetic gain than genome editing [31.29 vs. 21.81; (see Additional file 1: Table S1)] and 2.80 times more genetic gain than selection alone [31.29 vs. 11.16; (see Additional file 1: Table S1)]. Genome editing achieved 1.95 times more genetic gain than selection alone [21.81 vs. 11.16; (see Additional file 1: Table S1)].

Changes in allele frequency

Gene drives amplify the rate of increase in the frequency of favourable alleles at the QTN with the largest effect brought about by genome editing. This is shown in Fig. 3, which plots the average allele frequencies of the

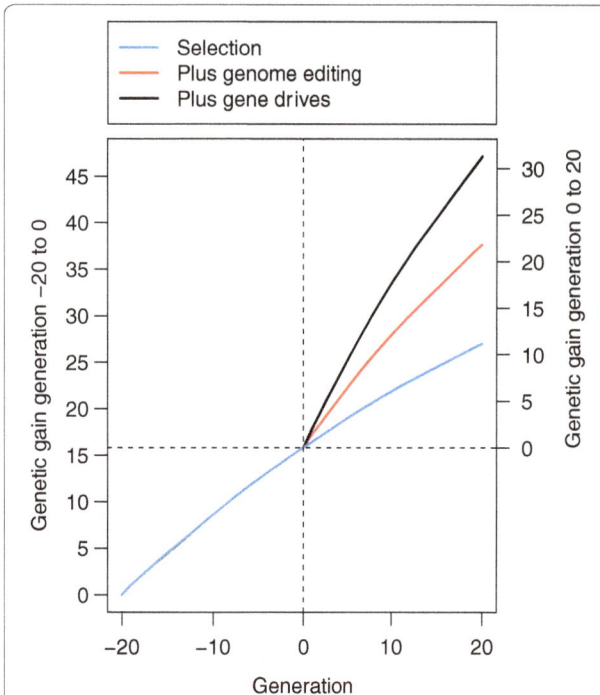

Fig. 2 Genetic gain using selection (*blue line*), selection with standard genome editing (*red line*) or selection, genome editing with gene drives (*black line*). The figure represents the scenario when all 25 sires in a given generation were edited at 20 QTN each

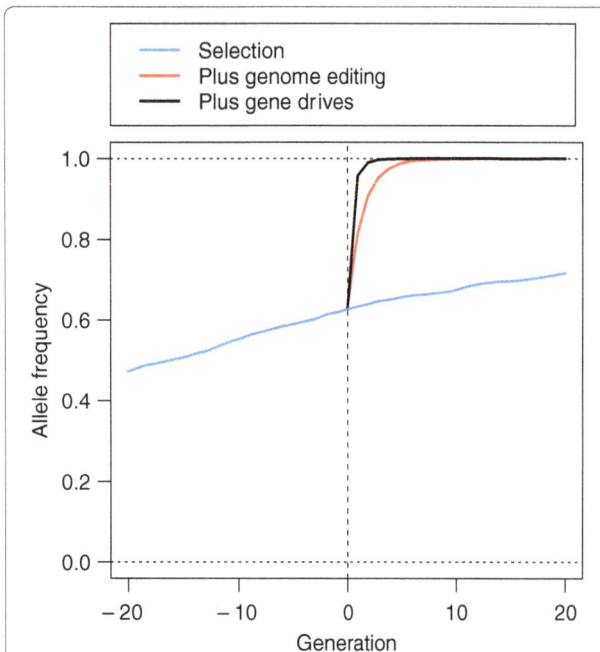

Fig. 3 Allele frequency in each generation of the 20 QTN with the largest effect using selection (*blue line*), selection and genome editing (*red line*), or selection and genome editing with gene drives (*black line*). The figure represents the scenario when all 25 sires in a given generation were edited at 20 QTN each

favourable alleles of the 20 QTN with the largest effect against time in generations −20 to 20. In the first generation of editing, gene drives produced nearly twice the increase in the frequency of favourable alleles at the 20 QTN with the largest effect than genome editing (increase of 0.33 vs. 0.18). This increase in frequency using gene drives was 47 times greater than that produced by selection alone, which produced an increase in the frequency of favourable alleles by 0.007 in the first generation and by 0.09 across all 20 generations.

Figure 3 also shows that gene drives fix the 20 QTN with the largest effect more quickly than genome editing alone. Gene drives achieve an asymptote of allele frequency higher than 0.99 in generation 2, whereas genome editing achieves it in generation 6. Selection without editing achieves a maximum frequency of approximately 0.72 across all 20 generations of future breeding. The rapid fixation of the 20 QTN with the largest effect when using gene drives would mean that QTN with lesser effect can be targeted for genome editing sooner.

Gene drives reduce the time required to target QTN with lesser effect and increase the frequency of their favourable alleles more quickly. This is shown in Fig. 4, which plots the average allele frequencies of favourable alleles in three categories of QTN against time in generations −20 to 20. The three categories of QTN were: (1) the 20 QTN with the largest effects; (2) the 20 QTN with effect sizes ranked from 101 to 120; and (3) the 20 QTN with effect sizes ranked from 201 to 220.

Figure 4 shows that the slope of the lines for all three QTN categories are much steeper and occur at earlier generations when using gene drives. Selection without genome editing resulted in very small increases in allele frequencies for all three QTN categories across all 20 generations. Gene drives caused the shift in allele frequency to occur two times earlier than genome editing alone for both QTN category (2) (generation 2 vs. 4) and QTN category (3) (generation 5 vs. 9). Gene drives also reduced the time taken to reach allele frequencies higher than 0.95 by a half for both QTN category (2) (two vs. four generations) and QTN category (3) (two vs. five generations). This reduction in the time required to shift allele frequencies when using gene drives could have additional benefits in the maintenance and fixation of favourable alleles.

Figure 4 also shows that gene drives can result in the rapid fixation of favourable alleles at QTN with lesser effect, which would probably never become fixed and may even be lost using selection or genome editing alone. When using gene drives, an asymptote of average allele frequency higher than 0.99 was achieved for QTN categories (1), (2) and (3) in generations 2, 5 and 8, respectively. When using genome editing alone, this asymptote

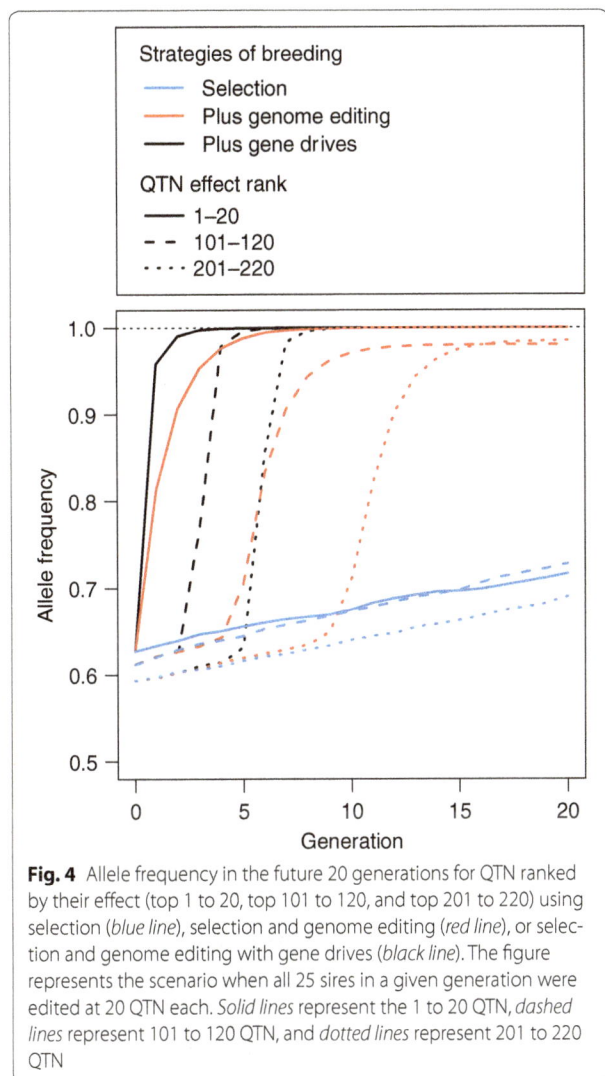

Fig. 4 Allele frequency in the future 20 generations for QTN ranked by their effect (top 1 to 20, top 101 to 120, and top 201 to 220) using selection (*blue line*), selection and genome editing (*red line*), or selection and genome editing with gene drives (*black line*). The figure represents the scenario when all 25 sires in a given generation were edited at 20 QTN each. *Solid lines* represent the 1 to 20 QTN, *dashed lines* represent 101 to 120 QTN, and *dotted lines* represent 201 to 220 QTN

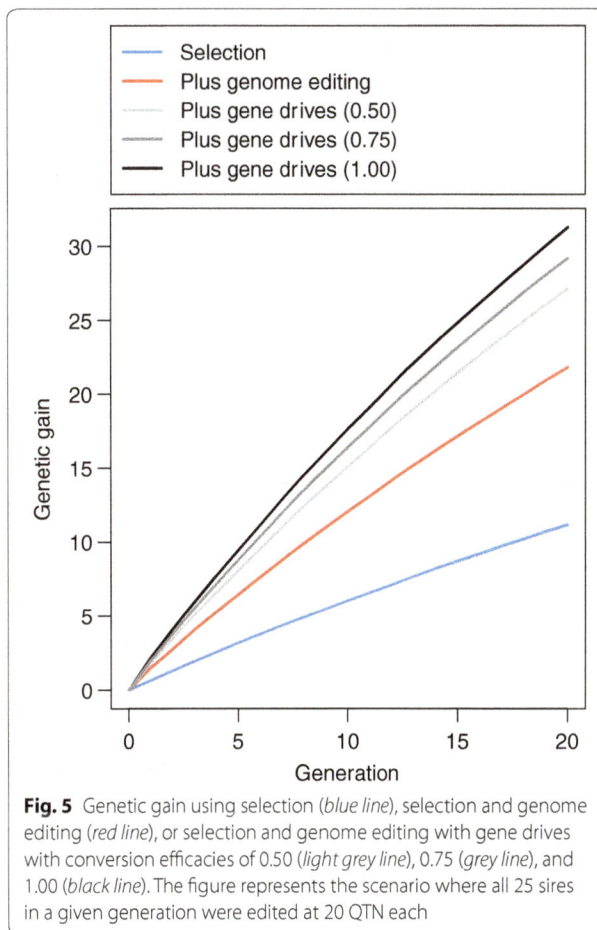

Fig. 5 Genetic gain using selection (*blue line*), selection and genome editing (*red line*), or selection and genome editing with gene drives with conversion efficacies of 0.50 (*light grey line*), 0.75 (*grey line*), and 1.00 (*black line*). The figure represents the scenario where all 25 sires in a given generation were edited at 20 QTN each

was reached only for category (1) while categories (2) and (3) had an asymptote of 0.98, which was reached in generations 14 and 17, respectively. This asymptote of 0.98 was caused by a loss of an average of four favourable alleles from the population before they could be targeted for genome editing. When using only selection, the maximum allele frequency reached for any category of QTN was approximately 0.72.

Effect of gene drive efficacy on genetic gain

Reducing the conversion efficacy of the gene drive mechanism reduces the genetic gain. This is shown in Fig. 5, which is a plot of the genetic gain against time in generations 0 to 20. At the three gene drive conversion efficacies that we tested, 1.00, 0.75 and 0.50, genetic gain was monotonically related to conversion efficacy. Gene drives with complete efficacy resulted in 1.07 times more genetic gain than gene drives with a conversion efficacy of 0.75 (31.29

vs. 29.22), and 1.15 times more genetic gain than gene drives with a conversion efficacy of 0.50 (31.29 vs. 27.16).

Gene drives with low conversion efficacies still substantially amplify the increase in genetic gain brought about by genome editing. Gene drives with a conversion efficacy of 0.50 resulted in 1.25 times more genetic gain than genome editing alone (27.16 vs. 21.81) and 2.43 times more genetic gain than selection alone (27.16 vs. 11.16).

Focusing editing resources on a subset of sires: genetic gain

Genetic gain was higher when editing a subset of the sires than when editing all 25 sires. This is shown in Fig. 6a, which plots the genetic gain against time in generations 0 to 20. Figure 6a shows scenarios in which either all 25 sires were edited at 20 QTN or the top 5 sires were edited at 100 QTN (both scenarios performed a total of 500 edits per generation).

Figure 6a shows that editing the top 5 sires resulted in more genetic gain than editing all 25 sires. This was the case with and without gene drives. With gene drives, editing the top 5 sires resulted in 2.25 times more genetic

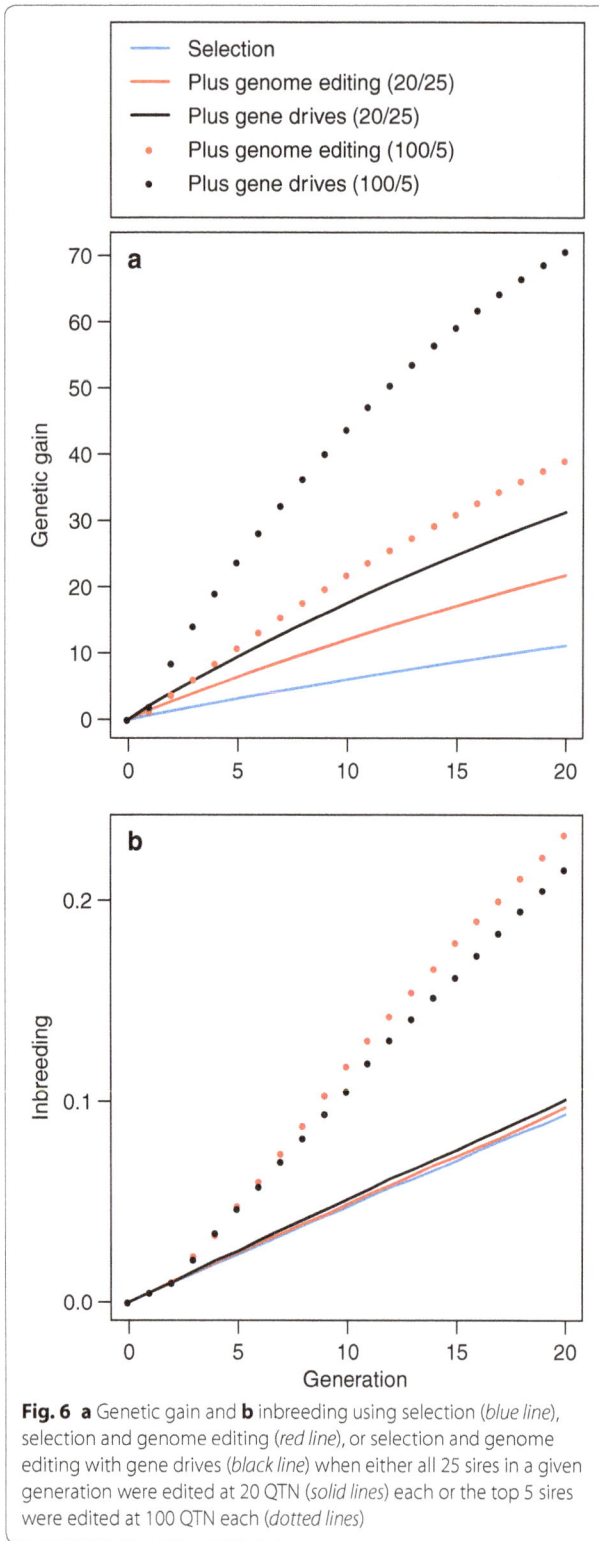

Fig. 6 **a** Genetic gain and **b** inbreeding using selection (*blue line*), selection and genome editing (*red line*), or selection and genome editing with gene drives (*black line*) when either all 25 sires in a given generation were edited at 20 QTN (*solid lines*) each or the top 5 sires were edited at 100 QTN each (*dotted lines*)

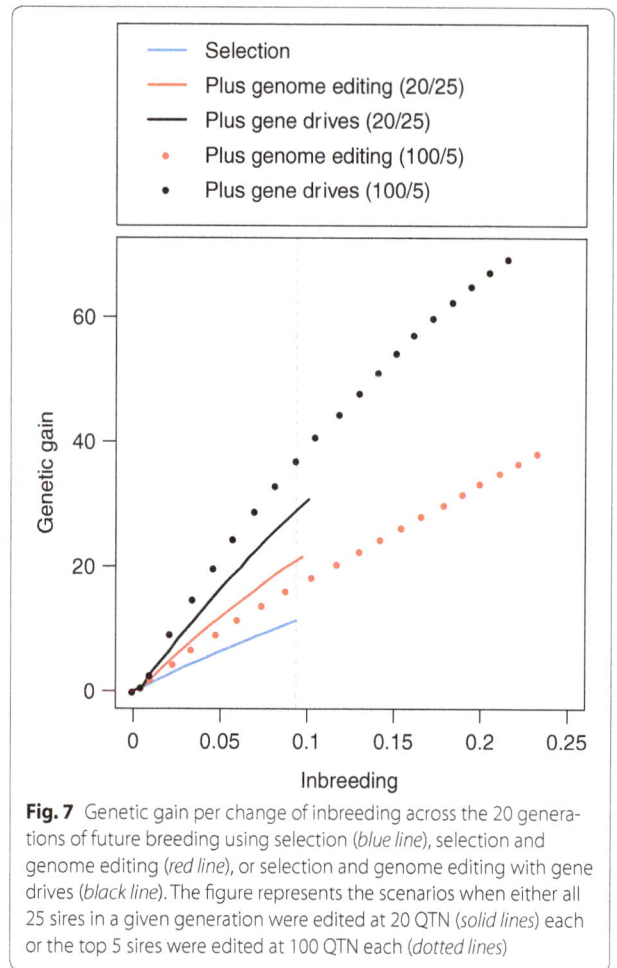

Fig. 7 Genetic gain per change of inbreeding across the 20 generations of future breeding using selection (*blue line*), selection and genome editing (*red line*), or selection and genome editing with gene drives (*black line*). The figure represents the scenarios when either all 25 sires in a given generation were edited at 20 QTN (*solid lines*) each or the top 5 sires were edited at 100 QTN each (*dotted lines*)

Figure 6a also shows that gene drives when editing the top 5 sires was the best strategy for maximising the genetic gain achieved. The lowest genetic gain was achieved when using selection alone. Editing the top 5 sires with gene drives resulted in 1.80 times more genetic gain than genome editing (70.66 vs. 39.17) and 6.33 times more genetic gain than selection alone (70.66 vs. 11.16). The second highest increase in genetic gain was achieved when editing the top 5 sires without gene drives. Editing the top 5 sires without gene drives resulted in 3.51 times more genetic gain than selection alone (39.17 vs. 11.16).

Focusing editing resources on a subset of sires: inbreeding
Inbreeding levels were higher when editing a subset of the sires than when editing all 25 sires. This is shown in Fig. 6b, which plots the genetic gain against time in generations 0 to 20. Figure 6b shows scenarios in which either all 25 sires were edited at 20 QTN or the top 5 sires were edited at 100 QTN (i.e., both scenarios performed a total of 500 edits per generation). Editing the top 5 sires doubled the final maximum level of inbreeding observed

gain than editing all 25 sires (70.66 vs. 31.29). With genome editing alone, editing the top 5 sires resulted in 1.80 times more genetic gain than editing all 25 sires (39.17 vs. 21.81).

with selection alone and when editing all 25 sires (~0.23 vs. ~0.10). The maximum level of inbreeding observed with selection alone and when editing all 25 sires was reached in half the time when editing the top 5 sires (generation 10 vs. generation 20).

Figure 6b also shows that the level of inbreeding attained when editing the top 5 sires was lower when gene drives were included. Figure 6b shows that at later generations, the reduction in inbreeding achieved with gene drives when editing the top 5 sires was more pronounced than in earlier generations. When editing the top 5 sires, the level of inbreeding attained with and without gene drives was equal across generations 0 to 5. By generation 20, the level of inbreeding reached without gene drives was 1.05 times higher than editing with gene drives (0.23 vs. 0.22).

Efficiency of converting genetic variation into genetic gain

Gene drives increase the efficiency of genome editing at converting genetic variation (measured by inbreeding) into genetic gain. This is shown in Fig. 7, which is a plot of the genetic gain against the inbreeding for generations 0 to 20. Figure 7 shows scenarios in which either all 25 sires were edited at 20 QTN or the top 5 sires were edited at 100 QTN (i.e., both scenarios performed a total of 500 edits per generation).

The two most efficient strategies were those including gene drives. The most efficient strategy was when the top 5 sires were edited with gene drives. The second most efficient strategy was when all 25 sires were edited with gene drives. The least efficient strategy was selection alone. By generation 20, the maximum level of inbreeding attained using selection alone was 0.0936 (indicated by the grey dashed vertical line in Fig. 7). At this level of inbreeding, editing the top 5 sires with gene drives achieved 3.92 times more genetic gain than selection alone (43.79 vs. 11.16). Editing all 25 sires with gene drives achieved 2.80 times more genetic gain than selection (31.29 vs. 11.16). Editing all 25 sires without gene drives achieved 1.95 times more genetic gain than

selection alone (21.81 vs. 11.16). Editing the top 5 sires without gene drives achieved 1.78 times more genetic gain than selection alone (19.82 vs. 11.16).

The number of sires edited influences efficiency differently, depending on whether or not gene drives were incorporated. With gene drives, editing the top 5 sires was more efficient than editing all 25 sires. Without gene drives, editing all 25 sires was more efficient than editing the top 5 sires. At the maximum level of inbreeding attained using selection alone, editing the top 5 sires with gene drives resulted in 1.40 times more genetic gain than editing all 25 sires with gene drives (43.79 vs. 31.29). In comparison, editing all 25 sires without gene drives resulted in 1.10 times more genetic gain than editing the top 5 sires with gene drives (21.81 vs. 19.82).

The efficiency of turning genetic variation into genetic gain was higher when inbreeding was lower (i.e., in early generations) compared to when inbreeding was higher (i.e., later generations). This pattern was consistent across all scenarios. This is shown in Table 1 as the ratio between the genetic gain and the change in inbreeding from generation 0 to generations 4, 8, 12, 16 and 20.

For selection, the efficiency of turning genetic variation into genetic gain in generation 4 was 1.18 times higher than the efficiency in generation 20 (1.34 vs. 1.14). For genome editing, the efficiency in generation 4 was 1.23 times higher than the efficiency in generation 20 when all 25 sires were edited (2.65 vs. 2.16), and 1.54 times higher when the top 5 sires were edited (2.12 vs. 1.38). For genome editing with gene drives, the efficiency in generation 4 was 1.24 times higher than the efficiency in generation 20 when all 25 sires were edited (3.66 vs. 2.95), and 1.75 times higher when the top 5 sires were edited (4.83 vs. 2.76).

The reduction in efficiency across generations was greater when using gene drives than without. Table 1 shows that the decay in efficiency from generation 4 to 20 was larger when using gene drives than without. When editing all 25 sires with gene drives, the reduction in efficiency was 3.55 times greater than with selection alone

Table 1 **Efficiency of turning genetic variation into genetic gain in generations 4, 8, 12, 16, and 20 of future breeding**

Editing strategy	Number of edited sires	Efficiency of turning genetic variation into genetic gain (95% CI)				
		Gen 4	Gen 8	Gen 12	Gen 16	Gen 20
Selection	0	1.34 (1.24–1.43)	1.27 (1.21–1.32)	1.21 (1.16–1.27)	1.18 (1.14–1.22)	1.14 (1.10–1.17)
Genome editing alone	25	2.65 (2.42–2.87)	2.51 (2.39–2.62)	2.38 (2.25–2.51)	2.27 (2.15–2.38)	2.16 (2.07–2.25)
With gene drives	25	3.66 (3.34–3.99)	3.43 (3.26–3.61)	3.25 (3.14–3.36)	3.10 (2.98–3.22)	2.95 (2.85–3.05)
Genome editing alone	5	2.12 (1.98–2.27)	1.73 (1.65–1.81)	1.52 (1.44–1.61)	1.44 (1.35–1.52)	1.38 (1.30–1.45)
With gene drives	5	4.83 (4.29–5.37)	3.84 (3.59–4.09)	3.38 (3.10–3.67)	3.06 (2.83–3.28)	2.76 (2.59–2.93)

CI confidence interval, *Gen* generation

(0.71 vs. 0.20). When editing the top 5 sires with gene drives, the reduction in efficiency was 10.35 times greater than with selection alone (2.07 vs. 0.20). Despite this, the use of gene drives was always more efficient than not using gene drives for all five generations tested.

Effect of gene drives on the number of distinct QTN edited

Across all generations, gene drives enable the editing of a larger number of distinct QTN. This is shown in Table 2, which gives the average number of distinct QTN edited across the 20 future generations. The table gives all scenarios when either all 25 sires were edited at 20 QTN or the top 5 sires were edited at 100 QTN (i.e., both scenarios performed a total of 10,000 edits across all 20 generations). Table 2 shows that when editing all 25 sires, gene drives resulted in 1.89 times more distinct QTN being edited than with genome editing alone (656.3 vs. 346.7). When editing the top 5 sires, gene drives resulted in 2.21 times more distinct QTN being edited than with genome editing alone (2612.9 vs. 1179.7).

While the use of gene drives enabled the targeting of more QTN for editing across all 20 generations, the rapid fixation of favourable alleles means that within a given generation, gene drives focus the editing resources on a smaller number of QTN. This is shown in Table 3, which gives the average number of distinct QTN edited per generation. The table gives all scenarios when either all 25 sires were edited at 20 QTN or the top 5 sires were

edited at 100 QTN (i.e., both scenarios performed a total of 500 edits per generation).

Gene drives resulted in fewer distinct QTN edited per generation. This pattern was consistent when editing either all 25 sires or the top 5 sires. When editing all 25 sires, genome editing alone resulted in 1.03 times more distinct QTN being edited than with gene drives (61.1 vs. 59.1). When editing the top 5 sires, genome editing alone resulted in 1.17 times more distinct QTN being edited than with gene drives (203.0 vs. 172.8).

Discussion

Our results highlight four main points for discussion, specifically: (1) the benefits of gene drives; (2) gene drives and editing strategies in livestock breeding; (3) gene drive risks and management strategies in livestock breeding; and (4) the assumptions made by the study and their effects on the application of gene drives in different settings.

Benefits of gene drives

Our simulations show that gene drives could amplify the benefits of genome editing in livestock breeding. The main benefit of genome editing is that it increases short-, medium- and long-term genetic gain [1]. This increase is brought about by: (1) increasing the frequency of favourable alleles at QTN; (2) reducing the time to fix favourable alleles at the largest effect QTN; and (3) minimising

Table 2 Average number of distinct QTN edited across all 20 generations of future breeding using genome editing alone or with gene drives of different conversion efficacies

Gene drive conversion efficiency	Average number of distinct QTN edited (95% CI)	
	25 Sires edited at 20 QTN each	5 Sires edited at 100 QTN each
Genome editing alone	346.7 (340.4–353.0)	1179.7 (1168.6–1190.8)
0.50	508.2 (498.2–518.2)	1943.5 (1923.7–1963.3)
0.75	582.4 (576.9–587.9)	2264.9 (2244.6–2285.2)
1.00	656.3 (648.7–663.9)	2612.9 (2593.1–2632.7)

CI confidence interval

Table 3 Average number of distinct QTN edited per generation using genome editing alone or with gene drives of different conversion efficacies

Gene drive conversion efficiency	Average number of distinct QTN edited (95% CI)	
	25 Sires edited at 20 QTN each	5 Sires edited at 100 QTN each
Genome editing alone	61.1 (58.6–63.6)	203.0 (198.1–207.9)
0.50	60.0 (58.2–61.9)	184.2 (181.2–187.2)
0.75	59.9 (57.2–62.5)	180.2 (174.3–186.2)
1.00	59.1 (57.3–61.0)	172.8 (168.7–177.0)

CI confidence interval

the chance of loss of favourable alleles at QTN with lesser effect by genetic drift.

Although genome editing alone results in large increases in genetic gain, the time taken to fix favourable alleles at the QTN with the largest effect could be up to six generations ([1] and our results). This reduces the chance of fixing the favourable alleles of QTN with lesser effect, since they may never become targets for genome editing or the favourable allele may be lost by genetic drift before it becomes a target for editing or for selection.

For livestock species with large generation intervals, the six generations would mean that fixing the favourable alleles at only the QTN with the largest effect could require a decade or more. Fixing only these QTN with large effect may not be enough for the return on investment if most of the traits that form parts of breeding goals are quantitative and are influenced by many QTN, all with small effect.

Gene drives can overcome these limitations by reducing the time to fix favourable alleles at the QTN with the largest effect. This enables the targeting of QTN with lesser effect for editing at earlier generations. This means that favourable alleles at QTN with lesser effect can be maintained in the population, are less prone to loss by genetic drift and are much more likely to reach fixation within a shorter time frame. Our simulations show that gene drives can achieve 1.5 times the genetic gain achieved with genome editing and can achieve 3 times that achieved with selection.

Gene drives and editing strategies in livestock breeding

With advances in genome editing technologies, genome editing of major genes within livestock breeding is a reality. More than 300 edits have been reported in livestock and plant species in the past five years, including edits for "double muscling" in pigs, cattle and sheep [4], to confer resistance to porcine reproductive and respiratory syndrome virus (PRRS) and African swine fever virus (ASFV) in pigs [4, 6–8], and has recently been adapted for use in poultry [22].

In spite of these advances, the economic and practical implications of genome editing means that it is likely that editing will be restricted to individuals with the largest impact on the population. In species such as pigs and cattle, these are the best performing males that are chosen as sires for the next generation. Editing these sires ensures that they are homozygous for the favourable allele. However, Mendelian sampling of alleles of the unedited dams means that there is no guarantee that all the progeny of an edited individual will also be homozygous for the favourable allele.

Gene drives eliminate the effect of Mendelian sampling by ensuring that all the offspring of an edited individual will be homozygous for the favourable allele, regardless of the genotype of its dam. In addition, all offspring will be homozygous for the gene drive, thus ensuring homozygosity in all future descendants of an edited individual [9, 10].

The economic and practical feasibility of genome editing may mean that the breeder must further prioritise amongst the selected sires. In this context, prioritising the top best performing sires for editing is the best option, and can even result in larger genetic gains over editing all sires. This increase in gain by editing only the best sires can be amplified by gene drives. We show that editing the top 5 best performing out of the 25 selected sires with gene drives can achieve over 6 times more genetic gain than selection alone and 2 times more genetic gain than editing the top 5 sires without gene drives.

The higher genetic gain achieved when editing a subset of the sires in this study is likely caused by the assumption of a fixed number of edits in a given generation (i.e., 500 edits per generation). This assumption meant that, within a given generation, a larger number of edits can be performed on a given individual when editing a subset of the sires than when editing all sires (i.e., top 5 sires edited at 100 QTN or all 25 sires edited at 20 QTN).

Applying a larger number of edits per individual in a subset of the sires means that the offspring of the edited subset perform better than the offspring of unedited sires and thus are more likely to be selected as parents for the next generation. The benefit of this is that the increase in frequency of favourable alleles occurs more quickly and results in higher genetic gains. The consequence of editing only a subset of the sires is that the increase in genetic gain comes at the expense of an increased rate of inbreeding.

Although gene drives cannot eliminate the increase in inbreeding observed when editing a subset of the sires, they can reduce it. They do this by speeding up the rate of spread of the favourable allele in the population (by implicitly editing the genome of non-edited mates of edited sires on the formation of zygotes). This achieves faster uniformity in performance across all individuals and reduces the relative advantage of the progeny and descendants of edited individuals both within and across generations.

Furthermore, gene drives increase the efficiency of converting genetic variation into genetic gain. This means that, for a given level of inbreeding, breeders could achieve more genetic gain with gene drives than with genome editing or genome selection alone. We show that when using gene drives, breeding programs can be four times more efficient than using selection alone and more than two times more efficient than using genome editing alone.

Gene drive risks and management strategies in livestock breeding

The use of gene drives when editing livestock populations is novel and thus care should be taken to consider the potential risks involved in the design and use of such technology. The potential risks of gene drives in livestock breeding are: (1) incorrect identification of favourable alleles within a given generation; (2) accidental spread of gene drives from a farmed population to a natural population; and (3) mutation of gene drive elements. Careful use and design of gene drives could eliminate these risks. Although some of the necessary technologies and risk-alleviating techniques and strategies we mention below are in their infancy or not yet developed, the field of genome editing and gene drives is rapidly advancing, and we believe that such technologies will eventually be available. Once developed, these technologies will need to be tested both in silico and in vivo before they can be applied at a larger scale within livestock breeding programs.

Risk 1: incorrect identification of favourable allele

The gene drive mechanism is very powerful at quick dissemination of alleles through a population. If the alleles are favourable and remain favourable into the future, there would be no negative consequences. However, if the alleles are incorrectly identified as favourable due to bad allele choice driven by underpowered experiments and less dependable data, or become unfavourable due to a change in environment, breeding goals or changes in the genetic background (e.g., negative epistasis), the rapid spread of a particular allele through a population could be negative or even catastrophic [23].

To overcome this, the gene drive mechanism could be used to switch back to the alternative allele in future generations. Alternatively, an additional gene drive could be introduced to deactivate and eliminate the initial gene drive from the genomes of future generations [12, 16, 24, 25]. This would be possible by combining the gene drive with a mechanism of underdominance, whereby individuals that are heterozygous for the deactivated gene drive would have a lower fitness than homozygous individuals [10, 26].

To further minimise the impact of incorrect allele identification, gene drives could be used to increase only the frequency of favourable alleles with proven effects. Those that appear to have favourable effects, but for which effects have not yet been proven, could be increased in frequency more conservatively using standard genome editing approaches.

Risk 2: accidental release of gene drives into wild populations

In livestock breeding schemes, the accidental introduction of the gene drive mechanism into a natural population could occur if a domesticated animal carrying a gene drive mates with an animal in a natural population from the same or related species. If an accidental introduction of the gene drive mechanism into a natural population did occur, it would result in the quick spread of the allele through the population. An allele that is considered favourable in farmed animals (e.g., double muscling) may be detrimental to the fitness of natural populations.

As a way of minimising this risk, physical containment strategies to reduce the likelihood of the gene drive escaping into natural populations could be used [24]. However, in some breeding programs physical containment may not be entirely possible. For example, the marine stage of the Atlantic salmon lifecycle in a breeding system takes place in seawater cages, where the possibility of escape and breeding with natural populations is quite high.

In such cases, alternative biological ways could be used to contain the gene drive system. These could involve attaching elements to the gene drive mechanism that control the number of times that the gene drive mechanism could act. A hypothetical example of this could involve adding five such elements and that each time the gene drive mechanism worked one of these elements was lost. Thus, the gene drive mechanism would only remain active in five descendant generations. To our knowledge no such mechanism has been developed, but the recently proposed 'daisy drive system' [27] bears some resemblance. The daisy drive system is identical in its effect to the normal gene drive system, but differs in its design. It involves a series of n unlinked gene drive elements that are unable to drive the spread of their own allele, but that control the spread of the gene drive element above it in the chain. Our results suggest that a gene drive mechanism with an element that enabled it to act for only two or three generations would convey all of the benefits of the efficiency of gene drives while removing the element of risk.

Risk 3: mutation of gene drive elements

Gene drives have two major components: (1) a guide RNA, which is the part of the gene editing mechanism used to recognise the specific target region of the genome where the gene drive will be incorporated; and (2) the Cas9 gene, whose protein product is responsible for cleavage of the targeted genomic region in order to initiate DNA repair. Without the *Cas9* gene, the gene drive mechanism is non-functional [28].

If the guide RNA mutates and is no longer able to specifically recognise the original targeted region, the gene drive mechanism could be incorporated into off-target regions of the genome. This would result in the uncontrolled and rapid spread of alleles at off-target regions with unknown consequences. Careful design of guide RNAs that require multiple mutations in order to target different genomic regions would minimise the probability of off-target incorporation of gene drives in future generations [24]. Alternatively, the guide RNA and the *Cas9* gene could be partitioned into separate cassettes. The genomic locations of the two cassettes could be carefully designed so that initial linkage between them ensures co-inheritance [15, 24, 29]. Recombination over a number of generations would break up this initial linkage, thus inactivating the gene drive mechanism in individuals who inherit only one of the cassettes and bypassing the problem of deleterious mutations accumulating in the gene drive over time.

Assumptions and applications

The benefits of gene drives are applicable in the context of some assumptions made in this study that are patently over simplified and technologically not possible currently. These include the genetic architecture of the trait of interest and the ability to discover many causal variants for quantitative traits, the absence of dominance, pleiotropy and epistasis, the ability to perform multiplexed genome editing, not accounting for the costs associated with each edit, and the absence of certainty that gene drives can be safely used (as discussed above in the section: "Gene drive risks and management strategies in livestock breeding"). We believe that the advances in genome sciences that will be made in the next decade or so will help to provide solutions to these simplifications, and provide some discussion around these assumptions below.

The impact of trait architecture and the discovery of causal variants on including gene drives

Potential targets for genome editing are already available in a variety of species for qualitative traits, but this is not always true for quantitative traits. The majority of traits forming breeding goals in livestock are quantitative, therefore it will be necessary to identify good targets in order to maximise the potential of this technology in livestock breeding. We chose to evaluate gene drives for a quantitative trait under the assumption that targets for editing were known and that the inheritance of the trait was additive.

In this study, QTN were prioritized for editing based on effect size. We do not believe that modest errors in the ordering of QTN would alter the results. Rather we believe that to use this technology, a breeding program would need to be able to find the approximately 500 to 600 of the QTN with the largest effect that control the genetic variation of the trait or selection index at some point over a 20-year time period.

We show that large positive impacts on genetic gain can be achieved with as little as 20 targets with large effect in any given generation. The identification of 20 or 30 targets for editing in the next few years is likely to be possible within large breeding schemes that routinely record and collect dense phenotypic and genomic data. However, the total number of targets that would need to be identified over several years is actually rather large (e.g., 500 to 600) and is more challenging. Our simulations show that these QTN would have to be discovered over a 20-year period. This may be possible given the huge advances in genome science that have been made in recent years and are likely to be made in the next two decades. Many breeding schemes are moving towards routine collection of sequence data, which will help in the precise identification and mapping of more QTN with large effect to target. Explicit approaches to discover genome editing targets will be needed. These approaches could make use of many different technologies including sequence enabled genome-wide association studies, genome annotation data, gene expression data, genome editing in vivo and in vitro and matings that are explicitly designed to enable allele-testing [30].

Impact of dominance, pleiotropy and epistasis on including gene drives

We also assume that the inheritance of the quantitative trait is fully additive. However, dominance and epistatic effects may exist, and could influence the number of edits required for a given individual and for a given QTN within a generation. For example, dominance of the favourable allele would mean that frequencies of favourable alleles need only be increased to ensure that individuals carry a minimum of one copy of the favourable allele, which would require fewer edits for a given QTN and may be done without the inclusion of gene drives.

In this study, a single trait controlled by 10,000 QTN each with additive effects sampled from a Gaussian distribution was simulated. This is a simplification, since most livestock breeding programs select for multiple traits. These traits have complex correlations with each other, caused by pleiotropy and linkage between alleles at QTN that affect different traits. However, we do not believe that the main conclusions from our results would be very different, since most livestock breeding programs select on an index. This index behaves like a single trait that is affected by many loci and thus our single-trait model could be seen as implicitly accounting for pleiotropic effects and complex genetic correlations between the component traits.

Negative epistasis of QTN may mean that editing multiple QTN for a given individual is required. We did not simulate epistasis because the data and theory suggest that epistasis has a minor contribution to total variation [31, 32]. However, if there are large epistatic effects, the value of genome editing and gene drives in livestock breeding would be significantly reduced. This is because on the one hand, the frequency of individual alleles would be shifted very rapidly by genome editing, resulting in these alleles being placed in different haplotypes that could have very different effects. On the other hand, this would also reveal epistatic effects that might otherwise be difficult to observe due to limited recombinations. The impact of epistasis is an open question that needs to be addressed with real data and populations.

Impact of multiplexed genome editing on including gene drives

The results of this study imply that multiplex editing of many alleles is needed to generate large increases in genetic gain in livestock. To our knowledge large multiplexing (e.g., 10 or more alleles) has not been successfully performed to date. However, genome editing techniques are improving rapidly and are an intensive area of research across all of the life sciences. We anticipate that multiplex genome editing will be possible in the future.

Impact of cost on including gene drives

The cost assumption made in this study was that a fixed editing resource of 500 edits was available within a given generation. These 500 edits could be distributed so that either all 25 selected sires were edited at 20 QTN each, or the top 5 sires were edited at 100 QTN each. If the cost of editing an individual is high, editing more QTN per individual enables a faster spread of favourable alleles across the population. In this context, gene drives will increase the rate of spread of favourable alleles throughout the population and reduce the impact of inbreeding. If the cost of a single edit is high, gene drives will be even more important for the fast dissemination of favourable alleles into the population. This is because within livestock breeding schemes where the majority of individuals are descendants of a few sires, editing with gene drives will mean that descendants of edited individuals will never require editing. Therefore the number of edits required in future generations for a given QTN is minimised.

Another assumption made was that gene drives do not constitute an additional edit by themselves. With the rapid fixation of the QTN with the largest effect with gene drives, this assumption meant that additional edits were available for QTN with lesser effect in future generations. If the gene drive is counted as an additional edit or if the cost of gene drives is too high, individual cost-benefit analyses would need to be conducted to evaluate the benefits of gene drives in the context of population size and structure and trait architecture.

Conclusions

Genome editing in livestock could be used to increase the frequency of favourable alleles at QTN with large effect. Gene drives could be used to increase the speed at which edited alleles are spread across livestock populations. They would do this by eliminating the effect of Mendelian sampling of alleles in unedited mates, resulting in complete homozygosity for the favourable allele amongst all descendants of an edited individual. Faster fixation of favourable alleles would mean that fewer edited founders would be required to fix the favourable allele for a given QTN. This would enable the targeting of more QTN and more effective distribution of editing resources across generations. Faster fixation of favourable alleles would also result in larger genetic gains in shorter time spans. In our simulations, we show that the larger genetic gains would come at no expense in inbreeding. In fact, gene drives could reduce the levels of inbreeding and increase the efficiency of the breeding program by resulting in higher genetic gains for a given unit of inbreeding. The magnitude of the benefits of gene drives would depend on three main factors: (1) the genetic architecture of the trait of interest, (2) the associated costs, and (3) risk management. Therefore, the additional benefits achieved with gene drives should be evaluated within individual breeding programs by conducting tailored cost-benefit analyses, taking into account population size and structure, trait architecture and the ability to successfully control the power of the technology.

Authors' contributions
JMH and CBAW conceived the study. JMH, GG, JJ and SG designed the study. SG wrote the software. JJ performed the analysis. SG and JMH wrote the first draft of the manuscript. CBAW, SG, GG, AJM and JJ helped to interpret the results and refine the manuscript. All authors read and approved the final manuscript.

Author details
[1] The Roslin Institute and Royal (Dick) School of Veterinary Studies, The University of Edinburgh, Easter Bush, Midlothian, Scotland, UK. [2] Genus plc, 1525 River Road, DeForest, WI 53532, USA.

Acknowledgements

The authors acknowledge the financial support from the BBSRC ISPG to The Roslin Institute with ISP numbers "BB/J004235/1" and "BB/J004316/1", from Genus PLC and from Grant Numbers "BB/M009254/1", "BB/L020726/1", "BB/N004736/1", "BB/N004728/1", "BB/L020467/1" and Medical Research Council (MRC) Grant Number "MR/M000370/1". This work made use of the resources provided by the Edinburgh Compute and Data Facility (ECDF) (http://www.ecdf.ed.ac.uk/). The authors thank Dr. Susan Cleveland (WI, USA) and Dr. Andrew Derrington (Parker Derrington Ltd.) for assistance in refining the manuscript and Dr. Maria C. Sanchez Perez for assistance in drawing Fig. 1 of the manuscript.

Competing interests

The authors declare that they have no competing interests.

References

1. Jenko J, Gorjanc G, Cleveland MA, Varshney RK, Whitelaw CB, Woolliams JA, et al. Potential of promotion of alleles by genome editing to improve quantitative traits in livestock breeding programs. Genet Sel Evol. 2015;47:55.
2. Carroll D, Charo RA. The societal opportunities and challenges of genome editing. Genome Biol. 2015;16:242.
3. Ainsworth C. Agriculture: a new breed of edits. Nature. 2015;528:S15–6.
4. Proudfoot C, Carlson DF, Huddart R, Long CR, Pryor JH, King TJ, et al. Genome edited sheep and cattle. Transgenic Res. 2014;24:147–53.
5. Tan W, Carlson DF, Lancto CA, Garbe JR, Webster DA, Hackett PB, et al. Efficient nonmeiotic allele introgression in livestock using custom endonucleases. Proc Natl Acad Sci USA. 2013;110:16526–31.
6. Lillico SG, Proudfoot C, King TJ, Tan W, Zhang L, Mardjuki R, et al. Mammalian interspecies substitution of immune modulatory alleles by genome editing. Sci Rep. 2016;6:21645.
7. Whitworth KM, Rowland RRR, Ewen CL, Trible BR, Kerrigan MA, Cino-Ozuna AG, et al. Gene-edited pigs are protected from porcine reproductive and respiratory syndrome virus. Nat Biotechnol. 2016;34:20–2.
8. Lillico SG, Proudfoot C, Carlson DF, Stverakova D, Neil C, Blain C, et al. Live pigs produced from genome edited zygotes. Sci Rep. 2013;3:2847.
9. Gould F. Broadening the application of evolutionarily based genetic pest management. Evolution. 2008;62:500–10.
10. Burt A, Trivers R. Genes in conflict: The biology of selfish genetic elements. Cambridge: Belknap Press; 2006.
11. Clark JB, Kidwell MG. A phylogenetic perspective on P transposable element evolution in Drosophila. Proc Natl Acad Sci USA. 1997;94:11428–33.
12. Burt A. Site-specific selfish genes as tools for the control and genetic engineering of natural populations. Proc Soc Biol Sci. 2003;270:921–8.
13. Ishino Y, Shinagawa H, Makino K, Amemura M, Nakata A. Nucleotide sequence of the iap gene, responsible for alkaline phosphatase isozyme conversion in Escherichia coli, and identification of the gene product. J Bacteriol. 1987;169:5429–33.
14. Barrangou R, Fremaux C, Deveau H, Richards M, Boyaval P, Moineau S, et al. CRISPR provides acquired resistance against viruses in prokaryotes. Science. 2007;315:1709–12.
15. Gantz VM, Bier E. The mutagenic chain reaction: a method for converting heterozygous to homozygous mutations. Science. 2015;348:442–4.
16. Gantz VM, Jasinskiene N, Tatarenkova O, Fazekas A, Macias VM, Bier E, et al. Highly efficient Cas9-mediated gene drive for population modification of the malaria vector mosquito Anopheles stephensi. Proc Natl Acad Sci USA. 2015;112:E6736–43.
17. Chen GK, Marjoram P, Wall JD. Fast and flexible simulation of DNA sequence data. Genome Res. 2009;19:136–42.
18. Faux A-M, Gorjanc G, Gaynor RC, Battagin M, Edwards SM, Wilson D, et al. AlphaSim: software for breeding program simulation. Plant Genome. 2016;. doi:10.3835/plantgenome2016.02.0013.
19. Hickey JM, Gorjanc G. Simulated data for genomic selection and genome-wide association studies using a combination of coalescent and gene drop methods. G3 (Bethesda). 2012;2:425–7.
20. Villa-Angulo R, Matukumalli LK, Gill CA, Choi J, Tassell CPV, Grefenstette JJ. High-resolution haplotype block structure in the cattle genome. BMC Genet. 2009;10:19.
21. Pérez-Enciso M. Use of the uncertain relationship matrix to compute effective population size. J Anim Breed Genet. 1995;112:327–32.
22. Bai Y, He L, Li P, Xu K, Shao S, Ren C, et al. Efficient genome editing in chicken DF-1 cells using the CRISPR/Cas9 system. G3 (Bethesda). 2016;6:917–23.
23. Unckless RL, Messer PW, Connallon T, Clark AG. Modeling the manipulation of natural populations by the mutagenic chain reaction. Genetics. 2015;201:425–31.
24. DiCarlo JE, Chavez A, Dietz SL, Esvelt KM, Church GM. Safeguarding CRISPR-Cas9 gene drives in yeast. Nat Biotechnol. 2015;33:1250–5.
25. Esvelt KM, Smidler AL, Catteruccia F, Church GM. Concerning RNA-guided gene drives for the alteration of wild populations. Elife. 2014;3:e03401.
26. Burt A. Heritable strategies for controlling insect vectors of disease. Philos Trans R Soc Lond B Biol Sci. 2014;369:20130432.
27. Noble C, Min J, Olejarz J, Buchthal J, Chavez A, Smidler AL, et al. Daisy-chain gene drives for the alteration of local populations. bioRxiv. 2016;057307. doi:10.1101/057307.
28. Jinek M, East A, Cheng A, Lin S, Ma E, Doudna J. RNA-programmed genome editing in human cells. Elife. 2013;2:e00471.
29. Akbari OS, Bellen HJ, Bier E, Bullock SL, Burt A, Church GM, et al. BIOSAFETY. Safeguarding gene drive experiments in the laboratory. Science. 2015;349:927–9.
30. Hickey JM, Bruce C, Whitelaw A, Gorjanc G. Promotion of alleles by genome editing in livestock breeding programmes. J Anim Breed Genet. 2016;133:83–4.
31. Hill WG, Goddard ME, Visscher PM. Data and theory point to mainly additive genetic variance for complex traits. PLoS Genet. 2008;4:e1000008.
32. Mäki-Tanila A, Hill WG. Influence of gene interaction on complex trait variation with multilocus models. Genetics. 2014;198:355–67.

PERMISSIONS

All chapters in this book were first published in GSE, by BioMed Central; hereby published with permission under the Creative Commons Attribution License or equivalent. Every chapter published in this book has been scrutinized by our experts. Their significance has been extensively debated. The topics covered herein carry significant findings which will fuel the growth of the discipline. They may even be implemented as practical applications or may be referred to as a beginning point for another development.

The contributors of this book come from diverse backgrounds, making this book a truly international effort. This book will bring forth new frontiers with its revolutionizing research information and detailed analysis of the nascent developments around the world.

We would like to thank all the contributing authors for lending their expertise to make the book truly unique. They have played a crucial role in the development of this book. Without their invaluable contributions this book wouldn't have been possible. They have made vital efforts to compile up to date information on the varied aspects of this subject to make this book a valuable addition to the collection of many professionals and students.

This book was conceptualized with the vision of imparting up-to-date information and advanced data in this field. To ensure the same, a matchless editorial board was set up. Every individual on the board went through rigorous rounds of assessment to prove their worth. After which they invested a large part of their time researching and compiling the most relevant data for our readers.

The editorial board has been involved in producing this book since its inception. They have spent rigorous hours researching and exploring the diverse topics which have resulted in the successful publishing of this book. They have passed on their knowledge of decades through this book. To expedite this challenging task, the publisher supported the team at every step. A small team of assistant editors was also appointed to further simplify the editing procedure and attain best results for the readers.

Apart from the editorial board, the designing team has also invested a significant amount of their time in understanding the subject and creating the most relevant covers. They scrutinized every image to scout for the most suitable representation of the subject and create an appropriate cover for the book.

The publishing team has been an ardent support to the editorial, designing and production team. Their endless efforts to recruit the best for this project, has resulted in the accomplishment of this book. They are a veteran in the field of academics and their pool of knowledge is as vast as their experience in printing. Their expertise and guidance has proved useful at every step. Their uncompromising quality standards have made this book an exceptional effort. Their encouragement from time to time has been an inspiration for everyone.

The publisher and the editorial board hope that this book will prove to be a valuable piece of knowledge for researchers, students, practitioners and scholars across the globe.

LIST OF CONTRIBUTORS

Ricardo V. Ventura
Centre for Genetic Improvement of Livestock, University of Guelph, Guelph, ON N1G2W1, Canada
Beef Improvement Opportunities, Guelph, ON N1K1E5, Canada

Stephen P. Miller
Centre for Genetic Improvement of Livestock, University of Guelph, Guelph, ON N1G2W1, Canada
Invermay Agricultural Centre, AgResearch Limited, Mosgiel 9053, New Zealand

Ken G. Dodds, Matthew Bixley, Shannon M. Clarke and John C. McEwan
Invermay Agricultural Centre, AgResearch Limited, Mosgiel 9053, New Zealand

Benoit Auvray and Michael Lee
Department of Mathematics and Statistics, University of Otago, Dunedin 9016, New Zealand

Yuanmei Guo, Yixuan Huang, Lijuan Hou, Junwu Ma, Congying Chen, Huashui Ai, Lusheng Huang and Jun Ren
State Key Laboratory of Pig Genetic Improvement and Production Technology, Jiangxi Agricultural University, Nanchang 330045, China

Timothy J. Byrne and Peter R. Amer
AbacusBio Limited, PO Box 5585, Dunedin 9058, New Zealand

Bruno F. S. Santos
AbacusBio Limited, PO Box 5585, Dunedin 9058, New Zealand
School of Environmental and Rural Science, University of New England, Armidale, NSW, Australia

John P. Gibson
School of Environmental and Rural Science, University of New England, Armidale, NSW, Australia

Julius H. J. van der Werf
Cooperative Research Centre for Sheep Industry Innovation, Armidale, NSW, Australia

Vanessa Lutz, Patrick Stratz, Siegfried Preuß, Michael A. Grashorn, Werner Bessei and Jörn Bennewitz
Institute of Animal Science, University of Hohenheim, 70599 Stuttgart, Germany

Jens Tetens
Division of Functional Breeding, Department of Animal Sciences, Georg-August-University Göttingen, 37077 Göttingen, Germany

Mehdi Momen, Ahmad Ayatollahi Mehrgardi and Masood Asadi Fozi
Department of Animal Science, Faculty of Agriculture, Shahid Bahonar University of Kerman (SBUK), Kerman, Iran

Ali Esmailizadeh
Department of Animal Science, Faculty of Agriculture, Shahid Bahonar University of Kerman (SBUK), Kerman, Iran
State Key Laboratory of Genetic Resources and Evolution, Yunnan Laboratory of Molecular Biology of Domestic Animals, Kunming Institute of Zoology, Chinese Academy of Sciences, Kunming 650223, China

Ayoub Sheikhy
Department of Statistical, Faculty of Mathematic and Computer Science, Shahid Bahonar University of Kerman (SBUK), Kerman, Iran

Andreas Kranis
Roslin Institute, University of Edinburgh, Midlothian, UK

Bruno D. Valente
Department of Animal Sciences, University of Wisconsin, Madison, WI, USA

Guilherme J. M. Rosa
Department of Animal Sciences, University of Wisconsin, Madison, WI, USA
Department of Biostatistics and Medical Informatics, University of Wisconsin, Madison, WI, USA

Daniel Gianola
Department of Animal Sciences, University of Wisconsin, Madison, WI, USA

Department of Biostatistics and Medical Informatics, University of Wisconsin, Madison, WI, USA
Department of Dairy Science, University of Wisconsin, Madison, WI, USA

Guiyan Ni, Anna Fangmann and Henner Simianer
Animal Breeding and Genetics Group, Georg-August-Universität, Göttingen, Germany

Malena Erbe
Animal Breeding and Genetics Group, Georg-August-Universität, Göttingen, Germany
Institute for Animal Breeding, Bavarian State Research Centre for Agriculture, Grub, Germany

David Cavero
Lohmann Tierzucht GmbH, Cuxhaven, Germany

Arne Johan Jensen, Bjørn Ove Johnsen and Sten Karlsson
Norwegian Institute for Nature Research (NINA), P.O. Box 5685, 7485 Sluppen, Trondheim, Norway

Lars Petter Hansen
Norwegian Institute for Nature Research (NINA), Gaustadalléen 21, 0349 Oslo, Norway

Ingrid David, Hervé Garreau and Laurianne Canario
GenPhySE, INRA, INPT, ENVT, Université de Toulouse, 31326 Castanet-Tolosan, France

Elodie Balmisse
Pectoul, INRA, 31326 Castanet-Tolosan, France

Yvon Billon
GenESI, INRA, 17700 Surgères, France

Roberto Antolín, Gregor Gorjanc, Daniel Money and John M. Hickey
The Roslin Institute and Royal (Dick) School of Veterinary Studies, The University of Edinburgh, Easter Bush Research Centre, Midlothian EH25 9RG, Scotland, UK

Carl Nettelblad
Division of Scientific Computing, Department of Information Technology, Science for Life Laboratory, Uppsala University, Lägerhyddsvägen 2, Box 337, 751 05 Uppsala, Sweden

Gary A. Rohrer and Dan J. Nonneman
U.S. Meat Animal Research Center, USDA, Agricultural Research Service, Clay Center, NE, USA

Laura Plieschke, Christian Edel, Eduardo C. G. Pimentel, Reiner Emmerling and Kay-Uwe Götz
Bavarian State Research Center for Agriculture, Institute of Animal Breeding, Prof.-Dürrwaechter Platz 1, 85586 Poing-Grub, Germany.

Jörn Bennewitz
Institute of Animal Science, University Hohenheim, Garbenstraße 17, 70599 Stuttgart, Germany

Hubert Pausch and Iona M. MacLeod
Agriculture Victoria, AgriBio, Centre for AgriBiosciences, Bundoora, VIC 3083, Australia

Phil J. Bowman and Hans D. Daetwyler
Agriculture Victoria, AgriBio, Centre for AgriBiosciences, Bundoora, VIC 3083, Australia
School of Applied Systems Biology, La Trobe University, Bundoora, VIC 3083, Australia

Michael E. Goddard
Agriculture Victoria, AgriBio, Centre for AgriBiosciences, Bundoora, VIC 3083, Australia
Faculty of Veterinary and Agricultural Sciences, University of Melbourne, Melbourne, VIC 3010, Australia

Ruedi Fries
Chair of Animal Breeding, Technische Universitaet Muenchen, 85354 Freising, Germany

Reiner Emmerling
Institute of Animal Breeding, Bavarian State Research Center for Agriculture, 85586 Grub, Germany

Hossein Mehrban
Department of Animal Science, Shahrekord University, P.O. Box 115, Shahrekord 88186-34141, Iran

Deuk Hwan Lee
Department of Animal Life and Environment Science, Hankyong National University, Jungang-ro 327, Anseong-si, Gyeonggi-do 456-749, Korea

Mohammad Hossein Moradi
Department of Animal Science, Faculty of Agriculture and Natural Resources, Arak University, Arāk 38156-8-8349, Iran

Chung IlCho
Hanwoo Improvement Center, National Agricultural Cooperative Federation, Haeun-ro 691, Unsan-myeon, Seosan-si, Chungnam-do 356-831, Korea

Masoumeh Naserkheil
Department of Animal Science, University College of Agriculture and Natural Resources, University of Tehran, P.O. Box 4111, Karaj 31587-11167, Iran

Noelia Ibáñez-Escriche
The Roslin Institute, Royal (Dick) School of Veterinary Studies, University of Edinburgh, Roslin, UK

Joonho Lee, Rohan Fernando and Jack Dekkers
Department of Animal Science, Iowa State University, Ames, IA 50011,USA

Hao Cheng
Department of Animal Science, Iowa State University, Ames, IA 50011,USA
Department of Statistics, Iowa State University, Ames, IA 50011, USA

Dorian Garrick
Department of Animal Science, Iowa State University, Ames, IA 50011,USA
Institute of Veterinary, Animal and Biomedical Sciences, Massey University, Palmerston North, New Zealand
ThetaSolutions, LLC, Atascadero, CA, USA

Bruce Golden
ThetaSolutions, LLC, Atascadero, CA, USA

Kyungdo Park
Department of Animal Biotechnology, Chonbuk National University, Chonju, Jeollabuk-do, South Korea

Deukhwan Lee
Department of Animal Science, Hankyong National University, Anseong, Gyeonggi-do, South Korea

Katharina Correa and José M. Yáñez
Facultad de Ciencias Veterinarias y Pecuarias, Universidad de Chile, Av Santa Rosa 11735, La Pintana, Chile
Aquainnovo, Cardonal s/n Lote B, Puerto Montt, Chile

Rama Bangera, René Figueroa and Jean P. Lhorente
Aquainnovo, Cardonal s/n Lote B, Puerto Montt, Chile

Serap Gonen, Janez Jenko, Gregor Gorjanc, C. Bruce A. Whitelaw and John M. Hickey
The Roslin Institute and Royal (Dick) School of Veterinary Studies, The University of Edinburgh, Easter Bush, Midlothian, Scotland, UK

Alan J. Mileham
Genus plc, 1525 River Road, DeForest, WI 53532, USA

Index